W9-CSA-899

Physics, Technology
and the Nuclear Arms Race
(APS Baltimore—1983)

AIP Conference Proceedings
Series Editor: Hugh C. Wolfe
Number 104

Physics, Technology and the Nuclear Arms Race

(APS Baltimore—1983)

Edited by
David W. Hafemeister
California Polytechnic State University
and
Dietrich Schroeer
University of North Carolina at Chapel Hill

American Institute of Physics

New York   1983

SEP IAE

PHYS

69665606

Copying fees: The code at the bottom of the first page of each article in this volume gives the fee for each copy of the article made beyond the free copying permitted under the 1978 US Copyright Law. (See also the statement following "Copyright" below). This fee can be paid to the American Institute of Physics through the Copyright Clearance Center, Inc., Box 765, Schenectady, N.Y. 12301.

Copyright © 1983 American Institute of Physics

Individual readers of this volume and non-profit libraries, acting for them, are permitted to make fair use of the material in it, such as copying an article for use in teaching or research. Permission is granted to quote from this volume in scientific work with the customary acknowledgment of the source. To reprint a figure, table or other excerpt requires the consent of one of the original authors and notification to AIP. Republication or systematic or multiple reproduction of any material in this volume is permitted only under license from AIP. Address inquiries to Series Editor, AIP Conference Proceedings, AIP, 335 E. 45th St., New York, N. Y. 10017

L.C. Catalog Card No. 83-72533
ISBN 0–88318–203–3
DOE CONF- 830463

U 264
P 578
1983
PHYS

v

PREFACE

The issues of war and peace in the nuclear age are of the highest importance for all of us. Since the arms race is a mixture of science, technology, political science, history, and ethics, it is difficult for any one group to claim that it understands this issue better than others. However, some physicists have felt that they have a special responsibility in this area. Physics has at times been at the leading edge of the next technological wave of arms developments; hence the public expects them to know a great deal about nuclear weaponry and its effects. Science is seen as containing some real truths; hence physicists have at times been seen as more thruthful even in non-technical matters. However, within the physics community, physicists are expected to conform to certain professional norms, and to focus on materials related to their expertise. Physicists may have made personal commitments to contribute in a useful way to solving the problem of the arms race, driven perhaps by a feeling that physicists still bear a special responsibility for having created nuclear weapons, or from judging that scientific rationality has been overrruled too often by political emotionalism. A sense of urgency about the arms race is reflected in the resolution concerning nuclear warfare adopted by the Executive Council of the American Physical Society on January 31, 1983, asking for efforts to control some nuclear weapons technologies.

It is our feeling that a better understanding of technical aspects of the nuclear arms race is a minimum requirement imposed on physicists, and that such improved understanding would contribute to alleviating the arms race. For this reason we have organized two short courses on the arms race to help physicists who either plan to teach about the arms race, or who want to study arms-race issues more deeply. The intent of the courses has been to provide technically-expert information on this topic, within a professional-physics context. Political topics have been dealt with only as they are either consequences of the technologies, or as they affect the technologies that have been developed. The goal has been to inform physicsts representing all parts of the political spectrum. This approach may indeed have left out the most-important political aspects of the arms-race debate, but it did bring together physicsts with diverse political persuasions. We have the belief that as physicsts we owe the world technological expertise first; political activism is a personal action that should be professionally secondary to making our technical knowledge available to the public. As physicsts, our first obligation is to be informed; hence the two informational short courses. This volume is the proceedings of the second short course on the nuclear arms race that was held on April 17, 1983 at the Baltimore Meeting of the American Physical Society. Like the first short course of January 24, 1982 in San Francisco, it was cosponsored by the Forum on Physics and Society of the APS and by the American Association of Physics Teachers. The invited talks presented at the course have been supplemented by several invited technical papers and appendices in order to broaden the coverage of the nuclear arms race.

The papers deal with the following issues, listed in more detail in the Table of Contents:

Chapter 1: Effects of Nuclear War;
Chapters 2-4: Technologies of the Arms Race;
Chapters 5-6: Technical Issues of Verification;
Chapters 7-10: Teaching Courses on the Arms Race;
Chapters 11-15: Policy Issues on the Arms Race;
Appendices A-I: Bibliography; Chronology;
   Agreements and Laws; Data Tables; Findings of OTA
   Studies; Equations and Calculations; and
   Biographical Notes on the Authors.

We would like to acknowledge the efforts by all the speakers and authors for their excellent contributions to this Proceedings. Space limitation does not allow us to include the proceedings of the first short course; therefore, we would like to thank the speakers who are not represented in this AIP volume: Michael Callaham (Carnegie-Mellon University), Barry M. Casper (Carleton College), Gloria Duffy (Stanford University), Marvin Goldberger (California Institute of Technology), Henry Kelly (Office of Technology Assessment), and Robin Staffin (Lawrence Livermore Laboratory). In addition, we would like to thank Bill Havens (APS), Jack Wilson (AAPT), and Hugh Wolfe (AIP) for helping make all the arrangements flow smoothly and pleasantly.

David Hafemeister   Dietrich Schroeer

San Luis Obispo, CA   Chapel Hill, NC

June 22, 1983

TABLE OF CONTENTS

APPENDICES

CHAPTER 1

The Effects of Nuclear War

Frank von Hippel

Woodrow Wilson School of Public and International Affairs
and
Center for Energy and Environmental Studies

Princeton University
Princeton, N.J. 08544

Table of Contents

I.   Nuclear Explosions

Heat: The release of nuclear energy converts the nuclear warhead
into an ionized plasma with a temperature of tens of millions of
degrees (see Appendix A [1]). At this temperature, most of the
energy is in the form of X-ray photons and, as a result, if air were
transparent to X-rays, the "X-ray fireball" would expand with the

0094-243X/83/1040001-46 $3.00 Copyright 1983 American Institute of Physics

speed of light. Since air is not transparent to X-rays, however, they are absorbed within about 10 meters - after which the fireball grows by "radiation diffusion" until, about one millisecond after the nuclear energy release, it has incorporated enough air to cool it to a temperature on the order of 100,000 degrees Kelvin. At this temperature, the rate of thermal radiation is low enough so that mechanical expansion begins to dominate (see Fig. I.1 [2, Fig. 3]). In Appendix A, the corresponding radius is is shown to be approximately

$$R_1 = 100*W^{0.3} \text{ meters} \qquad\qquad (I.1)$$

where W is the "yield" (energy release) in megatons TNT equivalent (one megaton = 10^{15} calories).

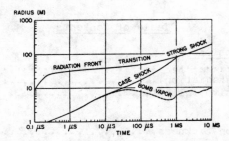

Fig. I.1 Space-time plot of early phase fireball (1 MT-sea level-air burst).

For the following second or so, the bubble of air which has been heated by radiation diffusion expands until its internal pressure has dropped to the ambient level. For a burst near sea level, this corresponds to an approximately fivefold increase in radius and therefore a drop in density to approximately 1 percent of atmospheric (see Fig. I.2 [2, Fig. 5]).

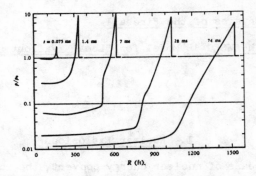

Fig. I.2 Fireball density versus radius at early times in the fireball history (1 MT-surface burst).

The fireball radiates a large fraction of its heat as visible and infrared radiation. For a surface facing the fireball at a distance R in km from the explosion, the incident heat is

$$7800 * f_T * W * T_A / R^2 \quad cal/cm^2 \qquad (I.2)$$

where the factor T_A takes into account the transmittance of the
atmosphere [3, Fig. 7.98], f_T is the fraction of the warhead's
energy which is radiated as heat and W is in megatons. For a low
altitude airburst, this fraction is about 35 percent [3, Table 7.88].
Because the surface area of the fireball grows less rapidly than the
yield, it takes longer for the fireballs from higher yield warheads
to radiate away their heat but, even for a one megaton warhead, half
of the thermal radiation is emitted within the first two seconds (see
Fig. I.3 [3, Fig. 7.87]).

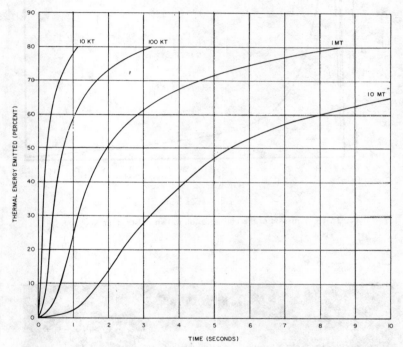

Fig. I.3 Percentage of thermal energy emitted as a function of time for air bursts of
various yields.

The thermal radiation pulse from a nuclear fireball has a
characteristic "spike plus a hump" shape (see Fig. I.4 [3, Fig.
2.39]). The drop in intensity after the first pulse is due to the
temporary masking of the fireball by various absorbing atomic and
molecular species created as a result of the pulse of ionizing
radiation emitted in the first fraction of a millisecond.

Shock Wave: Before the expansion of the fireball halts, the air
that has been compressed and accelerated outward by the fireball's

explosive expansion, separates and propagates outward as a shock wave (see Fig. I.5 [3, Fig. 2.32]).   The strength of this shock wave at the earth's surface depends upon both the distance from the explosion, and its height above the earth's surface.

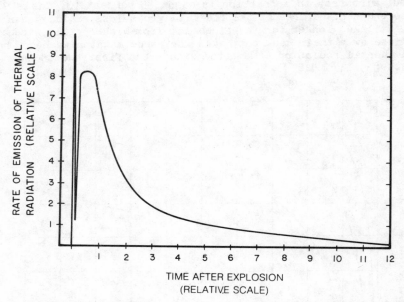

Fig. I.4   Emission of thermal radiation in two pulses in an air burst.

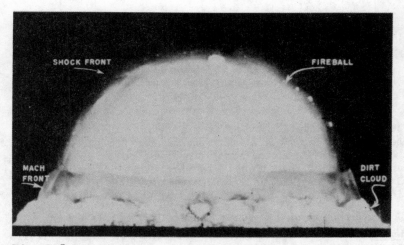

Fig. I.5 The faintly luminous shock front seen just ahead of the fireball soon after breakaway (see § 2.120).

A given peak overpressure occurs at a distance from the explosion (R) that scales with yield as

$$R = R_1 * W^{1/3} \text{ kilometers,}$$ (I.3)

where (R_1) is the distance for an explosion with a yield of one megaton. This scaling law can be derived by dimensional reasoning if one makes the approximation that the original energy release occurred at a point. The only dimensional quantities involved are then the yield, the speed of sound, and the density of the atmosphere [4]. For a one megaton ground-burst, the peak overpressure declines with distance as [3]

$$p = 6.4/(R_1)^3 + 2.2/(R_1)^{1.5} \text{ atmospheres.}$$ (I.4)

The overpressure at the earth's surface a given distance from ground zero can be maximized by exploding the warhead at an height where the shock wave from the explosion and the shock wave reflected from the earth's surface add "optimally" for this distance [3, Fig. 3.26]. Figs. I.6 [3, Figs. 3.73] show peak overpressures as a function of height-of-burst and distance from ground-zero for a one megaton explosion. Overpressures for other yields can be obtained by scaling these distances according to Eq. (I.3).

Fig. I6A ONE MEGATON - LOW PRESSURE REGION

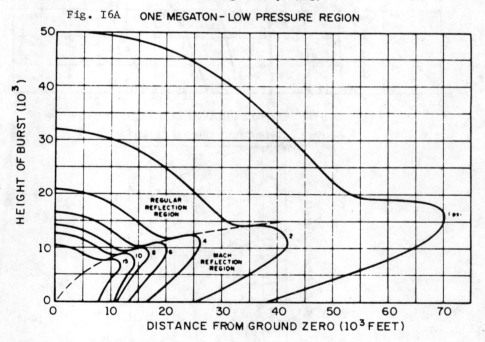

Nuclear Radiation: One of the ways in which the nuclear character of the energy release manifests itself is in the neutrons and gamma rays which are radiated by the nuclear explosion. Most of the neutrons (over 99 percent) are emitted during the fission events themselves.

(The remainder boil off of the excited fission products after fission has occurred.) In contrast, on average only about one half of the initial gamma energy radiated by a nuclear explosion comes from the fission events themselves. The remainder come from excited fission products and neutron reactions in the air surrounding the nuclear explosion (see Fig. I.7 [3, Fig. 8.14]).

ONE MEGATON - INTERMEDIATE PRESSURE REGION
HEIGHT OF BURST (10³ FEET)
REGULAR REFLECTION REGION
MACH REFLECTION REGION
10 psi
15
20
30
50
100
200
DISTANCE FROM GROUND ZERO (10³ FEET)

Fig. I.6B

ONE MEGATON-HIGH PRESSURE REGION
HEIGHT OF BURST (10³ FEET)
100 psi
200
REGULAR REFLECTION REGION
MACH REFLECTION REGION
500
1000
2000
5000
10,000
DISTANCE FROM GROUND ZERO (10³FEET)

Fig. I.6C

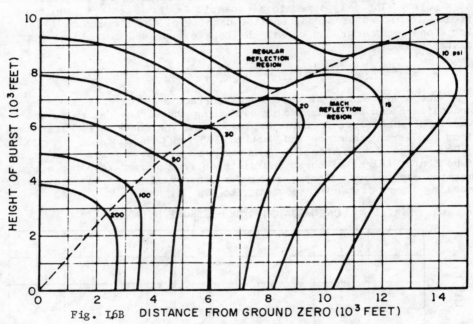

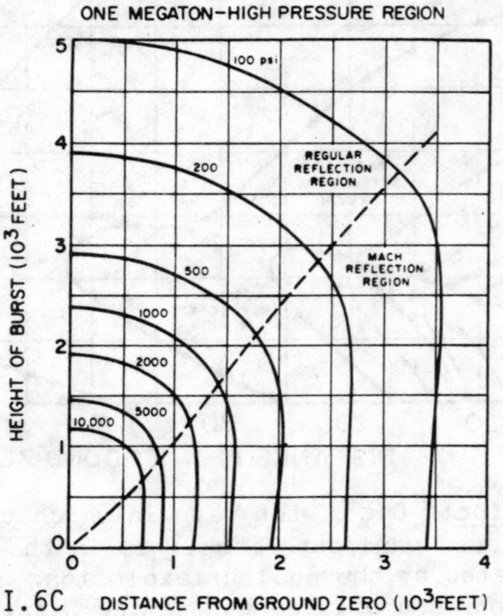

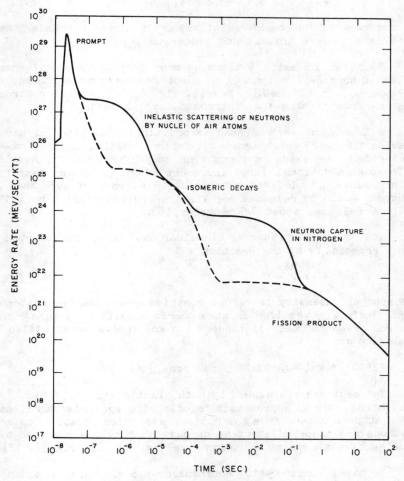

Fig. I.7 Calculated time dependence of the gamma-ray energy output per kiloton energy yield from a hypothetical nuclear explosion. The dashed line refers to an explosion at very high altitude.

Neutrons: Each fission in a fast-fission chain-reaction releases on average about 180 MeV. of kinetic energy and about 2.5 neutrons [3, pp. 12, 17]. Each megaton of fission therefore releases about $3.6*10^{26}$ neutrons.

Fusion fueled by lithium and deuterium does not release any net neutrons since, for each neutron released by the principal energy-releasing reaction,

$$D + T \longrightarrow n + He^4 + 17.6 \text{ MeV}, \qquad (I.5)$$

another must be absorbed in the tritium-generating reaction,

$$n + Li^6 \longrightarrow T + He^4 + 4.8 \text{ MeV.} \tag{I.6}$$

In the case of "enhanced-radiation" or "neutron" warheads, the fusion fuel is therefore tritium and deuterium.

Reaction I.5 emits 4 times as many neutrons per unit energy released as does fission and the neutrons carry more kinetic energy (14 versus about 2 MeV.). See [3, Fig. 8.116] for the neutron energy spectra from fission and "thermonuclear" warheads.

A "thermonuclear" warhead is ordinarily described as having a fission "trigger" which ignites a fusion reaction in lithium-deuteride. The extra neutrons from these first two "stages" then release an additional large increment of energy by the fast fission of uranium-238.[*] In such a warhead, an average of more than 2.5 neutrons would be released per fission because fission by a 14 MeV. neutron releases about 4.5 neutrons [6].

The neutrons emitted by a nuclear explosion are absorbed in air - especially by the reaction

$$n + N^{14} \longrightarrow p + C^{14}. \tag{I.7}$$

The neutron intensity therefore contains an exponential attenuation term. Ref. 2 gives (for an atmospheric density of 1 kg/m^3) the neutron "fluence" with distance R (in km) from an unspecified nuclear explosion as

$$W*2*10^{15}*\exp(-4.2*R)/R^2 \quad \text{neutrons/cm}^2. \tag{I.8}$$

The neutrons are slowed by both elastic and inelastic collisions, but an approximate "equilibrium spectrum" develops within a few hundred meters from the nuclear explosion because low energy neutrons are absorbed about as quickly as they are produced by the slowing of high energy neutrons (see Fig. I.8 [3, Figs 8.117]).[**]

It takes about $4.4*10^8$ neutrons/cm^2 to give a dose of one rad (100 ergs deposited per gram of soft tissue) [2] and the threshold for lethality is about 200 rads [3, p.583]. According to [3, Fig. 8.64], the distance at which a typical one kiloton fission explosion will deliver a 200 rad dose is about 1 km. Because of the exponential attenuation of the neutrons, this distance is only doubled for a thousandfold increase in yield. Since the radius for a given blast overpressure grows much more quickly, the relative importance of radiation effects declines with increasing yield.

----

[*] This picture is largely based on the fact that almost one half of the total energy released by nuclear weapons in atmospheric tests prior to 1962 (193 out of 511 megatons) was due to fission [5].
[**] Note that in [3, Figs. 117] the neutron fluences shown are about one order of magnitude higher than can be calculated from (I.8).

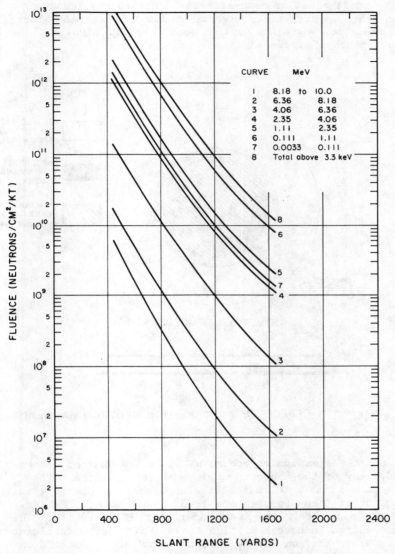

Fig. I.8 Neutron fluence per kiloton energy yield incident on a target located on or
near the earth's surface.

Consider a peak overpressure of 0.35 atmospheres (5 psi) which
occurs at a radius of 1 km from a 10 kiloton ground-burst, for
example. Ordinary residential structures will be completely
destroyed by blast at this overpressure [3, pp.178-189]. At
Hiroshima the area within this radius was also burned over by a
firestorm [3, pp.36, 300]. A neutron bomb with this yield would

hardly have small blast or heat effects. This is why "neutron bombs," in order to have their advertised low "collateral" blast damage at distances where the neutron dose is still high, must have yields on the order of one kiloton (see Fig. I.9 [7]).

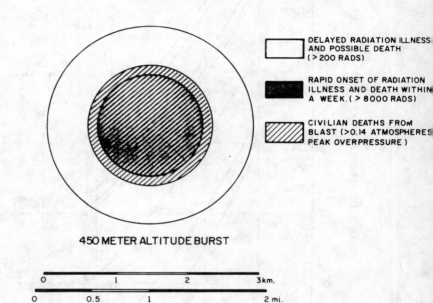

450 METER ALTITUDE BURST

DELAYED RADIATION ILLNESS AND POSSIBLE DEATH (> 200 RADS)

RAPID ONSET OF RADIATION ILLNESS AND DEATH WITHIN A WEEK.(> 8000 RADS)

CIVILIAN DEATHS FROM BLAST (>0.14 ATMOSPHERES PEAK OVERPRESSURE)

| 0 | | 1 | | 2 | | 3km. |
| 0 | | 0.5 | | 1 | | 2 mi. |

Fig. I.9   EFFECTS OF A ONE KILOTON NEUTRON WARHEAD

When a nuclear explosion occurs within a few hundred meters of the ground, many of the neutrons will be absorbed in the soil converting some atoms into radioactive species. It appears that the most important of these species are Al-28 (0.038 hour half-life), Mn-56 (2.58 hour half-life), and Na-24 (15 hour half-life) [8]. The resulting "neutron-induced radiation field" can be of significant intensity. Neutrons from the 74 kiloton "Hood" shot, set off on 5 July 1957 at a height of approximately 450 meters above the Nevada desert (high enough to avoid local fallout), induced a radiation field which had an intensity of over 100 gamma-rads per hour at ground zero one hour after the explosion (see Fig. I.10 [9]).

Gamma Rays: Gamma rays are about as numerous, carry about the same amount of energy and have about the same mean free path in air and water as neutrons. They therefore give comparable radiation doses to neutrons for warheads with yields below about 100 kilotons. For higher yield warheads, however, the large low density fireball allows the gamma rays emitted after the expansion of the fireball to travel

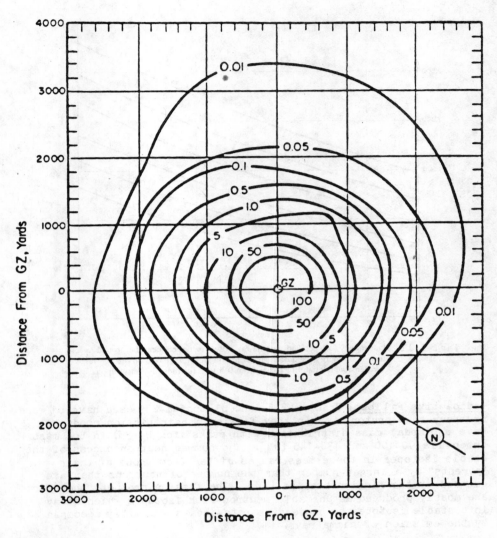

Figure I.10 Dose contours from neutron induced radioactivity in rads/hr one hour after the Hood explosion (74 kt at 450 meters).

out to greater radii with the result that lethal gamma ray doses extend to a significantly larger radius than lethal neutron doses. (See Fig. I.11 [3, Fig. 8.33].) The lethal radius of blast and heat effects would be still larger, however. At 3 km. from a one megaton ground-burst where the gamma dose drops through the threshold for lethality (200 rad), for example, the peak overpressure would be more than double the 0.35 atmospheres (5 psi) which defined the approximate radius of complete destruction at Hiroshima.

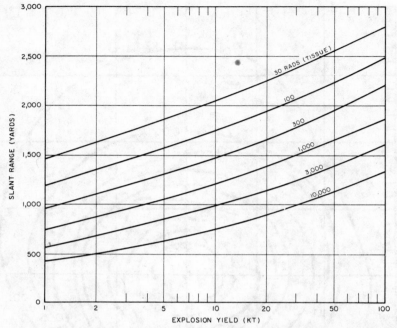

Fig. I.11 Slant ranges for specified gamma-ray doses for targets near the ground as a function of energy yield of air-burst fission weapons, based on 0.9 sea-level air density. (Reliability factor from 0.5 to 2 for most fission weapons.)

Radioactive Fallout: Uranium-235 and plutonium-239 each have 51 more neutrons than protons. When they fission, their fission products have a bimodal mass distribution with peaks (for U-235) in the mass intervals 90-100 and 130-145 [10]. The summed neutron excess of the stable isotopes in these mass ranges differ from those of their "parents" by a larger number than the number of neutrons that are emitted by the fission process [10]. The fission products therefore are mostly produced as unstable neutron-rich isotopes which decay into stable isotopes by successive beta-decays - usually accompanied by the emission of gamma rays.

About one hour after a nuclear explosion, the fission product residue is emitting gamma-rays with an average energy of about 0.7 MeV at a rate of about 530 gamma-megacuries per kiloton of fission yield [3, p.453]. The decline in the intensity of this radioactivity with time (t), as the shorter-lived fission products decay away, can be well approximated out to about six months by:

Gamma emission rate

$$= 2*10^{19}*t^{-1.2} \text{ gammas/sec/fission-kiloton.} \quad (I.9)$$

Here (t) is measured in hours (see Fig. I.12 [3, Fig. 9.16a]).

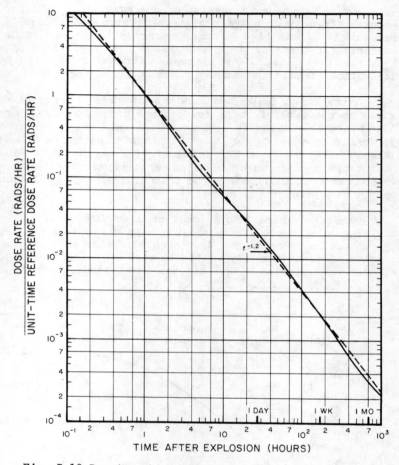

Fig. I.12 Dependence of dose rate from early fallout upon time after explosion.

If the fission products from a one kiloton fission weapon were spread uniformly over a square kilometer, one hour after the explosion, the flux of gammas integrated over all directions at a distance (H) above the surface would be

gamma flux

$$= -10^9*\text{Ei}(-H/L) \longrightarrow 10^9*\ln[L/(1.78*H)] \text{ gammas/sec/cm}^2, \quad (I.10)$$
$$(\text{when } H/L \ll 1)$$

where

$$\text{Ei}(x) = -\int_x^\infty dt*[\exp(-t)/t]$$

and L is the gamma energy absorption length in air, L = 32 gm/cm^2.

For an atmospheric density of one kg/m^3, L = 320m for gamma rays emitted directly from the nuclear explosion [2]. Since L is relatively insensitive to gamma energy [11, Fig. 9.1] and, when expressed in terms of gm/cm^3, is about the same for water and air [11, Fig. 2.11], Eq. I.10 will be used below to describe the energy absorption of fallout gammas in both air and the human body.

Using Eqs. I.9 and I.10, and a mean gamma energy of 0.7 MeV. gives an estimate of the average gamma dose rate of 6500 rads/hr one hour after the explosion at a point one meter above an area covered uniformly to a density of one kiloton of fission products per km.^{2} This is quite close to the value of 7200 rads/hr given in [3, p.454]. Radioactivity induced by non-fission absorption of neutrons in the weapons materials increases the intensity of this radiation field by only a few percent [3, p.454].

Fallout is more than a local problem because of the buoyancy of the huge hole in the atmosphere which is created by the heat released by the nuclear explosion. As a result of this buoyancy, most of the fission products from any explosion with a yield of over about 40 kilotons are carried within minutes to altitudes over 10 km [12] (see Fig. I.13 [3, Fig. 2.12]).

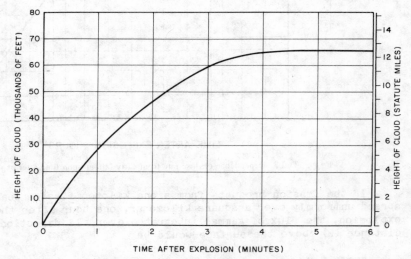

Fig. I.13   Height of cloud top above burst height at various times after a 1-megaton explosion for a moderately low air burst.

If the height-of-burst (H$_B$) is low enough so that the fireball touches the ground [3, p.71],

$$H_B < 0.87*W^{0.4} \text{ km,} \qquad\qquad (I.11)$$

then a great deal of surface material will be sucked up with the fireball and become contaminated with fission products. Much of this material will subsequently coalesce into particles large enough so that they will return to the earth again with a fallout time averaging on the order of ten hours.

During this time (t), the particles will be transported downwind by a distance

$$\underline{y} = \underline{W}_F * t \qquad\qquad (I.12)$$

where W_F is the (vector) time average of the winds experienced by the particle during its fall. Most of the particles are small enough so that their terminal velocity can be calculated by Stoke's Law which predicts that the terminal velocity is relatively insensitive to the density of the air and therefore the altitude. This is consistent with the approximate linearity of the curves in Fig. I.14 [3, Fig. 9.164] that show the time of fall as a function of initial altitude and particle diameter. The effective "fallout wind," W_F is therefore approximately the vector average of the wind velocities between the surface and the initial height of the cloud of fallout particles.

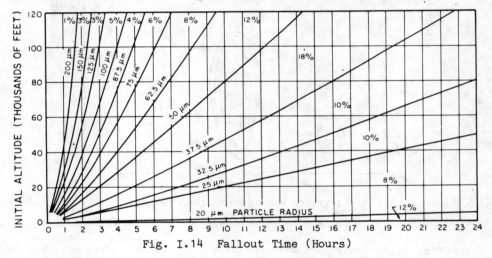

Fig. I.14 Fallout Time (Hours)

The width of the downwind fallout pattern close to ground zero is determined by the initial radius of the cloud of fallout. Further downwind, however, it is increased by "wind shear" as the different wind directions at different altitudes give particles which start at different heights in the cloud different average wind histories. The effect of wind shear is largest for the particles which fall most slowly, i.e. the particles which are carried furthest downwind. This is the principal reason why the fallout pattern broadens as one moves downwind from ground zero (see Fig. I.15 [3, Fig. 9.86a]).

In Appendix B, we give some of the basic formulae which are used in the "WSEG-10" general purpose fallout model originally developed in 1959 by the US Department of Defense's Weapons System Evaluation Group.[13]

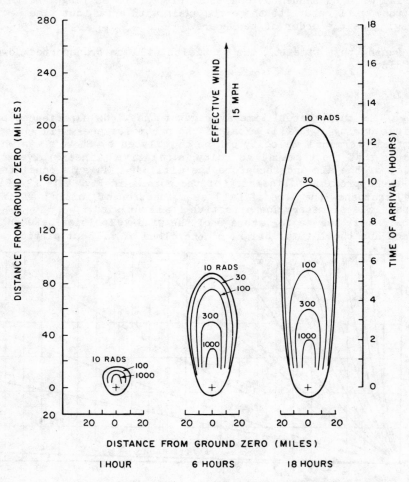

Fig. I.15   Total-dose contours from early fallout at 1, 6, and 18 hours after a surface
burst with a total yield of 2 megatons and 1-megaton fission yield (15 mph
effective wind speed).

II. <u>Effects</u>

Fortunately, the only experience of the intentional use of
nuclear weapons against human beings is still the original use at
Hiroshima and Nagasaki. The consequences of these uses were so
horrible, however, that the resulting worldwide revulsion against
nuclear weapons may be the principal reason why nuclear weapons have
not been used since.

The Hiroshima bomb had a yield of about 15 kilotons and the

Nagasaki bomb about 20 [14], [3, pp.36-37]. Both were exploded at an altitude of about 0.5 km. According to Eq. I.13, this height-of-burst was high enough to prevent significant local fallout. The one hundred thousand or so fatalities at each city [15, Table 10.11] were therefore due to prompt nuclear radiation, thermal radiation, and blast. We will discuss principally the consequences at Hiroshima, because that city is quite flat. Interpretation of the consequences at Nagasaki is complicated by the fact that there the bomb was exploded over a quite deep valley.

Radiation Sickness: Fig. II.1 [7] shows the approximate percentages of an irradiated population which will suffer nonfatal and fatal radiation illness as a function of absorbed dose.

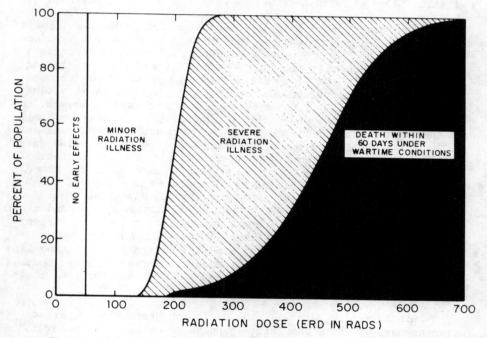

Figure II.1 Probability of Radiation Illness or Death

Because of the relatively large amount of material surrounding the nuclear cores of the Hiroshima and Nagasaki weapons [1], the neutron doses which they delivered were much smaller than their gamma doses [14]. At Hiroshima, the gamma dose was above 400 rads out to about 1 km from ground zero and therefore lethal to a large fraction of the population within that distance [14].

The effects of this gamma radiation was clearly manifested in the fate of a group of 168 workmen from villages outside Hiroshima who were standing outdoors about one km from ground zero at the time of the explosion. Six of these workmen were crushed by collapsing

buildings and three received fatal burns from exposure to the
fireball. The remaining 159 were shielded from the direct thermal
radiation of the fireball and were able to return by foot to their
homes 25 km away. Ninety-three (about 60 percent) of these
"survivors" died, however, of radiation illness 20-38 days after the
explosion [16, pp. 32-33]. Others as close as 0.25 survived because
of the radiation shielding afforded by reinforced concrete buildings.

The critical organ of the body least able to withstand the
effects of ionizing radiation is the blood-cell-producing bone
marrow. Consequently, in the lowest dose range where radiation
illness occurs, its most important symptoms and characteristic times
are associated with the decline in the populations of various types
of blood cells. Ref. 3 [pp. 583-585] describes the development
radiation illness in this (200-1000 rad) dose range as follows:

> "The initial symptoms are...nausea, vomiting, diarrhea, loss of
> appetite and malaise...After the first day or two the symptoms
> disappear and there may be a latent period of several days to 2
> weeks...Subsequently there is a return of symptoms, including
> diarrhea and a steplike rise in temperature which may be due to
> accompanying infection...Commencing about 2 to 3 weeks after
> exposure, there is a tendency to bleed into various organs, and
> small hemorrhages under the skin...are observed...Particularly
> common are spontaneous bleeding in the mouth and from the lining
> of the intestinal tract. There may be blood in the urine due to
> bleeding in the kidney...[These effects are due to] depletion of
> the platelets in the blood, resulting in defects in the blood-
> clotting mechanism...Loss of hair, which is a prominent
> consequence of radiation exposure, also starts after about 2
> weeks... Susceptibility to infection of wounds, burns, and other
> lesions, can be a serious complicating factor [due to] loss of
> the white blood cells and a marked depression in the body's
> immunological processes. For example, ulceration about the lips
> may commence after the latent period and spread from the mouth
> through the entire gastrointestinal tract in the terminal stage
> of the sickness..."

At Hiroshima and Nagasaki, the radiation doses were received
instantaneously. For a population exposed to radioactive fallout,
the doses would accumulate over a significant period of time during
which some offsetting biological repair could take place. The
ordinary assumption made concerning the efficiency of this repair is
that 90 percent of the damage done by ionizing radiation is
repairable and will be repaired in the surviving population with an
exponential time constant of 30 days (720 = 1/0.00139 hours) [13].
This results in the following prescription for the "equivalent
residual dose" (ERD) at any given time T:

$$ERD = R_{H+1} * \int_{T_A}^{T} dt * t^{-1.2} * [0.1 + 0.9 * \exp(-0.00139 * (T-t))]. \quad (II.1)$$

Here R_{H+1} would be the dose rate at the specified location one hour after the nuclear explosion(s) if all the fallout had arrived by then and T_A is the actual average time of arrival of the fallout. T_a, t, and T are all measured in hours. Of course, in a "real life" situation, one would have to put a time-dependent shielding factor in Eq. II.1 to take into account the fact that people would be moving around. In the absence of such a time-dependent shielding factor, the equivalent residual dose would peak at about 5 days, if the fallout arrived one hour after the explosion, and at about 11 days, if the fallout arrived after twenty hours.

Radiation illness is only the most immediate effect of radiation exposure. In the longer term, the Hiroshima-Nagasaki survivors have suffered, among other adverse effects: an increased incidence of cancers (a few percent excess per 100 rads dose) and chromosomal defects, and a high incidence of microcephaly (lack of full brain development) among offspring who were in utero at the time of the radiation exposure [15].

Burns: Another well-documented population in Hiroshima consisted of 17,000 students - 4000 outside working on fire-breaks and 13,000 inside wooden school buildings at the time of the explosion. Fig. II.2 [16, Fig. 3.15] shows that the mortality of those indoors dropped sharply beyond about one km from ground zero - approximately the distance at which the nuclear radiation dose dropped to nonlethal levels. The mortality rate for those outdoors stayed very high out to distances of about 2 km, however - apparently because of flash burns resulting from their direct exposure to the explosion.

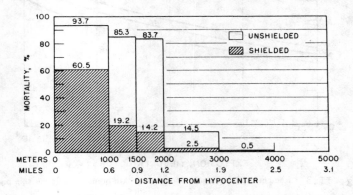

Fig. II.2 Mortality in shielded and unshielded students in relation to distance, Hiroshima.

According to Eq. 1.2, if one sets the fraction of the explosive energy which goes into thermal radiation (f_T) equal to 0.35 and assumes an 85 % atmospheric transmittance (T_A), a 15 kiloton explosion will deliver about 10 cal/cm^2 at a distance of 2 km. According to Fig. II.3 [3, Fig. 12.64], this is enough to deliver third degree burns to uncovered skin. It is also enough to ignite "black cotton rags" [3, Table 7.40].

20

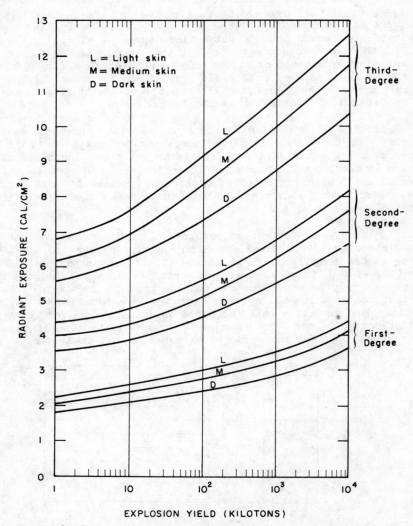

L = Light skin
M = Medium skin
D = Dark skin

RADIANT EXPOSURE (CAL/CM²)

EXPLOSION YIELD (KILOTONS)

Fig. II.3   Radiant exposure required to produce skin burns for different skin pigmentations.

Blast Injuries: The peak blast overpressure one km. from ground zero at Hiroshima would have been almost one atmosphere (see Fig. I.6). By 2 km the peak overpressure would have dropped to about 0.2 atmospheres (3psi). Fig. II.4 [16, Fig. 3.20] shows that, even at 4 km, where the peak overpressure was on the order of 0.07 atmospheres (1 psi), there was a significant incidence of blast-related injuries among those alive 20 days after the explosion. According to [16],

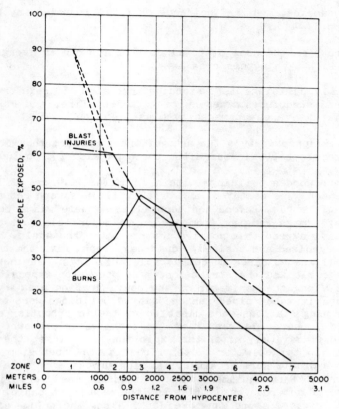

Fig. II.4 Incidence of blast injuries and burns by distance, Hiroshima. ----, probable incidence in the innermost zones, assuming that all those killed were injured by blast and radiant heat. See note in legend for Fig. 3.16.

"direct blast injuries were few... ...Almost all injuries due to blast were of the indirect type, being caused by building collapse or flying missiles...Blast injuries are least frequent among people in the open, well away from buildings and walls, which are likely to be more of a risk than a protection, especially at close range...Even in well-built concrete and steel frame buildings, which retain their structural integrity, most of the occupants may be injured by flying debris or falling partitions or ceilings. In buildings with a large amount of window space and plaster ceilings and walls, lacerations and contusions may be very numerous. In contrast, when brick or wooden buildings collapse, fatal wounds from beams, roof tiles, and bricks are frequent; also many occupants are pinned in the wreckage and burned to death...Most of the fatal injuries from blast may be expected to occur in the area of complete destruction. Beyond this area, however, there may be a large region in which a great number of buildings are destroyed or badly damaged, and the incidence of fatal injuries may be high [16, pp.37-41]."

At Hiroshima the area of total destruction had an average radius of about 2 km. Within this area

"the only buildings left standing were about 50 reinforced-concrete structures...These were gutted by fire, and most of them were severely damaged by blast [16, pp.14-15]."

This was approximately the area of the firestorm that developed in Hiroshima about 20 minutes after the explosion [3, pp.300-304].

"All the wooden buildings within 1.5 miles [2.4 km] of the hypocenter were destoyed, and brick buildings were reduced to piles of rubble...Beyond the central area there was a very large zone of damage ranging from severe to superficial...It is generally agreed that no fire department in the world, however well organized and trained, could have coped with the thousands of simultaneous fires in combustible buildings. Furthermore wreckage blocked the streets and made operation impossible except along the perimeter of the conflagration area. Innumerable water pipes inside damaged buildings were broken, and pumping stations were disabled by the interruption of electric power. The result was that the water pressure dropped to zero immediately after the explosion, and it was therefore impossible to get water to the fires in the upper stories. Street hydrants were broken by falling wreckage or were buried in debris. Hand equipment in buildings was used very little because of numerous casualties and the resultant panic, but bucket brigades saved many dwellings along the fringe of the conflagration area. Fire walls and doors did little to prevent firespread within buildings. Interior fire walls were blown down or damaged. Many fire doors were open at the time of the explosion, and many of those that were closed were bent or displaced. Fire shutters were blown off by the blast wave, exposing interiors to heat and burning debris from nearby buildings [16, pp. 17-20]."

Fires were ignited throughout Hiroshima by both the thermal radiation from the fireball and as a result of blast effects creating other ignition sources - especially by turning over cooking stoves.

Equivalent Areas of Death and Casualties: Fig. II.5 [16, Fig. 3.10] shows the estimated total mortality and casualty rates as a function of distance from ground zero at Hiroshima. By integrating under the curves, one obtains circular "equivalent areas of death and casualties" with radii of 1.5 and 2.5 km respectively (areas of 6.6 and 22 km^2 respectively).

If one makes the "cookie cutter" approximation that there is a 100 percent fatality rate inside a certain radius and that there are no fatalities further out, the equivalent areas of death and casualties correspond to the areas subjected to peak overpressures greater than 0.35 atmospheres (5 psi) and 0.14 atmospheres (2 psi) respectively. This fact is often used to extrapolate the equivalent areas of death and casualties to higher yields. This is a very rough approximation, however, since the casualties from flash burns and fire would not scale with distance from ground zero and height of burst in exactly the same way as those from blast. At lower yields the scaling breaks down entirely as the effects of direct nuclear radiation become increasingly dominant.

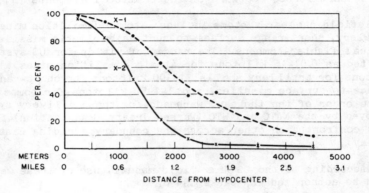

Fig. II.5 Relation of mortality and casualty rates to distance, Hiroshima. ●, total casualty rate. X, mortality rate. At 0.62 mile from hypocenter: X-1, mortality from burns among people in the open; X-2, mortality from ionizing radiation among people with shielding.

III. Nuclear War

In theory, at least, possible nuclear wars range from the use of only one or a few warheads in a Vietnam-style war against a nonnuclear weapons state to the deliberate attempt by the superpowers to destroy each other as completely as possible - what Herman Kahn has called "Wargasm." Below, we discuss the civilian fatalities which would result from the use of nuclear weapons for several cases which have been analysed in some depth:

o Battlefield use in the Germanies (East and West) against conventional forces such as tanks;

o Use on the "theatre" level - once again in the Germanies - against medium- and intermediate-range nuclear delivery systems and nuclear warhead storage depots;

o Use against missile silos, bomber bases, nuclear naval support facilities, and command and control systems in the US; and

o Use against the cities of the superpowers.

<u>Nuclear</u> <u>Weapons</u> <u>on</u> <u>the</u> <u>Battlefield</u>: Any major conflict on the
conventional level between the superpowers would bring with it the
danger that the losing side would resort to battlefield nuclear
weapons. In Europe, this threat of "escalation" is implicit in the
deployment of thousands of delivery systems designed for use against

> "committed enemy units, reserves, lead elements of second
> echelon forces, enemy nuclear systems, field artillery, air
> defense artillery, selected command and control elements, [and]
> support forces rearward of the committed elements [17]."

Battlefield nuclear systems include: atomic demolition mines, nuclear
artillery, short range surface-to-surface ballistic missiles, and
tactical fighter-bombers. The ranges of yields for US systems are
reported as 0.01-15 kilotons for atomic demolition mines, 1-12
kilotons for artillery shells, 1-100 kilotons for short-range
surface-to-surface missiles, and 1-1450 kilotons for bombs [1].
On the order of ten thousand warheads for these delivery systems are
deployed by the NATO and WTO (Warsaw Treaty Organization) forces
which confront each other across the boundary dividing East and West
Germany.

According to the US Army field manual, use of these weapons
would be authorized in "packages" of

> "nuclear weapons of <u>specific</u> <u>yields</u> for employment in a
> <u>specified</u> area, within a <u>limited</u> <u>timeframe</u> to support a <u>tactical</u>
> <u>contingency</u>. Sufficient nuclear weapons should be planned in
> each package to alter the tactical situation decisively, and to
> insure accomplishment of the the assigned mission [17, emphasis
> in original]."

The field manual gives as an example a package consisting of: 2
atomic demolition mines (ADM), 30 rounds of nuclear artillery, 10
surface-to-surface missiles, and 5 air-delivered bombs (see Figure
III.1 [17]).

Efforts would, of course, be made in planning and targeting such
a package to minimize "collateral damage" to populated areas. This
would be very difficult, however, in an area as densely populated as
central Europe. The average population density of the two Germanies
is 200 persons per square kilometer and they contain an average of
one populated place per four square kilometers (hence the well known
lament of nuclear weapons officers: "The towns and villages in
Germany are only one to two kilotons apart [7])." Additional
difficulties would flow from the fact that roads naturally pass
through towns and, therefore, so would many of the military units
that would be the targets of battlefield nuclear weapons. In many
cases, the roads between towns would be crowded by fleeing civilians.
Finally, attacking military forces might be expected to use urban and
refugee "hugging" tactics so as to discourage the use of battlefield
nuclear weapons against themselves.

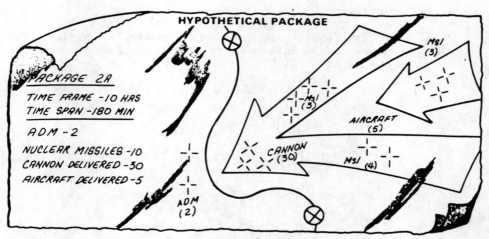

EMPLOYMENT

Figure III.1 An Hypothetical "Package" of Battlefield Nuclear Weapons (from the US Army Field Manual).

In 1955 NATO conducted a war game, "Carte Blanche," which resulted in the simulated use of 335 nuclear weapons (80 percent in West Germany) over a period of two days. Since the yield of the typical "battlefield" nuclear weapon at the time was of the same order as that of the Hiroshima bomb, it is not surprising that "prompt" West German casualties (those due to blast and thermal radiation) were estimated at 1.5-1.7 million dead and 3.5 million wounded [18]. Even a "small" one kiloton yield neutron bomb, exploded over an area of average population density in the Germanies would cause about 1000 deaths due to radiation illness, however, (see Fig. I.9) and it would require more than one thousand such explosions, resulting in on the order of a million deaths, to immobilize a significant fraction of the 20,000 tanks which might be involved in a full-scale battle between NATO and WTO forces in the Germanies [7].

Unfortunately, it is not clear that nuclear weapons planners understand the horrendous carnage which would result from the use of even the lowest yield nuclear warheads on the battlefield. According to Paul Braken of the Hudson Institute, much of their planning is based on computerized war games which predict thousands not millions of of civilian fatalities from nuclear war on the battlefield. According the Bracken, the computer programs assume that there are no refugees on the roads and

"treat Soviet forces as automata who cross into West Germany and advance directly into [unpopulated] NATO nuclear killing zones. Here they are detected and destroyed by the lowest yield nuclear weapon capable of doing the job. Command and control difficulties, confusion, false targeting and other problems are simply assumed away...

As Bracken points out:

> "What actually prevents either side from getting too close to the 4000 towns and cities in Germany is a collection of Fortran statements [18]."

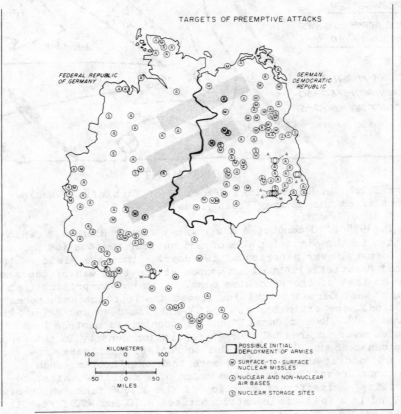

TARGETS OF PREEMPTIVE ATTACKS

FEDERAL REPUBLIC
OF GERMANY

GERMAN
DEMOCRATIC
REPUBLIC

KILOMETERS
100   0   100

50   0   50
MILES

☐ POSSIBLE INITIAL
  DEPLOYMENT OF ARMIES

Ⓜ SURFACE-TO-SURFACE
  NUCLEAR MISSLES

Ⓐ NUCLEAR AND NON-NUCLEAR
  AIR BASES

Ⓢ NUCLEAR STORAGE SITES

Fig. III.2   The locations of 92 military airbases in the Germanies are indicated by the circles containing the symbol "A", 56-surface-to-surface missile sites by a circle with an "M", and 24 known and inferred nuclear weapons storage depots by a circle with an "S". These are the assumed targets of the preemptive attacks discussed in the text. The area of each circle is 180 square kilometers, the estimated equivalent "area of death" below a 200 kiloton warhead exploded at an altitude of 2 kilometers.
The shaded rectangles show the sizes of the areas which might be covered in a confrontation along the border by three Soviet Armies and the opposing NATO forces. Battlefield nuclear weapons would be used against the forces in such areas.

Regional: It is difficult to believe that, if the circumstances appeared so desperate as to require the use of nuclear weapons, the US and USSR could subsequently successfully cooperate to limit their use short of all-out nuclear war. Indeed, a whole family of "medium" and "intermediate range" nuclear weapons have been deployed in Europe to make more credible the danger that nuclear "escalation" would occur. There are approximately 2000 warheads on land- and submarine-based missiles with ranges greater than 150 km in and around Europe plus an estimated 2500 nuclear-capable fighter-bombers and medium-

range bombers.   The warheads for these systems range in yield
from about one kiloton to more than one megaton [7].

These "theatre" nuclear weapons delivery systems have among
their designated targets:

"IRBM/MRBM [Intermediate and Medium Range Ballistic Missile]
sites; naval bases; nuclear and chemical storage sites;
airbases; command, control, and communication centers;
headquarters complexes; surface-to-air missile sites; munitions
and petroleum storage areas and transfer facilities; ground
forces installations; choke points; troop concentrations; and
bridges [19]."

This list corresponds to over one thousand potential targets in the
Germanies alone [7].

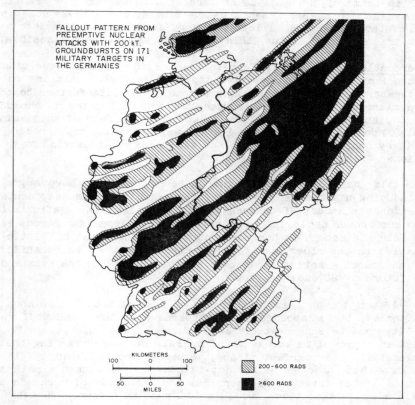

FALLOUT PATTERN FROM
PREEMPTIVE NUCLEAR
ATTACKS WITH 200 kT.
GROUNDBURSTS ON 171
MILITARY TARGETS IN
THE GERMANIES

KILOMETERS
100   0   100

50   0   50
MILES

200-600 RADS

>600 RADS

Figure III.3 The fallout pattern, given "typical June winds", for an
attack with surface-burst 200 kiloton warheads against the targets
shown in Fig. III.2.

Both sides would attach the highest priority to destroying as
many of each other's nuclear weapons as possible before they could be

used. In [7], therefore, a scenario was considered in which 200 kiloton warheads carried by intermediate range delivery vehicles were targeted against 171 surface-to-surface missile sites, military airbases and nuclear weapons storage depots in East and West Germany (see fig. III.2 [7]). The resulting civilian casualties in the two Germanies were estimated to range from a low value of 2 million deaths (7 million total casualties) to a high value of about 10 million deaths (25 million total casualties). The low values were obtained by assuming that all the warheads were exploded at an altitude of 2 km - too high to cause local fallout - while the high value assumed attacks with both an air-burst and a ground-burst on each target. The combined population of the two Germanies is 76 million. Figure III.3 [7] shows the projected fallout pattern from the groundbursts using "typical June winds."

The attacks envisioned in this scenario are very restrained - both in terms of the classes of targets attacked and the fraction of the available nuclear arsenal used. Because the attacks would put all of the land-based nuclear delivery systems in Europe into a "use them or lose them" status, this restraint is not very plausible.

Intercontinental - Counterforce: Leaders of both superpowers have stated (and the US has reassured its NATO partners) that the use of theater nuclear weapons would probably escalate further to the use of nuclear weapons against "strategic" targets located in the US and the Soviet Union. Many believe that it is the fear of triggering a global nuclear conflagration in this way which has "deterred" military conflicts between the two alliances of industrialized nations during the nuclear era.

This "balance of terror" cannot be relied upon, however, without it inducing nightmares. Recently, for example, there has been a considerable amount of concern about the destabilizing effects of each superpower deploying ballistic missile warheads numerous enough and accurate enough to destroy most of the land-based ballistic missiles on the other side. Indeed, the "window of vulnerability" of US land-based missiles has become a rallying cry in the debate over the future of US nuclear weapons policy.

Although even a small increase of risk of catastrophe should not be ignored, it is important that concerns over the "window of vulnerability" not be exaggerated. Even if one of the superpowers could mount an attack with thousands of nuclear warheads reaching their targets with near perfect timing, accuracy, and reliability, it could not know that the other would not launch its threatened missiles on warning of the attack. Furthermore, the thousands of warheads that would survive the attack, would still pose an enormous threat.

The submarines which the US has hidden under the ocean at any one time, for example, by themselves carry over 2500 warheads which could cover an area of over 50,000 km^2 with peak overpressures in excess of 0.35 atmospheres (5psi) [20]. (Figs. III.4 [3, Figs. 5.55, 5.57] show a before and after sequence of an ordinary house exposed

Fig. III.4a Wood-frame house before a nuclear explosion, Nevada Test Site.

Fig. III.4b Wood-frame house after a nuclear explosion (5 psi peak overpressure).

to a blast wave with 0.35 atmospheres peak overpressure.) This is three times the area of the land on which the Soviet urban population (50 percent of the total Soviet population) lives (see Fig. III.5 [21, Fig. 1]). Perhaps the greatest danger associated with the "window of vulnerability" is that constant discussions of counterforce strategies by nuclear policy makers will make such attacks more thinkable.

30

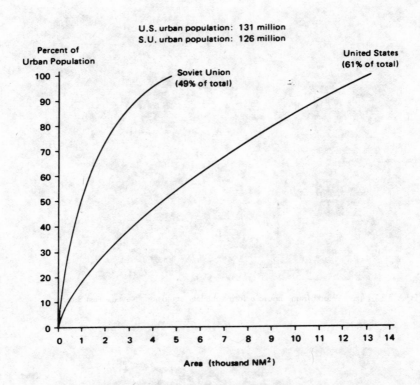

Figure III.5 Cumulative Soviet and US Populations as a function of urban land area.

Nevertheless, if a strategic nuclear "exchange" between the superpowers should occur, it would be directed, at least initially, at each side's instruments for waging war. While such targets include conventional military installations, such as ports, airfields, barracks, and depots as well as nuclear installations, the most "time urgent" targets would be the other side's strategic nuclear forces and their associated early warning, command, communications, and strategic defense facilities.

Attacks on ICBM Fields: The bulk of the warheads involved in an attack against US strategic nuclear forces would be thrown against the US land-based intercontinental ballistic missile (ICBM) force which consists of 1000 Minutemen missiles, several dozen Titan II missiles and their associated local launch-control facilities (one for each ten Minuteman missiles and one for each few Titan missiles. These missiles and launch-control facilities are protected in hardened, underground capsules that are widely distributed on the US Great Plains in six major and three minor missile fields.

To have any reasonable chance of destroying a high percentage (better than 90 percent) of the US ICBM force, it would be necessary to use two or more accurate ICBM warheads against each US missile silo. The only Soviet warheads which are both sufficiently accurate and numerous to accomplish such an attack are the multiple warheads on the Soviet SS-17, -18, and -19 missiles. These warheads have yields which are commonly estimated as falling in the range between 500 and 1000 kt.

The launch-control facilities could be targeted by similar or higher yield warheads. It appears that in the past US ICBM launch control facilities may have been targeted with the single large (estimated 10-20 megaton yield) warheads carried by SS-9 missiles [22]. Since the SS-18 missile which has replaced the SS-9 has been tested with a similar large warhead, they may still be so targeted. In that case the total megatonnage exploded over the launch-control centers in a counterforce attack could be greater than that exploded over the ten times more numerous missile silos.

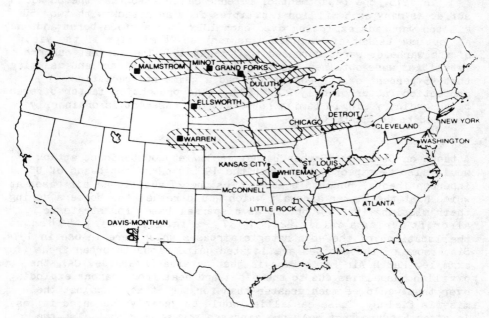

FATALITIES

- ▨ GREATER THAN 50%, INDOORS ABOVE GROUND
- ▨ GREATER THEN 50%, OUTDOORS
- ■ MINUTEMAN FIELDS
- □ TITAN FIELDS

ASSUMPTIONS

- o 2-ONE MEGATON WARHEADS ON EACH SILO
- o 50% FISSION YIELD
- o SURFACE BURSTS
- o TYPICAL MARCH WINDS

Fig. III.6 Fallout from an attack on US missile silos.

Since US missile fields are located in relatively sparsely populated areas, the blast and heat of Soviet warheads exploding over them would cause relatively few civilian casualties. The missile silos and launch-control centers are so hardened, however, that, in order to maximize the likelihood of their destruction, at least one warhead would have to be surface-burst at each silo. As a result, hundreds of thousands of square miles, including densely populated urban areas, would be covered with lethal levels of radioactive fallout. Fig. III.6 (adapted from [23, p. 51]) shows that, given "typical March winds," the overlap of the fallout patterns from individual missile silos would result in the areas of lethal fallout extending to a distance on the order of one thousand miles downwind from the Minuteman fields. Within the shaded area doses indoors (assumed to be one third of those outdoors) would exceed 450 rads. According to Fig. II.1, such a dose could be expected to result in a fatality rate from radiation illness on the order of 50 percent.

In 1975, the Department of Defense calculated that the US would suffer as many as 5 million fatalities from an attack with two 550 kiloton warheads exploding over each ICBM (one surface-burst and one at "optimum height-of-burst) and as many as 18 million if the yield of the warheads were increased to 3 megatons. The largest numbers of fatalities were obtained assuming "typical March winds" and assuming that 45 percent of the population had a gamma ray radiation protection factor of only 2.25 [23]. (The protection factor offered by an ordinary single family house without basement is ordinarily assumed to be about 3.)

Attacks on Strategic Bomber Bases: A Soviet counterforce attack would also be expected to target the 19 [1] permanent bases of US intercontinental nuclear bombers plus the 27 additional air bases at which host the tanker aircraft which would refuel the bombers during their missions (9 [24]) or act as dispersal bases for strategic aircraft during a crisis.[22, p. 137] Fig. III.7 [23, p. 50] shows the locations of the 46 strategic airbases which were in use in 1974. Since some of these bases are located quite close to urban areas (for example, March Air Force Base is just outside Riverside, CA), the civilian casualties due to the blast and heat from weapons exploding over them would be much greater than for the attacks against the missile fields. These casualties would be greatly increased if, as is often assumed, not only the airbases themselves, but also the areas around them were barraged in order to try to destroy in the air as many as possible of the bombers and tankers which had taken off on warning of attack. In the DOD's 1975 analysis, for example, it was assumed that "pattern attacks" would destroy all aircraft flying within 8 nautical miles of the 46 Strategic Air Command bases [23].

Attacks on Submarine Bases: In peacetime, nearly half of US ballistic missile submarines and therefore over 2000 US strategic warheads are in four ports: Charleston, South Carolina; Kings Bay,

U.S. TARGET STRUCTURE

Fig. III.7 SAC bomber, tanker, and dispersal bases (1975).

Georgia; Bremerton, Washington; and Groton, Connecticut.[1] Other
potential counterforce targets would be US naval bases hosting other
naval ships carrying nuclear weapons - many designed for attack
against Soviet submarines (e.g. weapons carried on attack submarines)
or that could be used for attacking the Soviet Union itself (e.g.
bombs carried by aircraft carriers and sea-launched cruise missiles).
This would increase the counterforce target list by six naval bases
in the US and five abroad [1]. Attacks on most of these bases would
result in substantial numbers of fatalities in nearby urban areas.

Attacks Against Strategic Warning, Communications, Command, and
Defense Facilities: In order to minimize the number of surviving
bombers and the chances of launch-on-warning of US ballistic
missiles, the highest priority targets of a Soviet counterforce
attack would be US early warning systems, the control centers which
would issue the instructions for US nuclear weapons use, and the
communications attennae which would transmit these instructions.
Presumably, an attempt would also be made to destroy US strategic
defensive systems. As a result of such considerations, [22] lists,
in addition to nuclear delivery systems, the following "nuclear
threat targets" for Soviet intercontinental forces:

o 60 National Command Authority Centers;

o 5 airbases for airborne command posts;

o 60 transmitters for communicating with ballistic missile
   submarines;

o 132 radars;

o 28 fighter-interceptor sites; and

o 1 ABM test site.

Ref. [25] suggests that, in addition, a number of transmitter links for communication to early warning, navigational and military communication and meteorological satellites would be targeted.

A forthcoming article [26] will present improved estimates of the casualties which would result from an all-out Soviet counterforce attack on the US. It is already evident from the above discussion, however, that the number of US deaths would probably number in the tens of millions. With such horrendous consequences, retaliation against urban/industrial targets would become quite credible.

Intercontinental - Counter-City and Industry:

Attacks on Soviet Cities: In 1968, Secretary of Defense McNamara tried to quantify the amount of destruction which could be done to the cities and industry of the Soviet Union by nuclear blast as a function of the "equivalent megatonnage"* of nuclear warheads used.[27]   An interpolation of McNamara's table gives the curves shown in Fig. III.8 [28]. No explanation was given about the derivation of these curves. In the case of population, however, it is instructive to compare the curve in Fig. III.8, showing the cumulative Soviet fatalities as a function of equivalent megatons used, with the curve in Fig. III.5, showing the cumulative Soviet urban population as a function of urban land area. One finds from Fig. III.5 that the 25 percent of the total Soviet population (50 percent of the total urban population) which lives in the most densely populated urban areas of the Soviet Union lives on about 1000 nm^2 or about 3500 km^2. According to Fig. III.8, this many people could be killed by 270 equivalent megatons. This gives an equivalent area of death of about 13 km^2 per equivalent megaton.

This equivalent area of death corresponds to the area of a circle approximately 2 km in radius. Fig. I.6b shows this to be the area which would be subjected to an overpressure greater than about 2 atmospheres (30 psi) for an airburst at optimum height-of-burst (about 1.5 atmospheres for a ground-burst). As was discussed in Section II, however, the "equivalent area of death" at Hiroshima was approximately equal to the area subjected to an peak overpressures

----

*Since the area subjected to more than a given peak blast overpressure scales as the two-thirds power of the megatonnage (Y) of a warhead, its megatonnage equivalent (MTE) is defined as:

$$MTE = (Y)^{2/3}.$$

greater than 0.35 atmospheres (5 psi). If this criterion had been used in the calculations done for McNamara, the megatonnages shown along the horizontal axis of Fig. III.8 would be lower by a factor of 5 (ground bursts) to 12 (optimum height of burst). Corrections due to the overlapping of circles of death with each other and with the edges of urban population areas would reduce these factors somewhat. The addition of fallout effects and secondary effects such as illness and starvation among the survivors of the direct effects of the nuclear explosions, would offset these reductions, however.

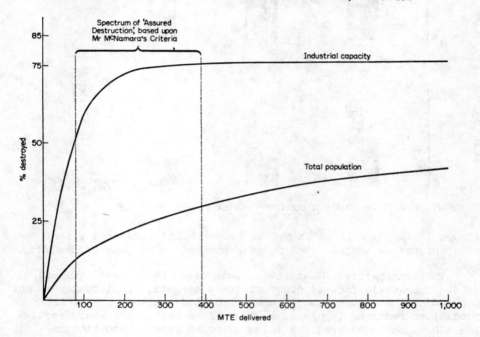

Fig. III.8 Soviet Population and Industrial Capacity Destroyed

Fig. III.9 [20] shows that in 1983, even after a first strike, both the US and USSR would have thousands of equivalent megatons in their nuclear arsenals. A comparison with Fig. III.8 shows that, even without taking into account the conservatism in the calculations done for McNamara, this explosive power is well into the "overkill" region for both the cities and industry of the USSR.

More recent studies have given this conclusion more detail. The Arms Control and Disarmament Agency (ACDA) concluded in 1978, for example, that US strategic bombers and ballistic missile submarines surviving after a Soviet first strike could destroy [subject to more than 0.7 atmospheres (10 psi)] 65 to 90 percent of "key Soviet production capacity" (primary metals, petroleum products, electric power generation, etc.) and would destroy 60 to 80 percent of the remaining (non-targeted) Soviet production capacity by "collateral damage [21]."

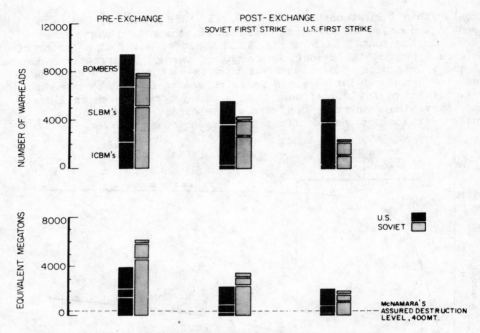

CALCULATED RESULTS OF STRATEGIC COUNTERFORCE EXCHANGES; 1982 FORCES
(BOTH SIDES ON GENERATED ALERT)

Figure III.9. Levels of US and Soviet overkill remaining after counterforce exchanges (1982 forces, both sides on generated alert).

The hypothetical US attack in this case was directed against Soviet "strategic forces, other military targets, and industry" - not population. Nevertheless, it was calculated that, if the Soviet population remained in place, 80-95 million fatalities would result. The ACDA also considered the value of prior urban evacuation in reducing Soviet fatalities and concluded that the number of fatalities could be reduced to 23-34 million if

"80% of the urban population evacuates the cities to ranges up to 150 km and the remaining 20 percent take protection in the best available shelters. The evacuated people are located with the rural population, and both the evacuees and rural people go to the best available rural shelters and build hasty shelters."

The report comments that

"This posture represents an immense civil defense effort and no analysis was made to determine the feasibility of implementing such a posture. "

The report also adds that, even with this civil defense posture, if "residual weapons [were used] to directly target the evacuated population," the number of Soviet fatalities could be increased back up to 54-65 million [21].

None of the ACDA fatality numbers include indirect and long-term deaths due to exposure, starvation, epidemics, etc. With the destruction of the Soviet economic base, it is likely that the death rate from all of these causes would be enormous.

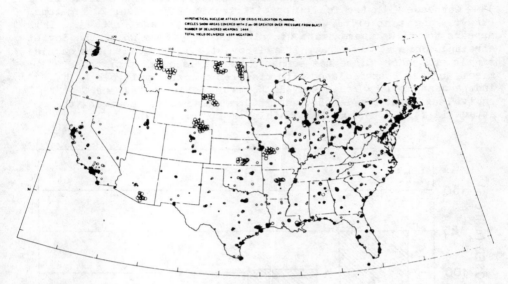

Fig. III.10 The Federal Emergency Management Administration's projection of the targets of an all-out Soviet attack on the US.

Attacks on the US: In a 1979 report, the US Federal Emergency Management Agency designated certain areas of the US as "high risk areas" for civil nuclear defense planning purposes [29]. According to this report:

"Potential target values were developed ... based on the following criteria listed in descending priority order:

a. U.S. military installations.

b. Military supporting industrial, transportation and logistics facilities.

c. Other basic industries and facilities which contribute significantly to the maintenance of the U.S. economy.

d. Population concentrations of 50,000 or greater...

[Then, after taking into account] projections of Soviet capabilities (circa 1980)...envelopes were plotted...to depict areas subject to a 50% or greater probability of receiving blast overpressures of 2 psi or more."

38

An apparently identical attack is discussed in a report published by Oak Ridge National Laboratory in 1976 (see Fig. III.10 [30]). There it is described as being associated with a specific attack scenario with the designation, CRP-2B, involving a total of 1444 warheads with the following distribution of yields: 20 megatons (241), 3 megatons (176), 2 megatons (184), and 1 megaton (843). This appears to be the approximate distribution of yields which the Soviet strategic arsenal would have if all Soviet ballistic missiles carried single warheads. Since the scenario was devised, many of these missiles have been replaced with missiles carrying multiple independent reentry vehicles. In terms of Soviet total and equivalent megatonnage (6560 and 3300 respectively), however, the attack is still physically possible (see Fig. III.9).

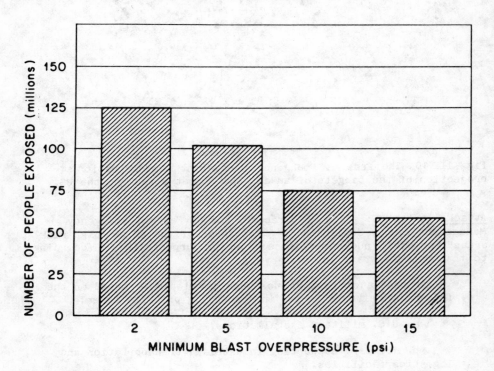

Fig. III.11 Exposure of the US population to blast - given the attack postulated by FEMA and assuming no evacuation.

Fig. III.11 [30, Fig. 3.2] shows the estimate in the Oak Ridge report of the distribution of overpressures to which the US population would be subject in the absence of urban evacuation. Fig. III.12 [30, Fig. 4.4] shows the corresponding distribution of radiation doses from fallout for an unsheltered population with and without urban evacuation. Here it has been assumed that 77 percent of the megatonnage in the attack is ground-burst on military and industrial targets and that the winds are blowing due east at 25 mph.

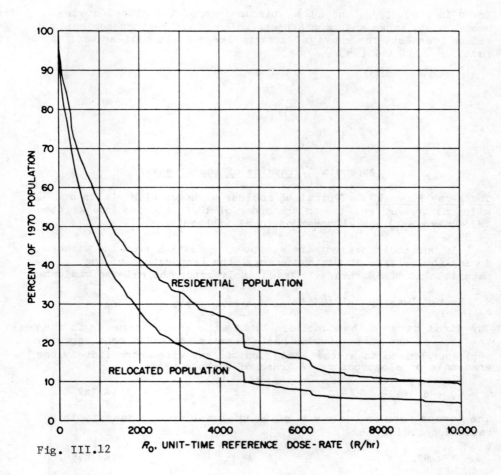

Fig. III.12

The "unit-time reference doses" shown in Fig. III.12 are of the same order as the peak equivalent residual doses which would be calculated by integrating the associated dose rate from the time of arrival of the fallout (see eqn. II.1). Either Fig. III.11 (in combination with our definition of "equivalent areas of death" in Section II) or Fig. III.12 (in combination with Fig. II.1) would therefore result in estimates of on the order of 100 million US dead from this hypothetical attack. In fact, the ACDA report estimated 105-131 million US fatalities from a comprehensive Soviet attack in the absence of urban evacuation and 69-91 million assuming urban evacuation [21].

Once again, fatalities due to the indirect effects of nuclear attack such as exposure, starvation, epidemics, cancer and the breakdown of the institutions of society have not been included in the numbers discussed here. Discussions of these effects can be

found in [30], [31], and [32]. Discussions of the global-scale
ozone, dust, smoke and smog effects which would further seriously
reduce prospects for survival after a large scale nuclear wars can be
found in [33] and [34].

Appendix A. Physics of the Fireball

The X-ray Fireball: Typical US nuclear warheads with yields of 100
kilotons or more release on the order of 10^{12} calories (one kiloton
TNT equivalent) per kilogram of total warhead mass.[1]

If one could neglect the escape of radiation from the warhead
during the release of nuclear energy, the temperature of the
materials in the warhead (T) immediately after the release would be

$$\text{Energy Density} = 10^{16} \text{ j/m}^3 = C_V*d*T + a*T^4. \qquad (A.1)$$

The first term on the r.h.s. of (A.1) is the thermal energy in the
the mass of the warhead which has a density (d) and a heat capacity
(C_V) appropriate to a completely ionized gas with approximately one
gram-mole of electrons per 2 grams of warhead mass:

$$C_V = 6.2*10^3 \text{ j/}^{o}\text{K-kg.} \qquad (A.1a)$$

The second term on the r.h.s. of (A.1) is the energy density in a
radiation field. Here

$$a = 4*s/c \qquad (A.1b)$$

where c is the speed of light and s is the Stefan-Boltzmann constant,

$$s = 5.7*10^{-8} \text{ W/m}^2\text{-}^{o}\text{K}^4. \qquad (A.1c)$$

For a warhead with a yield of one kiloton per kilogram and a density
(d) of about $2.5*10^3$ kg/m^3, (A.1) gives an initial temperature of
approximately 60 million ^{o}K. More than 90 percent of the thermal
energy would be in the radiation field.

The First Millisecond - Radiation Diffusion: Because most of the
energy in this initial "X-ray fireball" would be in the form of
photons, the flow of radiation out of its surface would be enormous.
The mean free path of the X-rays would be relatively short, however,
so that the energy would not be dissipated in a flash. Instead the
fireball would grow initially by "radiation diffusion" (see Fig. I.1
[2,Fig. 3]).

If we model the fireball during this stage by a uniform sphere with a sharp boundary at a radius of R meters, the total power of the black-body radiation leaving the surface of the sphere would be

$$P = 4*\pi*R^2*s*T^4 \text{ Watts,} \tag{A.2}$$

where (s) is once again the Stefan-Boltzmann constant.

As the fireball expanded, the temperature would fall rapidly to the point where virtually all the energy would be carried by the electrons in the ionized air which had been heated by the radiation diffusion. Thereafter, neglecting the second term in (A.1) and assuming a uniform sphere of air with a density of one kg/m^3, the temperature for a weapon with a yield of W megatons would decline with increasing (R) as

$$T = 1.6*10^{11}*W/R^3. \tag{A.3}$$

The rate of radiation described by (A.2) would decline still faster as

$$P = 4.9*10^{38}*W^4/R^{10}. \tag{A.4}$$

If one divides (A.4) into the energy of the explosion,

$$E = 4.2*10^{15}*W \text{ joules,} \tag{A.5}$$

the characteristic radiation time of the fireball is

$$E/P = 8.6*10^{-24}*R^{10}/W^3 \text{ seconds.} \tag{A.6}$$

This time constant has grown to one millisecond when

$$R = R_1 = 100*W^{0.3} \text{ meters.} \tag{A.7}$$

This is a long enough time constant so that physical expansion of the hot gases would begin to dominate (see Fig. I.2 [2, Fig. 5]).

The Next Second - Adiabatic Expansion: If one assumes that, starting at a radius, R_1, the fireball expands adiabatically until its internal pressure drops to atmospheric pressure, it is possible to obtain a reasonable estimate of the final fireball diameter.

According to (A.3), the temperature associated with R_1 is

$$T = 15*10^4*W^{0.1} \text{ °K.} \tag{A.8}$$

According to the ideal gas law, the corresponding pressure when R = R_1 would then be (approximating the heat capacity of the isothermal sphere by that of fully ionized air with a mass density of 1 kg/m^3):

$$p_1 = 6*10^3*W^{0.1} \text{ atmospheres.} \tag{A.9}$$

The pressure would subsequently fall during an adiabatic expansion,if
we use the constant volume heat capacity given by (A.1a), as

$$p = p_1*(R_1/R)^5 \qquad\qquad\qquad (A.10)$$

and would reach atmospheric pressure when

$$R = 6*R_1*W^{0.02} = 600*W^{0.32} \text{ meters.} \qquad\qquad (A.11)$$

The average mass density when the fireball reached this radius would
be about one percent of the original atmospheric density and the
average temperature would be a few thousand degrees.[2]

Eqn. (A.11) gives a prediction which is in surprisingly good
agreement with the observed size of the fireball of a 20 kiloton
explosion at 1 second [3,Fig. 2.121]. The scaling with yield in the
corresponding empirical formula is, however, given in ref. 3 as $W^{0.4}$
[3, p.70].

Appendix B. The WSEG-10 Model for "Local" Fallout

In the WSEG (Weapons System Evaluation Group) model for the
distribution of "local" fallout, the ultimate altitude to which the
fission product cloud from a low-altitude air-burst rises is given as
a function of yield as [13, p.3]

$$H_C = 13.4 + 1.9*\ln W + 0.062*\ln(11.2W)*\text{Abs}[\ln(11.2W)] \text{ km.} \quad (B.1)$$

Here W is the yield of the nuclear warhead measured in megatons. By
the time the fission products reached the altitude H_C, the mixing
which occured during the rise and cooling of the fireball would
result in the debris being distributed with a vertical rms radius of
[13, p. 5]

$$R_V = 0.18*H_C \qquad\qquad\qquad (B.2)$$

and a horizontal rms radius of comparable size [3, Fig. 2.16].

If the height-of-burst (H_B in km) were low enough so that the
fireball touched the ground, i.e.

$$f = H_B/(0.87*W^{0.4}) < 1, \qquad\qquad\qquad (B.3)$$

then a great deal of surface material would be sucked up with the
fireball and become contaminated with fission products. Much of this
material would subsequently coalesce into particles large enough so
that these particles would fall to earth again within a few hundred
miles in what is called "local" fallout. Ref. [13, p. 7] gives a
formula for the fraction of the total fission products deposited in
this local fallout pattern, when inequality (B.3) is satisfied, as

$$F_L = (1 - f)^2*(1 - f/2). \qquad\qquad\qquad (B.4)$$

In the WSEG-10 model, it is assumed that the particle size distribution in the cloud associated with a surface burst is such that the time distribution with which the local fallout arrives on the surface has an exponential form

$$f_t(t) = \exp[-t/T_F]/T_F \tag{B.5}$$

where the characteristic fallout time

$$T_F = 1.06*[0.66*H_C-0.0075*H_C^2]*[1.0-0.5 \exp(-0.017*H_C^2)] \text{ hours.} \tag{B.6}$$

If we assume, for simplicity, that the wind does not change with time or downwind distance, then (B.5) can be converted into a distribution as a function of the downwind distance (x) by making the substitution

$$t = x/W_F \tag{B.7}$$

where W_F is the magnitude of the average wind velocity seen by the fallout particles during their descent. In the actual WSEG model, the resulting distribution is rounded near x = 0 because of the finite initial size of the cloud [13, p. 4]. Since most of the particles are small enough so that their terminal velocity can be calculated from Stokes Law, the terminal velocity in air will be relatively insensitive to the air density and therefore the altitude (see Fig. I.14 [3, Fig. 9.164]). W_F is therefore approximately the vector average of the winds between the surface and H_C.

The width of the downwind fallout pattern close to ground zero is determined by the initial radius of the cloud. Further downwind, however, it depends upon the transverse wind "shear" at the altitude of the cloud. This tranverse wind shear blows different layers of the cloud in slightly different directions. At a downwind distance where the initial size of the cloud can be neglected, the resulting transverse half-width of the fallout distribution becomes [13, p.5]

$$S_{TR} = x*R_V*W_{TRS}/W_F \tag{B.8}$$

where W_{TRS} is the transverse wind shear measured in km/hr/(km altitude).

Given a wind distribution which does not change with either time or distance and neglecting the original transverse dimensions of the cloud, the normalized local fallout distribution f(x,y) may be written in terms of the variables and constants defined above as

$$f(x,y) = [\exp(-x/W_F*T_F)/(W_F*T_F)]*\exp(-0.5*y^2/S_{TR}^2)/[2*\pi*S_{TR}]^{1/2} \tag{B.9}$$

where y is the distance transverse to the downwind direction.

References

1. Thomas B. Cochran, William M. Arkin, and Milton M. Hoenig, Nuclear Weapons Databook, Vol. I: US Nuclear Weapons and Forces, (Cambridge, Mass: Ballinger, to be published).

2. Harold L. Brode, "Review of Nuclear Weapons Effects," Annual Reviews of Nuclear Science, 18 (1968), p. 153.

3. US Departments of Defense and Energy, The Effects of Nuclear Weapons, 3rd edition, Samuel Glasstone and Philip J. Dolan, eds. (1977).

4. Gilbert Ford Kinney, Explosive Shocks in Air, (New York, Macmillan, 1962).

5. C.L. Comar, Fallout from Nuclear Tests, (US Atomic Energy Commission, 1964), p. 12.

6. John R. Lamarsh, Introduction to Nuclear Reactor Theory, (Reading, Mass: Addison-Wesley, 1972), Table 3-5.

7. William M. Arkin, Frank von Hippel, and Barbara G. Levi, "The Effects of a `Limited' Nuclear War in East and West Germany," Ambio Vol.XI, Number 2-3, 1982, p.163; and "Addendum," Vol. XII, Number 1, 1983, p. 57. A popular version was published in Spektrum der Wissensschaft, the German language outlet for Scientific American, March 1983, p. 18.

8. David Albright, "Gamma Radiation from Activation Products Induced in the Soil by Neutrons from Atmospheric Nuclear Explosions" (1982, to be published).

9. Compilation of Local Fallout Data from Test Detonations 1945-1962 Extracted from DASA 1251, Volume I - Continental U.S. Tests, (Washington, DC: Defense Nuclear Agency, DNA 1251-1- EX, 1979), p. 281.

10. Seymour Katcoff, "Fission-Product Yields from Neutron-Induced Fission," Nucleonics, November 1960, p. 201.

11. Samuel Glasstone and Alexander Sesonske, Nuclear Reactor Engineering (New York: Van Nostrand, 1967).

12. Kendall R. Peterson, "An Empirical Model for Estimating World-Wide Deposition from Atmospheric Nuclear Detonations," Health Physics 18 (1970), p. 357.

13. Leo A. Schmidt, Jr., Methodology of Fallout-Risk Assessment, (Arlington, Va: Institute for Defense Analyses, Paper P-1065, 1975).

14. W.E. Loewe and E. Mendelsohn, "Neutron and Gamma Doses at

Hiroshima and Nagasaki," <u>Nuclear Science and Engineering</u> <u>81</u> (1982), p. 325.

15. The [Japanese] Committee for the Compilation of Materials on Damage Caused by the Atomic Bombs in Hiroshima and Nagasaki, <u>Hiroshima and Nagasaki: The Physical, Medical, and Social Effects of the Atomic Bombings,</u> (New York: Basic Books, 1981).

16. <u>Medical Effects of the Atomic Bomb in Japan</u>, Ashley W. Oughterson and Shields Warren, eds. (New York: McGraw-Hill, 1956).

17. US Army Field Manual, <u>Operations,</u> (FM 100-5 1976).

18. Paul Bracken, <u>On Theater Warfare</u>, (Croton-on-Hudson, NY: Hudson Institute, Report # HI-3036-P, 1979).

19. <u>Fiscal Year 1981 Arms Control Impact Statements</u>, printed jointly by the US Senate Foreign Relations Committee and the House Committee on Foreign Relations, May 1980, p. 243.

20. Blast and fallout areas for the US and Soviet arsenals are given in Harold A. Feiveson and Frank von Hippel, "The Freeze and the Counterforce Race," <u>Physics Today</u>, January 1983, p. 36. To convert the blast areas for 0.25 atmospheres peak overpressure to those for 0.35 atmospheres, multiply by a factor of approximately 0.6.

21. US Arms Control and Disarmament Agency, <u>An Analysis of Civil Defense in Nuclear War</u>, December 1978. [Reprinted in Arthur M. Katz, <u>Life After Nuclear War</u> (Cambridge, Mass: Ballinger, 1981), Fig. 10-1.]

22. Robert P. Berman and John C. Baker, <u>Soviet Strategic Forces: Requirements and Responses</u>, (Washington, DC: Brookings, 1982).

23. Subcommittee on Arms Control, International Organizations and Security Agreements of the Senate Committee on Foreign Relations, Committee Print, <u>Analyses of Effects of Limited Nuclear Warfare</u>, September 1975.

24. "Guide to USAF Bases at Home and Abroad," <u>Air Force Magazine,</u> May 1982, p. 188.

25. Desmond Ball, <u>Can Nuclear War be Controlled?</u> (London: International Institute for Strategic Studies, Adelphi Paper #169, 1981).

26. Frank von Hippel and John Duffield, <u>Short Term Effects of Thermonuclear Explosions: An Update</u>, invited paper to be presented at the annual meeting of the American Association for the Advancement of Science (AAAS), Detroit, May 28, 1983 and to be published subsequently in a AAAS Symposium volume.

27. Statement of Secretary of Defense Robert S. McNamara before the Senate Armed Services Committee on the Fiscal Year 1969-73 Defense Program and 1969 Defense Budget, Jan. 1968, p. 57.

28. Geoffrey Kemp, Nuclear Forces for Medium Powers Part I: Targets and Weapons Systems (London, International Institute for Strategic Studies, Adelphi Paper #106, 1974).

29. Federal Emergency Management Agency, High Risk Areas for Civil Defense Planning Purposes, (Report # TR-82, September 1979).

30. Carsten M. Haaland, Conrad V. Chester, and Eugene P. Wigner, Survival of the Relocated Population of the U.S. After a Nuclear Attack,, (Oak Ridge National Laboratory, ORNL-5041, 1976).

31. Arthur M. Katz, Life After Nuclear War (Cambridge, Mass: Ballinger, 1981).

32. Herbert L. Abrams, "Infection and Communicable Diseases," in The Final Epidemic: Physicians and Scientists on Nuclear War, Ruth Adams and Susan Cullen, eds. (Chicago: University of Chicago Press, 1981), p. 192.

33. US National Academy of Sciences, Long-Term Worldwide Effects of Multiple Nuclear-Weapons Detonations, (Washington DC, National Academy of Sciences, 1975).

34. Paul J. Crutzen and John W. Birks, "The Atmosphere After a Nuclear War: Twilight at Noon," Ambio, Vol. XI, Number 2-3, 1982, p.114.

CHAPTER 2

PHYSICS AND TECHNOLOGY OF THE ARMS RACE

R.L. Garwin
IBM Thomas J. Watson Research Center
P.O. Box 218, Yorktown Hts, NY   10598

ABSTRACT

Traditional military concepts of superiority and effectiveness (as embodied in Lanchester's law) have little relevance to thermonuclear weapons, with their enormous effectiveness in destruction of society. Few are needed to saturate their deterrent effect, but their military effectiveness is limited. The evolution and future of strategic nuclear forces is discussed, and their declining marginal utility emphasized. Some calculations relevant to the nuclear confrontation are presented (Lanchester's Law; skin effect of VLF and ELF signals to submarines; the rocket equation; simple radar-range equation) and recommendations presented for future strategic forces and arms control initiatives. Recommended programs include a silo-based 12-ton single-warhead missile (SICM), the development of buried-bomb defense of individual Minuteman silos, the completion of the deployment of air-launched cruise missiles on the B-52 fleet, and the development of small (1000-ton) submarines for basing ICBM-range missiles. Limiting the threat by arms control should include ratification of SALT II, followed by negotiation of a protocol to allow a SICM and dedicated silo to be deployed for each two SALT-II-allowed warheads given up; a ban on weapons in space and anti-satellite tests; and an eventual reduction to 1000 nuclear warheads in U.S. and Soviet inventories.

INTRODUCTION

We see the United States and the Soviet Union with 20-25,000 nuclear warheads each. Why? What do they plan to do with them? How does this situation come about?

History and the world are full of signs of past conflicts and the means taken to avoid them. Among these are cemeteries, scorched earth, hill towns, and castles with moats. The sequence of measures and countermeasures is well known: arrows, armor, crossbow, rifle, machine gun, tank, trench.

Throughout history, groups have sought the means to attack without peril, but those subject to attack have sought countermeasures, to which counter-countermeasures were found, and so on. Thus initially in World War I aircraft were used for observation, until some enterprising pilot found that he could drop something on somebody or shoot somebody, and aircraft rapidly became fitted with bombs and guns. In World War II, the art of anti-aircraft fire was advanced by the introduction of the proximity fuze, and the counter-air activity of aircraft was extended to the attack of aircraft on airfields. It is this legacy of

0094-243X/83/1040047-18 $3.00 Copyright 1983 American Institute of Physics

weapons attacking weapons which is responsible, in part, for the vast number of nuclear weapons now extant. This presentation is organized as follows:

OUTLINE

-Weapons and countermeasures.
-Lanchester's law.
-Nuclear weapons are not just ordinary weapons.
-Types of deterrence.
-Evolution of nuclear deterrent force(s).
    --Survivability; penetrativity; "overkill"; marginal utility.
    --I.I. Rabi's "We meant well."
-Current forces and their prospects:
    --ICBMs; SLBMs; Bombers and ALCMs.
    --C^3I.
The future of defense against nuclear weapons.
    --A world without nuclear weapons. Can we have it? Do we want it?
-What follows from this analysis?
-Question-and-answer session.

LANCHESTER'S LAW

Lanchester's law expresses the fact that a weapon platform not only kills other weapon platforms, but also is subject to destruction itself. Under certain assumptions, (1) represents the rate of consumption of a weapon platform in a conflict, with the coefficients ε proportional to the unit effectiveness of the force.

$$dN_B/dt) = -\varepsilon_A N_A; \quad (dN_A/dt) = -\varepsilon_B N_B \tag{1}$$

From (2) we see that a constant of the motion emerges, the unit effectiveness times the <u>square</u> of the number of units.

$$-\varepsilon_B N_B (dN_B/dt) = -\varepsilon_A N_A (dN_A/dt); \quad \varepsilon_B N^2_B(t) - \varepsilon_A N^2_A(t) = \text{const.} \tag{2}$$

$$\text{"Force effectiveness"} \equiv \varepsilon N = \text{"F"}; \quad \text{"L-value"} = \varepsilon N^2 = F^2/\varepsilon! \tag{3}$$

In fact, we might as well call the quantity, the difference of which is the constant of the motion, the "L-value" for a force ("L" for Lanchester). This is quite distinct from the commonly imagined "force effectiveness", which is the effectiveness of a unit times the <u>number</u> of units. The L-value is equal to the force effectiveness (squared) <u>divided</u> by the unit effectiveness!

In the example (4), with equal unit effectiveness, a force of 5 units encountering a force of 3 units eats it up, and has 4 units surviving.

$$\varepsilon_B = \varepsilon_A = 1; \ N_B = 5; \ N_A = 3; \ \rightarrow N_B = 4; \ N_A = 0 \tag{4}$$

Of course, fluctuations would affect the outcome with such small numbers. However, fluctuations would be unimportant to a B-force of 500 units meeting an A-force of 300 units, destroying the other side and emerging with 400 units intact.

A B-force of 5 units can defeat a force of 6 units if it can contrive to fight 3 units at a time. The first engagement would then result in a residual of 4 units on the B-side, facing 3 units on the A-side. Further conflict will reduce the number of A-forces to 0, while leaving 2.6 B-forces. "Divide and conquer":

$$N_B = 5; \; N_A = 6 = 3+3; \; \rightarrow N_B = 4; \; N_A = 3 \rightarrow N_B = 2.6; \; N_A = 0 \tag{5}$$

Lanchester's law is a good indication of the value of local superiority, and there are many other lessons to be drawn. Unfortunately, when we studied practice engagements some years ago and interpreted the results in terms of Lanchester's law, some 7 out of 9 general officers had never heard of it!

But nuclear weapons don't (effectively) fight nuclear weapons, and the traditional military concepts and intuition of mass and force do not necessarily apply! The primary utility of nuclear weapons is for <u>deterrence,</u> and they are so much more effective in destroying value than in destroying other nuclear weapons that one side or the other could not resist this counter-value role.

TYPES OF DETERRENCE

Deterrence (affecting the decisions of the other side so as to keep them from doing something obnoxious) could be obtained through (1) a passive shield; or (2) through active defense; or (3) by threat of retaliation or punishment. No one is going to attack when his forces are simply going to be blunted and repelled; on the other hand, there would be very little penalty in trying. In any case, passive defense of society is infeasible against nuclear weapons of the numbers and yields existing in the world today.

Active defense against nuclear weapons must be leak-tight at the present number of nuclear weapons in the world. Between 20 and 100 nuclear weapons could destroy the US or the Soviet society, and against some 7-9000 strategic nuclear weapons, such a defense would have to be extraordinarily and unprecedentedly effective. So all agree (including President Reagan in his March 23, 1983 speech) that deterrence of attack by threat of retaliation <u>works.</u> Some are uncomfortable with it morally; some wonder whether there is not a total defense possible, so that one will not live in "daily terror" of attack by nuclear weapons. These questions are a central topic of this short course.

EVOLUTION OF STRATEGIC NUCLEAR FORCES

Three days ago, I was at Los Alamos for the 40th anniversary of the founding of that laboratory, and on December 2, 1982 I attended the 40th anniversary of the first chain reaction at the University of Chicago. The occasions recalled the environment of World War II-- of battle and conquest and occupation and deliberate genocide by the Nazis, the strategic bombing primarily by our side, and eventually after the war a review which indicated that strategic bombing is not really effective in disrupting war potential, except as it kills people. Hundreds of thousands of people died in strategic bombing-- in the Dresden fire storm, in the Tokyo fire bombing. On our side there was the fear of the Nazis getting the fission bomb (if such a device were possible), and we began a program (not I; I was introduced to this field in 1950) to develop atomic bombs. The purpose was to end the war in any way necessary.

However, the purpose of nuclear weapons when there is no war is different. It's to _prevent_ war, to deter war. The idea of deterrence through threat of punishment is old, even proportional deterrence. The Bible prescribes "an eye for an eye and a tooth for a tooth", in the expectation that a possible perpetrator is going to value _his_ eye more than the destruction of _your_ eye. But this assumes high probability of apprehension and punishment, and a rule of law. There is no "international law" enforceable against nations. You cannot enforce treaties except by the use of force. The possibility of destruction, to be used in retaliation if the other side attacked, could be used also for compellence, a term invented by Tom Schelling around 1950. If you are able to destroy the other fellow, and he can't do anything to you, then you threaten to destroy him unless he does what you want him to do-- compellence. At present we and the Soviet Union are far beyond that, with similar capabilities to destroy the other side. With some 9000 strategic nuclear weapons on our side, 7000 on the Soviet side; a megaton weapon will promptly kill about a half million people when dropped on a city. Of course after the first few 1-megaton weapons are used, it's not possible to continue to kill half a million people with each one. Those large numbers of weapons are not worth very much at the margin unless you've thrown away the first 95% or 99% of them, and that's what people often talk about doing when they argue the necessity for more or different nuclear weapons.

But back to deterrence-- deterrence of what, by what? Well, deterrence by threat of retaliation. President Reagan in his March 23 speech acknowledges that deterrence by threat of retaliation works-- deterrence of Soviet attack on the United States. In general, there are at least 3 kinds of deterrence. Deterrence is to affect the mind of the other side so as to keep him from doing what you don't want him to do. You can deter, for instance, by having a perfect shield. Then there's no sense his using his expensive forces because it won't do any good. You could deter by having an active defense, so that his forces are used up in combat with yours and he doesn't achieve his aims, so why start? And you can deter probably most reliably (and with nuclear weapons it turns out, so far, the only way) by threat of punishment elsewhere. People pretty much agree that it is effective, even those who want to change the present system because deterrence does not make destruction impossible, or who hold that deterrence is immoral, or whatever their reason for criticizing it.

Of course, the Soviet Union would work as hard as possible to overcome active defense against their weapons, and we have worked hard to avoid losing our deterrent to Soviet actions-- that is, to defenses against our bombers, to possible attack on the bombers before they could be launched in response to a Soviet strike, and so on. Some of these are technical means: against radar we have built jamming systems or decoys; we have divided the force to have more penetrating aircraft. We can jam fuzes so that even after our aircraft are acquired by Soviet radars and surface-to-air missiles launched, those missiles pass by without harm.

We had our bombers carry "Bullpups" in the old days-- subsonic cruise missiles of not very good accuracy or reliability, or short-range nuclear-armed attack missiles (SRAM), and so on-- and we made major changes in our forces.

For instance, in the 1950s, we had thousands of B-47 bombers and even a good many B-58 supersonic bombers. As the Soviet Union developed their surface-to-air missile systems, it became clear that we could no longer use a high-altitude aircraft, so we abandoned the B-70, a supersonic long-range aircraft, cancelled that program in a rare demonstration of rationality, because it wouldn't be able to penetrate Soviet SAM systems. We've concentrated on low-altitude penetration, even though it costs a good deal more in fuel and stress on the aircraft-- altitudes on the order of 100 meters or less, which require careful planning of the attack. The problem with bombers is that they can be seen by radar; they can be attacked by fighter aircraft and air-to-air missiles, or by surface-to-air missiles.

THE RADAR-RANGE EQUATION

Equation (6) calculates the energy received by a radar antenna illuminating an aircraft with a single pulse of rf energy.

$$E_r = (E_t G/4\pi R^2)\sigma(A/4\pi R^2), \text{ with } G = 4\pi A/\lambda^2 \tag{6}$$

$$E_r > fkT = (E_t \sigma A^2/4\pi R^4 \lambda^2); \tag{7}$$

for $R = 300$ km, $\sigma = A = 1$ m^2, $f = 20$; $\lambda = 10$ cm; $\rightarrow E_t = 100$ j.

Say 6 kw for one circular scan per second.

The returned energy is the transmitted energy (times the gain over isotropic) divided by the area of the sphere at the range R, times some scattering cross section (which is different by 4π, incidentally, from the cross section as defined in nuclear physics) times the receiving antenna area A, again over $4\pi R^2$. The gain of an antenna is related to its area as shown. You need at least that the returned energy exceed kT by a certain factor, typically 20 or so, in order that a threshold can be set to distinguish real echoes from thermal noise peaks. Putting all these things together, and assuming a very small cross section for an airplane of one square meter at 300 km radius, you find that you need 100 joules of radiated RF energy at 10-cm wavelength to see an airplane (with this very small antenna). But that's only to see an airplane, and if you want to find an airplane, you have to look where it isn't as well as where it is, and it takes perhaps 6 kilojoules to look out in all directions.

In the early 1950s when we had the nuclear weapon, and the Soviets had no effective numbers of them, we struggled to avoid being deterred by the Soviet Union. It is not comfortable to go from absolute security to survival dependent on threat of retaliation. But no matter what we did, short of nuclear attack on the Soviet Union, we could not defend against their bombers. I worked on such things for a couple of years, and I don't think we ever had the capability to shoot down more than 30% of their attacking bombers.

We were never able to find a way to defend against Soviet missile forces effectively, although we developed marvelous technology, which would in exercises show, as Kruschev put it, that one could hit a bullet with a bullet. We paid attention in this evolution of our nuclear deterrent forces to survivability, to penetrativity, and one way or another we developed what is characterized by unsympathetic people, as "overkill". We have many more nuclear weapons than are required to carry out the deterrent role, in part because when we organized these forces we could not guarantee for the future their survivability or the ability to penetrate. These forces, as I've indicated, have extremely little utility at the margin; the last 1000 or 5000 nuclear weapons don't contribute much to the ability to destroy Soviet industry or population or whatever it is you think that they value-- conventional reserve forces, Communist Party leadership, and so on. Some conclusions follow from this analysis, which I will point out at the end.

The possibility of missile attack on bomber bases made us uneasy, of course, especially when we were working on air defense and penetration, and the Soviets were working on missiles, and so we built ourselves some missiles.

Of course in order to have a force to be used for deterrence rather than compellence, you have to promise the other side that your force will destroy them after they have done everything that they can do to coerce you, destroy your force, and so on. That means that in addition to the weapons, the bombers, and their air-launched cruise missiles, the submarine-launched ballistic missiles, the intercontinental ballistic missiles, you need to have a command, control, communications, intelligence (C^3I), and warning system, which will let you observe what's happening, and make the right decision to destroy the Soviet Union-- only after the U.S. has been destroyed!

Now, that doesn't seem to make much sense, and people get to this point, particularly military people and political people and ordinary people, and say: "What good will it do to destroy the Soviet Union after they destroy us? And, by the way, how will we use all these nuclear weapons in detail when the time comes?" One enters in that way a morass of not knowing how nuclear weapons can be used, acknowledging that it's really more important to have live Americans than dead Russians, so why should we concentrate on the ability to destroy the Soviet Union?

The whole problem is given by the mood "What if?". The Soviet leaders have to understand that if they attack and destroy the United States, they will be destroyed. They don't have to worry about the details of how they will be destroyed, and neither do we. We just have to make sure that they understand that they will be. The arguments about our lack of planning for detailed conduct of nuclear war, our lack of ability to release the forces flexibly and responsively months after the war has started-- are all irrelevant to the deterrence of the Soviet Union.

So, I will talk about the future of defense against nuclear weapons, whether we can have a world without nuclear weapons, whether we want it, and some specific recommendations.

C^3I-- COMMUNICATIONS TO SUBMARINES

One would like to talk to the missile-launching submarines reliably. We can talk to them reliably now. We will be able to talk to them reliably in the future, no matter what. In order to do that talking, of course, one ought to have something to say. There is the same problem with the land-based missiles as it is with the submarines: the transmitters have to survive. The way we talk to submarines is by radio. Everybody knows that radio waves don't go through a conductor, and sea water is a conductor, but they do penetrate a little bit, and here are the skin-effect equations (Maxwell equations neglecting displacement current, which is the right thing to do).

$$\nabla xE = -dB/dt; \quad J = \sigma E; \quad \nabla xB = 4\pi J = 4\pi\sigma E \tag{8}$$
$$\rightarrow \quad -\nabla^2 B = -d^2B/dz^2 = -4\pi\sigma dB/dt \tag{9}$$
$$\text{if } B_x = B_0 \exp(i\omega t - \alpha z), \quad \alpha \equiv 1/Z = (1+i)(2\pi\sigma\omega)^{1/2} \tag{10}$$

For $\sigma = 0.03$ mho/cm, $f = 10$ kHz, $Z \sim 5$ m; and for $f = 50$ Hz, $Z \sim 70$ m. One has magnetic fields produced by currents; currents produced by electric fields; electric fields induced by change of flux; and you put all these things together to obtain the skin-effect equation, to find an exponential falloff with depth of the envelope of the magnetic field, or electric field.

You wonder how a submarine can patrol at 100-m depth and still communicate, when the radio waves penetrate only to a few meters from the surface. The submarines carry buoyant wires which they extend from the hull, and which then float on the surface. As submarines patrol at low speed, these wires sense the electric field, bring it down to the hull, so they are in constant receive-only communication. If they don't like to tow the wire, they can use a buoy, streamed on a cable, floating a few meters below the surface, which does the same sort of thing. Recently, the U.S. has begun operational experiments with extremely low frequency (instead of 10-kHz VLF, 50-Hz ELF) signals, and those penetrate further by the square root of the wavelength-- a characteristic falloff distance of 70 m or so instead of 5 m. Furthermore, the transmitters, which are soft (both VLF and ELF), are replaced in wartime by aircraft with bellies full of 10-kHz transmitting equipment, and those have been practiced for a long time.

ICBMs and SLBMs

Now how about missiles? How do they get there? What can they do? Well, originally ballistic missiles were unguided artillery. You pointed one in the right direction, you burned its motor until it got up to a certain velocity, and it landed where the shot would have fallen at that same speed in that same direction. That's not a very good way to strike a target at a great distance, and so the next generation was a radio system. Because the path is so great, it's not the uncertainty in position of the missile but the uncertainty in vector velocity at burnout which contributes most to the error. So, a good system would be a Doppler system in which one would have several receivers around, the missile could re-radiate the signal from a transmitter, and a computer on the ground or a programmer could program the velocity of the missile in 3 dimensions until it was right, making corrections (if you have a fancy-enough computer) if the missile gets off the track.

That's what we did, and followed that with the all-inertial navigation which has been so successful. But inertial isn't everything, and in fact it would be better now to have a mix of inertial navigation, self-contained, with radio aids. One of the best ways these days is the Navstar global positioning system (or ground beacon system) which can give in 0.1 s, 10-m accuracy and can provide very good velocity information also-- to the point that, although the existing CEP (circle of equal probability within which half of the shots will fall) is much smaller for ICBMs than for the submarine-launched ballistic missiles (primarily because the subs don't know well which way they're pointing or how fast they're going) the next generation or any retrofit of guidance systems could provide the same accuracy in submarine-launched missiles as in ICBMs.

Based on studies for the Defense Department, several reasonably respectable people have proposed to the over the years that one could and should base ballistic missiles of ICBM size on small submarines. The present sub-launched ballistic missiles are all carried vertically inside the submarine, so the biggest missile planned is 43 ft long-- the Trident II, 24 of which fit in an 18,000 ton submarine. But MX missiles which are 72 ft long would require a 100,000 ton submarine. We suggested an idea not new with us, but unpopular nevertheless, that if the missiles lie next to the submarines in the water, then one could have 2-4 MX missiles on a submarine which are below 1000 tons. I'll get to current status of understanding of that in a minute.

One can deliver weapons by aircraft or by rockets. There is a crossover in efficiency of delivery of munitions by one or the other at a particular range. For instance, if you want to go 1000 times around the earth, a rocket coasts in earth orbit once it has been launched, whereas an aircraft has to struggle all the way. So, that's a problem.

In (11) are derived the satellite velocity in low-earth orbit, 8 km/s, and the escape velocity of 11 km/s. If you launch east you get free, the rotation of the earth, 0.5 km per second, times the cosine of the latitude λ .

$$V^2/R = g; \quad V^2 = gR = gS/2\pi; \quad V_0 = [(9.8)(4\times10^7)/2\pi]^{1/2} = 7.9 \text{ km/s}; \quad (11)$$

$$V_e = 11.2 \text{ km/s}$$

launch East? $V_r = 4\times10^4/86,400 = 0.46\cos\lambda$ km/s.

Rocket propulsion with exhaust velocity V_r: $F = 0 = d(mv)/dt$

$$-V_r dm/dt = m(t)dv/dt; \quad (1/m)dm/dt = -(1/V_r)dv/dt \qquad (12)$$

$$\ln m/m_0 = -\Delta V/V_r; \quad m_0/m = e^{\Delta V/V_r} \qquad (13)$$

$V_r{}^2/2 = \varepsilon$ (internal energy per gram) if fully expanded in the rocket plenum and nozzle.

For $\varepsilon = 1000$ cal/g (x4.186×10^7 erg/cal),
$V_r = 2.9$km/s $\equiv gI_{sp} = gx296$ sec; so $I_{sp} = 296$ sec. $\qquad (14)$
So for LEO $m_0/m = 15$. To escape, $m_0 = 47$.

Orbital pellets (meteors):

$$\varepsilon' = (V_0 + V_t)^2/2 = (8+5)^2 \times 10^{10}/2 = 20 \text{ kcal/g}$$

(versus 1 kcal/g for high explosive).

Conservation of momentum applied to a system which expels an exhaust stream at velocity V_r and mass rate dm/dt gives the ratio of the initial mass to the final mass as an exponential (Eq. 13). The exhaust velocity of the fuel simply is obtained by converting in a combustion chamber the internal chemical energy, stored energy, into thermal energy, and then expanding that through a rocket nozzle to get directed velocity-- no internal energy-- and so 100% conversion into kinetic energy. Rocket propellant has some 1000 calories per gram internal energy ε, (as does high explosive). In (14) the "specific impulse" I_{sp} is calculated as 296 or 300 seconds.

That lets you understand that for low-earth orbit, that is a ΔV of 8 km/s, you need a mass ratio of 15, assuming that there is no excess mass in motors or things like that (continuous staging); to escape you need something like 50. And while we're at it, there are other things you can do with these numbers. You can ask, for instance, about the utility of pellets in orbit for destroying ballistic missiles or satellites. So you calculate the kinetic energy of a pellet in orbit, which is going at 8 km per second, in the rest frame of a ballistic missile going at 5 km per second. The result is that it has a kinetic energy of 20 kcal/g, versus about 1 kcal/g for high explosives. So, once one puts things into orbit and separates them a little bit, one has a very effective warhead against missiles.

CURRENT STRATEGIC FORCES

I've touched on the strategic forces. More specifically, our primary strategic forces now are a rapidly building number of 3000-1b air-launched cruise missiles, carried about 20 per B-52. We have several hundred of them now, are producing about 100 a month, and going to have force of some 3000 air-launched cruise missiles, with unprecedented accuracy of the order of 70 m or better. They obtain this accuracy by terrain comparison-- TERCOM. They look down with the radio altimeter, and at specific chosen regions in the Soviet Union, the terrain topography has been mapped and stored in the memory. Updates are made to the inertial guidance system on the ALCM by the difference in the expected trace of altitude versus time and the observed trace. The ALCM has the same warhead as the Minuteman.

We have had 31 Poseidon submarines, each with 16 missiles, each with 10 or 14 warheads. Those are being replaced by the Trident I. We have several shiploads of Trident I-- 16 missiles, 8 MIRVS each, 4000 nautical mile range, and considerably improved accuracy. On land we have the 550 Minuteman-III ICBMs with 3 MIRVs each, 200-m CEP maybe, submegaton yield.

Although Secretary Weinberger has said recently that we have not improved our land-based force since it was deployed in the 1950s, except by upgrading the warheads, that's not true at all. The only thing in common that the Minuteman III and the Minuteman I have are the name. The Minuteman III is a 1970s system. It has MIRV; it has improved accuracy; it has many features which the Minuteman I did not have. And we still have 450 Minuteman IIs with single warhead and megaton-class yield.

The Soviet Union has mostly ICBM warheads, in contrast to the vast number of submarine warheads on our side, including 300 troublesome SS-18s, monster missiles with as many as 10 MIRVs each, also megaton class, and CEP in the same order as Minuteman III.

For years, the next US ICBM has been called the MX; it simply stands for "missile experimental". It used to be that an MX would become an MY, would become an M-34, or whatever it is. I think the MX isn't going to become anything, and that will be good. It was to be a 100-ton missile with 10 warheads, largely because that was the biggest we could build under SALT II. It would have two objective features though. First, the paper missile would cure Minuteman vulnerability by the way it was based, not by any characteristic of the missile itself. And second, it would threaten Soviet silos and "drive the Soviet Union to the arms control bargaining table."

But consider the history of MIRV in the 1960s, the multiple independently-targeted re-entry vehicle that Secretary McNamara revealed in a Life magazine article interview in 1965. The Secretary of Defense wanted to deploy MIRV because the Soviets might sometime in the future build an ABM system, and it was better to have more warheads for penetrating than less. MIRVs would be 100% certain, whereas other kind of penetration aids, decoys, and so on, might be distinguished ("discriminated") by a sufficiently capable ABM system. However, the military wanted MIRVs simply because more warheads are better; you can destroy more targets with them. And when the ABM threat disappeared in 1972 with the SALT I Treaty, we had MIRVs. We did not really consider giving them up. We argued that the Soviets (you'll hear about this in the arms control portion of the discussion) would not accept the abandonment of MIRVs when we had tested and they hadn't. Similarly, if we persuade ourselves now to build the MX when it is admittedly without justification for either deterrence or military tasks, we may regret the response by the Soviet Union.

On the MX, the real question is: "Are there other options to the same ends, that is to curing Minuteman vulnerability, arising because the Soviets have so many large-yield re-entry vehicles with reasonable accuracy; and is there another way, maybe even cheaper and sooner, to threaten Soviet silos, if you think that's a good thing to do. And, by the way, does the MX cure Minuteman vulnerability?" You may even ask, do we need, can we have, survivable, enduring, land-based ICBMs? The first quick comment is that even if Minuteman is not survivable, it's still useful, and not subject to attack if the other strategic forces are survivable and capable of deterrence by assured destruction, and so on. Even if the whole force is vulnerable, we have a fall-back-- the doctrine, capability, and the will for launching our weapons under attack, with adequate command, control and intelligence.

FUTURE STRATEGIC FORCE OPTIONS

I talked to the Scowcroft Commission January 27, and told them the same things. In considering these options, one notes that future ICBMs are no longer unique in accuracy. MIRVs on ICBMs yield short flight time and responsiveness, but they will have to compete in cost and survivability with shorter flight-time submarine-launched missiles or other things. Besides, how can we make the ICBMs survive, for a given number of warheads? We could harden them; and the Soviets could counter with accuracy. We could make them mobile; the Soviets could counter by finding out where they were, or by barraging the entire deployment area. We could deceive by having more aimpoints than the Soviets have missiles; and the Soviets might try to detect where these missiles are. Some believe we could keep it secret indefinitely; some don't. We could de-MIRV; instead of having a force of 2000 missiles in 200 silos (if you went to a total fixed MX-system) we could have single warhead missiles, and have each in a small silo. And we could, if necessary, increase the number of warheads and the number of silos.

We could have defenses, especially specific defenses of silos-- buried nuclear weapons 1 km north of each of the silos, which, when a re-entry vehicle was a few seconds from striking the silo, would be exploded and 100 kilotons of reasonably clean nuclear energy release would throw 100 kilotons of earth into the air and bar access to the silo. Or, one could use a non-nuclear shotgun approach. The Soviet counter to that would be to throw more than one warhead against a silo. However, that causes a great deal of difficulty because the environment after nuclear weapons have exploded near a silo, especially with 100,000 tons of earth in the air, is not conducive to effective attack. So, they would have to wait, probably for 30 minutes, or an hour. Increasing the "price" of each of 1000 or 2000 silos (from a single Soviet RV required for destruction, to 2 or 3) would have a substantial impact on their force requirements.

De-MIRVing is a good idea, and the key is to have more aimpoints and missiles than the other side has reliable, effective, usable warheads. Under those circumstances, each Soviet warhead could destroy less than one US warhead, and the Soviet Union would disarm itself, if we had equal numbers of warheads, by using them against our warheads. De-MIRVing would help the Soviet Union if they moved also to these small ICBMs. Technically, they would be best deployed in silos or shelters; don't have to be very hard, just so that not more than one could be destroyed by a Soviet warhead. And the firing and monitoring would be by encrypted radio link-- a lot better than the Dense Pack system, which has to work after the very silos which are going to be used have been attacked, and tilted, and pushed up and down by nuclear weapons. One could increase the leverage by a single-silo buried-bomb defense, and that would allow us to guard against Soviet warhead members 2 or 3 times the number that we have in our land-based force. I don't believe that mobility, although a technical option, is a good choice.

A reasonable path to a silo-based small ICBM, is to develop a 10-12 ton small ICBM (SICM), carrying the Mk-21 MX warhead and deploy some of them-- 450 in the Minuteman II silos. That's allowable under SALT II. Ratify SALT II, and preferably negotiate a protocol after we ratify, allowing each of the 820 land-based MIRV missiles to which we are entitled under SALT II to be traded for skin-tight, new unique silos-- 4 of them for each of those 10 warheads to which we're entitled-- each containing one of the 12-ton SICMs. We should direct the START talks toward limits on numbers of warheads consistent with this kind of "freedom to mix," and stability. Is such a missile feasible? The answer is "yes," as much as anybody knows anything about any of these things. If you talk to the contractors, they say that missiles really cost so much per pound for a given level of technology. You could build the SICM in either 2 or 3 stages; two stages is probably cheaper, although a little bit bigger.

The contractors don't need to do any development of the rocket motor for a proof test. And the guidance which looms large in various criticisms of small ICBMs (or Midgetmen) because you have to have a guidance system per warhead, costs $0.5 million per warhead. Let's just add it in; never mind that we can probably get it for less, let's just pay that cost, and one can have the global positioning system: the Navstar adjunct, which now weighs less than 10 pounds, to provide a better accuracy with the small ICBM than would have been available with the MX.

How much are you allowed to spend on these things? Well, here are some quick numbers: Trident II, which is a sound and secure system, costs more than $15 million per warhead deployed; the smallsub undersea mobile system (SUM)-- the small submarine carrying an MX, which is also a technically-sound system, might cost somewhat less, but you can get an argument with the Navy about that-- its 10-year cost might be very similar to that of a Trident II, although it would carry more capable warheads; Dense Pack, according to the numbers given by the Administration, would cost $26 billion for 100 missiles with 1000 warheads, so $26 million per warhead deployed. This little ICBM, including its silo, would be an investment cost per deployed warhead of the order of $6 million. How can you afford not to buy?

But you say, "how about the development costs?" Well, the development cost is also so much per pound if it is totally within the MX technology, within the state of the art. First, each time you shoot, you only expend 12 tons of hardware instead of 96 tons of MX; second, you need fewer test flights because you don't have to test the MIRV-dispensing bus. You would be generous if you would plan for $1 billion, and more than generous for $2 billion development cost.

Eventually, we could build enough silos to eat up the Soviet warheads, no matter how many warheads they built. That could be an arms race; it wouldn't do us any good; it wouldn't do the Soviets any good. We would restore deterrence at a _higher_ cost. Better would be to have big reductions, and that's one reason to go to small ICBMs with single warheads, because if we reduce to what I think is a reasonable level of 1000 nuclear warheads total on either side, I would not like to have my 1000 warheads in 2 Trident submarines and on 10 MXs. I would far rather have them in 400 single-warhead ICBMs, and 100 submarines each carrying 4 single-warhead missiles, and 200 or 100 aircraft each with 2 air-launched cruise missiles.

The Scowcroft Commission has now reported, but the full report is not yet available. The excerpts which have been released put the question of Minuteman vulnerability in perspective. They agree, as Harold Brown said in his last annual report as Secretary of Defense, that so long as the strategic force is not overall vulnerable, Minuteman vulnerability is not a temptation to the Soviet Union to attack, and not a significant reduction in our deterrent capability. The Commission apparently endorses de-MIRVing by going to small, single re-entry-vehicle ICBMs (Midgetman); but they propose building 100 MX for Minuteman silos, and you have to ask: "Why?" From what I've seen of the report and the response of the commissioners to questions, there is no justification in the report for building MX and putting it in Minuteman silos, except that in the distant past, various people-- Presidents, Secretaries of Defense-- have said favorable things about MX, and it would be regarded as irresolute to abandon a bad idea _now_ just because it's a bad idea.

It is more than possible that the Soviet Union would much rather have us build the MX than not build the MX, and I just don't understand _why_ we should build the thing, which will assuredly impede the evolution of the _small_ ICBM, will make it as long-delayed and as costly as possible, just as when the Chief of Naval Operations, Elmo Zumwalt, tried to build a smaller sea-control ship while allowing the big aircraft carriers to go their way. The supporters of the big aircraft carriers not only impeded, but they totally scuttled the sea-control ship program. And that's what will happen here. The MX would be our one new missile under SALT II, which would be a _bad_ thing because the Scowcroft Commission really provides motivation to go to the single warhead, small ICBM. Furthermore, you cannot tell the difference between an MX development and production program leading to 100 MXs, and the first stage of one which leads to 600 or 800. It will be intentionally, in my opinion, confused, so as to threaten the Soviet Union more. Instead, if you really believe that threats to the Soviet Union, threats to their silos, (not to their missiles, incidentally, because they could also always launch their missiles under attack) have a beneficial affect on arms control, then you should have improved in 1979, and you still should, improve the accuracy of the Minuteman III. You could still do that sooner than you could get the MX deployed, and you can do it surely for less cost.

RECOMMENDATIONS

We should develop the small ICBM. I know lots of people feel we have plenty of weapons, but eventually somebody will want either the opportunity to build more, or we will have increased cost of maintenance of Minuteman. Sometime, we'll have to build a new missile. Now when there seems to be a consensus to build a useful missile, one which is better for arms control and security, one which is cheaper-- a small ICBM-- we ought to develop it right away, deploy 450 in the Minuteman II silos, without deciding how many more we're going to build, and whether we will be permitted to build small silos, or whether they'll have to be mobile. That's enough of a production run to get started.

In agreement with the Scowcroft Commission, which has finally come around to view the technical merit in a program which was anathema to the Air Force and unpopular with the Navy, we should begin the development of small submarines, because a reasonable number of warheads on Trident submarines with 24 missiles-- billions of dollars in a single submarine-- is not a sensible way to have submarines remain indefinitely invulnerable. Small subs carrying a few warheads are the way to go. We ought to strengthen command and control.

We ought, finally, to develop the buried-bomb defense of Minuteman silos. I first published this in 1976 after talking about it for some years, and in fact, the Army was directed to look at it, which they did; but they did not say anything about their conclusions, for instance, to the first Townes Panel on the MX in 1981. Finally when we had a full day of discussion (Charles Townes, Mike May, the Army, and I) the Army showed foils which were in reasonable agreement with mine about the effectiveness of buried-bomb defense of Minuteman silos, and explained that they had not presented those results to the Townes Panel because there wasn't any "near-term, high-level interest in a limited capability defense of Minuteman." Naturally not, when the idea is how you're going to deploy an MX, and a reduced vulnerability of Minuteman does not conduce toward that.

We ought to cancel the MX program now. We've spent enough money on it, and we've created enough myths. We have no justification for building MX either from the point of view of improving our security or scaring the Soviet Union.

Related to the question of Command, Control, Communication and Intelligence, even though we have deterred the Soviet Union from nuclear attack on the United States, and they us, war could start in various ways. War could start over our trying to exert our technological dominance in space, for instance, by attempting to deploy (or leave room for deploying) effective ABM systems there. These things deployed in space are extremely vulnerable. The Soviet Union would put space mines next to them; we would then destroy their space mines preemptively, or keep them from launching things into space which did not meet with our approval. If they got there first, they would do the same thing. The result would be to destroy all satellites which serve important military, as well as civil, functions in space. Worse, war in space is, in my opinion, very likely to spread to earth.

We ought to constrain the Soviet Union by ratifying the SALT II agreement, signed by the Soviets and signed by President Carter, and then further move from that as a base to allow either side to build small ICBMs in limited numbers, and to do other things which are in our security interest (and which the Soviet Union will regard as in their overall security interest, or of course, they won't sign a treaty).

I've given you my recommendations, which follow from this analysis. I don't believe that we can have a world without nuclear weapons, or a world in which nuclear weapons have been "negated and rendered impotent." Suppose we really did that? Where would we be? We would be facing the Soviet Union's conventional forces; by definition, we would have no nuclear deterrent. The President has asked for a world in which nuclear weapons have been negated, rendered impotent, and maybe faded away. But if no nuclear weapons existed in the world, we would be only a development program away from nuclear weapons again, unless we are considering some kind of world government-- maybe ours, maybe theirs, maybe some cooperative world government. Any crisis under those circumstances would lead to the re-building of nuclear weapons, with full knowledge this time of how to do it, with many tons of fissile material available in the world, both from the military program and from the civilian power reactor program. I think that we could build nuclear weapons from scratch in about 6 months-- not very good ones, but we could build them-- and so could the Soviet Union. I believe it's a more secure world in which we try hard to reduce to about 1000 nuclear weapons so we don't spin fantasies about how our 10,000th and 20,000th nuclear weapons will destroy more of their nuclear weapons than we will use. We need a regime in which the nuclear weapons are for deterrence, are not military, and don't fight one another at all. They're to be used for the threat of punishment thereby eliminating the possibility of intentional use of nuclear weapons by the Soviet Union or by us.

--- END OF PRESENTATION ---

QUESTIONS AND ANSWERS

Male voice: Thank you very much. Let me ask first for questions of Dr. Garwin as contrasted to expressions of contrary views or judgments.

Male voice: Just a quickie. You spoke in terms of Trident I. I understand, or have read, that there's a Trident II on the way. Can you give me some information?

RLG: Trident I is a real missile. We have I think about either 4 or 6 Poseidon submarines, retrofitted to carry this 4000-nmi range,

Trident I, with improved accuracy, and so on. Trident II was put into development by President Reagan in October of 1981. It's a 6000-nmi range missile with somewhat bigger warhead capability; it won't be around until 1990 or later, but it's committed to have silo-killing accuracy in order to compete with the MX.

Other male voice: Ok, implicit in your proposal of a small ICBM approach is a model of deterrence, which is almost identical with the hostage-holding of other people and cities. It's called "mutual assured destruction". What evidence do you have, in the history of Soviet weapons development, which suggests that they believe that doctrine?

RLG: Well, the question is: "I am proposing to deter the Soviet Union by threat of retaliation..., against cities." Well, then, why do I believe the Soviets are deterred, because Soviet weapons are "not compatible with their believing that." The Soviets are deterred as any human being is. Talk to anybody. Speak of your own views; go see a Russian. Anybody is deterred because he knows that if you destroy everything he holds valuable-- societal control by the Communist Party, nuclear weapons, conventional forces-- that is just not worth starting a war. What will they gain from a war? What will they lose from a war? They will lose everything that they have built up, military or civil, over these decades. And there's no question, even among those people who so dispute minimal deterrence-- Richard Pipes, and so on-- no question about the fact that the Soviet Union has been deterred. However, they don't accept the principle, the desirability, of deterrence. Now, I don't particularly accept the desirability of mutual assured destruction. I just want to maintain a capability to destroy the Soviet Union if they destroy us, and then I'm convinced. I don't know anybody who says: "I, myself, am not deterred under those circumstances." Now they sometimes say: "Maybe there are people who aren't deterred." Well, so long as they have a capability for destruction, you can always say that maybe they will have a government which is totally insane, and they will then destroy us because they have nuclear weapons. Maybe there will be war by accident. But, in my opinion, there is no evidence. And if the President, who is very much against this regime, says it's intolerable; and the Secretary of Defense who says he is confident, without doubt, that in the future we can have a leak-tight defense, not only against ballistic missiles, but against cruise missiles; all these people still appear to feel that deterrence by a threat of retaliation really works, and the President said so explicitly in his speech of March 23.

Male voice: If a nuclear freeze were to come about, would that make it more difficult or less...

RLG: The question is: "Would a nuclear freeze be helpful?" A nuclear freeze is orthogonal to everything that I am talking about here. A nuclear freeze in principle says: "Stop everything." It has been elaborated more in recent months to say: "Yes, you can change weapons. It should be possible under a freeze to build new weapons to replace old weapons, especially if the new weapons are less capable or less numerous." That's sensible, but I think by the time you flesh out a nuclear freeze, and you insist that it be mutual and verifiable, it will be a more difficult course and a less desirable one-- less secure for us, less secure for the world-- than what I'm proposing. I do believe that the freeze effort, all these people who are interested, finally, in reducing the threat of nuclear war, is a _very_ good thing, (and that the government is noticing this pressure even though they say things to the contrary) and I welcome the public pressure to reduce the threat of nuclear war (not necessarily by reducing the number of weapons-- that's not necessarily going to help). I think that's a good thing.

Male voice: I wonder if your proposal really is orthogonal to the freeze. In looking ahead a little bit at the political situation, I think what's coming is a confrontation between one approach which is the freeze, and a new concept which has gotten a lot of interest in the Senate, the guaranteed build-down idea, which is for every warhead which is deployed on a new system, you would destroy 2 warheads on an old system. I may be wrong, but if I see developing is a political confrontation between those advocates of the freeze, and another approach, which I think will undermine the freeze, which is to allow the building of new weapons systems of the sort you described. It seems to me, in other words, that you're going to end up testifying _against_ the freeze in favor of the guaranteed build-down idea.

RLG: Well, as Niels Bohr said: "It's very difficult to predict, particularly the future."

Male voice: How does Europe fit into this scheme of things?

RLG: The same way Europe has always fit into the deterrent equation. If we care enough about it, we can deter attack on Europe just as we deter attack on New York. I do believe that we ought to improve our conventional capabilities in Europe. That doesn't solve the fundamental problem of nuclear coercion against Europe and whether the United States would feel fully engaged. I think we _must,_ and I think the Soviet Union, under those circumstances, is deterred from attack from Western Europe as well. I'm in favor of removing the battlefield of nuclear weapons from Europe (they don't help), and _not_ deploying the Pershing II or the ground-launched cruise missiles. Strengthen the submarine deterrent. Maybe even deploy some of those Pershing IIs and ground-launched cruise missiles, if by doing so, we can get the Soviets to reduce their nuclear threat. But even if the Soviets had no SS-20 missiles facing Europe, the nuclear threat to Europe would not be changed significantly because their strategic weapons can do the same kind of job. The whole thing is very much overblown in my opinion.

Male voice: If 100 MX missiles were deployed, reducing the asymmetry between the US and the Soviet Union in land-versus sea-based missiles, would the US then possibly have a credible first-strike capability..."

RLG: No, 100 MX missiles are only 1000 warheads, and so you would get a much better first-strike capability by improving the Minuteman accuracy, which I suppose we would do in addition. There's no such thing as a "first-strike capability". I should have mentioned that. "First strike" is shorthand for the ability to disarm substantially the other side's strategic offensive force. And if our anti-submarine warfare were so good that we could destroy all their submarines at sea and in port, and we had enough warheads to destroy their silos, they would simply launch their missiles under attack; and therefore, we would not use our first-strike capability, and would not have had it in the first place.

Male voice: You suggested that the Soviet weapons buildup stopped at the end of SALT.

RLG: I suggested that there was no ballistic missile defense after 1972. In May 1972 the SALT agreements were signed; neither side could have an effective missile defense of the territory. Shortly thereafter, instead of two sites, each side could have just one site with 100 interceptors or fewer, and that is not a significant defense, no matter how they upgrade their radars, no matter how many surface-to-air missiles they have; that's not a defense which is significant versus our offense. They've continued to work at maybe the $1 B/yr level on research and development on BMD, and we have continued to work at a $0.5 B/yr level on BMD. We have no effective BMD. We mothballed ours the day we deployed it, and in my opinion, shouldn't have finished that $10 B expenditure, begun as a bargaining chip to get SALT I. Thank you.

Male voice: Is your contention that by not needing to worry about war fighting... and putting all our eggs in assured destruction, that any kind of Soviet attack on the United States should be met by an all-out attack on the Soviet Union?

RLG: That's not desirable, and it's not necessary. In fact, it would deter any attack if you could make up your mind to do that. However, if you consider that the Soviet Union might sometime consider intentional use of a nuclear weapon against us, they should feel that we might send 1 or 2 back in return. The idea of deterrence is that they care more about the values which they hold, than about destroying things over here, and they should understand that we will respond. We'll not respond precisely equally, but we will respond. That has nothing to do with the overall number of weapons that you want to buy. It has something a little bit to do with the nature of the force. Single-warhead missiles are better for that, with adequate command and control, and with decisions made beforehand.

--- END OF QUESTION-AND-ANSWER SESSION ---

CHAPTER 3

New Weapons and the Arms Race

Kosta Tsipis
Program in Science/Technology for International Security
Department of Physics
Massachusetts Institute of Technology
Cambridge, Massachusetts   02139

In speaking about technologies that could further animate the weapons competition between the United States and the U.S.S.R., it would be useful to distinguish between technologies that have already been incorporated into specific weapons systems, and new technologies that are of a generic nature, can be used in a variety of applications, and can best be described by the tasks that they can perform rather than any specific weapons' application. Let me begin with the latter class.

I.   NEW GENERIC WEAPONS TECHNOLOGIES.

1. Perhaps the most significant and important - in the long term - development has been the initiation, during the Carter Administration, of a new technology aimed at producing very large scale integrated circuits (VLSIC) for military use. This effort entails the basic research necessary for the development of the methods and techniques to manufacture large scale integrated logic circuits which then can be used in a wide variety of weapons systems, strategic, but mainly conventional battlefield systems. We can expect that as a direct result of this effort future weapons systems will be able to perform, unaided by their operator, a large variety of missions, with considerably improved but not perfect reliability under an expanded set of environmental conditions.

Controversy has surrounded this program mainly because it encourages complexity and multimission capabilities for conventional weapons and therefore would increase their cost. It has also been hailed as a panacea to remedy all our security ills, which is largely an overstatement. More probably this technology will allow the development of more effective conventional weapons but will result in true gains in cost/effectiveness only if applied and managed properly.

2. In addition to the VLSIC, electronics (both communications and control) have continued to increase in capability, sophistication and uses. Their advances seem

0094-243X/83/1040065-06 $3.00 Copyright 1983 American Institute of Physics

to be pacing the majority of defense R&D projects.

3. An important specific application of solid-state electronics has been successful development of large arrays (roughly $10^3 \times 10^3$ elements) of infra-red sensors. Based on the principle of photosensitive charged-coupled-devices, these i.r. sensors can _image_ targets that are hot with respect to their environment, transmitting instantaneously the imaging information in the form of electronic signals. As a consequence they can be used to acquire targets - tanks, aircraft, satellites, missiles - and either track them or be part of a homing guidance system that can guide a projectile onto the warm target. Thus, i.r. mosaic sensors can find wide applications in conventional weapons systems as well as the more exotic variety of futuristic space-related anti-satellite, anti-aircraft, and the recently contemplated anti-ballistic systems.

4. A new technology that combines inertially stabilized platforms and large aperture optics - visible or i.r. - is emerging in response to the perceived need to detect, track and aim against distant targets such as satellites in high orbits from the ground, and ballistic missiles during their boost phase, or low flying aircraft from space. Laboratory set-ups have achieved repeatable aiming accuracies of tens of nanoradians with useful settling times. The stability and scanning rate of these prototype systems is orders of magnitude away from what a space-borne antiballistic beam weapon would require, but it is characteristic of the direction in which the U.S. Air Force is proceding that these systems are being developed.

5. An interesting technology that may find widespread applications in military $C^3 I$ and control systems is the family of photonic circuits that utilizes photons and fiber optics rather than electrons and metallic conductors to manipulate and transmit information. Systems that operate with photons rather than electrons are immune to the EMP generated by nuclear detonations. Therefore, research into these photonic circuits has been encouraged by the perceived strategic need to be able to fight a relatively protracted nuclear war during which the $C^3 I$ facilities should remain unaffected by the EMP. Fiber optics have a multiplicity of useful applications in the civilian sector as well as in military systems and consequently their continuing development should not be interpreted as a sign of adoption of the doctrine of protracted nuclear war fighting.

6. Recent presidential pronouncements in the United

States have focused attention on the possible use of lasers as a space-based weapon. The current generation of high energy lasers that emit in the infra-red part of the spectrum are unsuitable for such applications. Emphasis, therefore, has been put into the development of shortwave lasers that emit in the visible, ultraviolet or even the x-ray part of the electromagnetic spectrum. Bound-electron laser systems such as the excimer lasers do not have at this time the energy densities needed in weapons applications and seem instrinsically inefficient. Free-electron lasers can in principle be quite efficient, but they are in a very early stage of development and they could prove to be a promising technology for producing intense short wave beams of coherent-radiation.

X-ray lasers that consist of a thin rod of a material with suitable energy levels - for example calcium - that can be pumped by a small nuclear explosive are, in principle, possible. An experimental prototype has been tested with marginal results. Such a laser would be unusable operationally because the nuclear detonation would destroy both its own platform, and by means of the EMP and electron flux would destroy adjacent similar weapons.

Incidentally, particle beams and focused microwave beams do not offer themselves to weapons uses: Their production and propagation properties make them unsuitable for weapons uses.

7. Increased interest in military operations in space has encouraged two parallel technologies: Space transportation and space-borne large scale systems. The space shuttle is the first step in the development of re-usable space-transportation systems. Although limited in its capabilities and not quite cost-effectively competitive with single-use boosters for payload transportation into Space, the Shuttle is, nonetheless, a prototype of future systems that will, no doubt, follow. The Shuttle has significant civilian applications and, therefore, cannot be considered a weapon anymore than a truck is a weapon.

Large scale installations in space (whether manned or not) are at a very early stage of development - mainly paper studies. But there is accelerated interest in a technology that would support the development of such structures.

II.   SPECIFIC WEAPONS SYSTEMS.

I now list briefly the specific weapons systems that have been recently developed or are still under development, testing and evaluation, that can contribute to the continuation if not exacerbation of the strategic nuclear arms race between the U.S. and U.S.S.R.

1.   The Miniature Homing Vehicle (MHV) is a U.S. antisatellite weapon capable of attacking satellites up to altitudes of 200-300 km from mobile bases on the surface of the Earth. The weapon consists of a sounding rocket fit with a small barrel-like object with its own small booster rocket and infra-red homing sensor. This vehicle is filled with steel rods able of being fired against an enemy satellite. The entire system is carried by an F-15 and released at altitudes of 40-50,000 feet towards its orbiting target which the MHV homes onto using its infra-red sensor. When within striking range, the barrel-like cannister with the steel rods launches towards the satellite and attacks it at close range with the steel rods destroying it on impact with the kinetic energy of the rods. The system will have its full first test sometime in the summer of 1983. It is expected to be more effective and much cheaper to use than the Soviet ASAT system that utilizes ballistically travelling boosters designed to place an explosives-laden satellite in the same orbit as the target satellite, and then maneuver it to approach its target within the kill radius of the explosive charge. Used in a multiple attack, MHV antisatellite weapons can attack successfully very large and well defined platforms.

2.   A new class of endoatmospheric weapons carriers such as bombers and cruise missiles are now under development taking advantage of the "Stealth technology". This is a generic term that encompasses a host of different technologies which, when applied in combination in the manufacturing of a flying vehicle, make it much less visible to radar. The principle technologies are three.
     a.   The use of graphite-epoxy bonded materials for the construction of key parts of the aircraft. Being non-metallic these composite materials have much lower reflection coefficients for radar-waves and, therefore, their return signals can be orders of magnitude weaker than those of metallic aircraft at the same radarwave frequencies.
     b.   The reflected signal can be further attenuated by the application of coats of non-reflecting material on the body of the aircraft. Even though these coatings are effective absorbers of radar energy only at given wavelength bands, a combination of the composite materials and several coatings that absorb optimally at different wavelengths can diminish the size of the reflected signal.
     c.   A third method of reducing the strength of the return signal is to configure the aircraft surface in such a fashion that its contours are not good reflectors of radar energy. Sharp angles, the openings in front and

the back of jet engines, sharp edges are all efficient
reflectors of radar radiation. By avoiding or masking
such features, the reflected radar signal from an
aircraft can be further diminished.

In combination these "Stealth" technologies can
reduce the radar visibility of an aircraft (piloted or
unmanned) by operationally significant amounts.

3. A new and significant development has taken
place in the guidance technology of land-based and
sea-based ballistic missiles. The nominal precision of
these missiles is now about one part in 10^5. Warheads of
MX and Trident II missiles will be capable of such
precision. A combination of improvements in guidance of
the missile itself and in the positioning accuracy of the
missile at launch have contributed to this improvement.
The Trident II, in addition, is taking advantage of a
newly developed star-tracker that enables the missile to
perform last-minute adjustments to its terminal velocity
before warhead release that improve the precision of the
re-entry vehicles.

4. Another system that can contribute substantially
to the accurate delivery of explosives onto their target
(in addition to assisting in the navigation of military
vehicles of all kinds) is the constellation of Ground
Positioning Satellites (GPS) also known as NAVSTAR.
Twenty-four satellites orbitted in six circular orbits of
4 at 22,000 km above the Earth carry exactly synchronized
radio transmitters that emit coded signals with a
simultaneity of 0.1 nanosecond. The difference in the
time of arrival of such signals from four different
satellites at a receiver on Earth or on an aircraft or
missile, enables the operator of the receiver to
determine his position to within 10 meters in 3
dimensions (50% of the time) and 30 meters (90% of the
time). The receiver, which is the size of a large book
and weighs about 10 pounds, can be carried by a missile -
ballistic, or cruise - aircraft, ship or land vehicle.
Thus GPS can facilitate enormously the accuracy of the
delivery of their munitions.

5. The development of powerful satellite-borne
synthetic aperture radar now under way can have
significant implications for the strategic balance
between the U.S. and U.S.S.R. Early results from a
similar radar carried by "Seasat", an experimental
ocean-surveillance satellite, have led to the belief that
the system could detect submarines submerged to
missile-launch or radio-contact depths by resolving the
surface effect of the bow-wave of a rapidly moving
submerged submarine. Little is known at this time about
this and other capabilities of such a radar, but if

adequate space-borne power supplies could be developed,
such a radar could improve the monitoring capabilities of
the United States but could also complicate to some
extent the operations of submarines as well as surface
ships which are clearly visible to such radar.

6. Perhaps the most profound but least visible
technical development that could have drastic effects on
the strategic arms race, the strategic stability during
crises, and the development of the doctrine of nuclear
warfighting, is research and development in the general
field of command, control, communications and
intelligence (C^3I).

The Reagan Administration has initiated a very large
scale program to develop a C^3I system able to survive and
operate during a protracted nuclear war. Confidence in
such a system could sometime in the future embolden a
political leadership to proceed with the initiation of
nuclear hostilities in the expectation (which could prove
optimistic) that its nuclear forces could be commanded
and controlled in a nuclear environment. Although it is
essential for avoidance of crisis instability and
accidental initiation of nuclear war, that the C^3I system
of a nuclear country can functon robustly immediately
before, during and immediately after a nuclear attack by
an opponent, this operational requirement is not the same
as those incorporated in a system designed to support a
protracted nuclear war. Therefore, the effort of the
Reagan government to install a nuclear-warfighting C^3I
capability must be seen as a very serious development.

CHAPTER 4

The Physics and Employment of
Neutron Weapons

Peter D. Zimmerman
Department of Physics and Astronomy
Louisiana State University
Baton Rouge, Louisiana 70803 USA

ABSTRACT

Well known physical relationships make it possible to estimate
the military effects of low yield nuclear weapons. Since weapons
effects tables already exist for fission explosives, it is possible
to calculate approximately the impact of thermonuclear and enhanced
radiation weapons (ERWs) on a tactical situation. The principal
physical problem is the calculation of the ratio of neutron
production in an ERW explosion to that in a fission explosion.

Our calculations indicate that the ERW is less effective
against tanks than is widely believed. The collateral damage for
an ERW with a reasonable ratio of fission to fusion yield is shown
to extend to a distance of about 75% of that of a fission weapon
with the same energy release.

While the engineering details of nuclear weapons must remain
secret, the underlying physical principles are well known. It is
possible to use those principles to estimate the performance of
nuclear weapons against specific targets, obtaining results which,
although approximate, are accurate enough to serve as guides for
reaching decisions on arms control, procurement and deployment
policies and, to some extent, on operational doctrine. This is a
case where physical reasoning can be of great value in reaching
conclusions on broad questions, even when the analysis is not based
upon classified data. While physics alone should not set policy,
scientific realities always act to constrain the feasible set of
policy alternatives.

This paper will focus particularly on those characteristics of
nuclear battlefield weapons proposed for use in the defense of
Western Europe, in particular those likely to be used on West
German territory. These weapons have relatively low yields,
typically on the order of one kiloton, in contrast to the 13
kilotons of the Hiroshima bomb. Because the Enhanced Radiation
Weapon (ERW), or "neutron bomb", is both a technical and political
issue, this paper will concentrate on the ERW and then look quickly
at other battlefield weapons in the nuclear stockpile.

0094-243X/83/1040071-14 $3.00 Copyright 1983 American Institute of Physics

As Sam Cohen, the "father of the neutron bomb", has sketched in his paper published by the Institute for Foreign Policy Analysis,[1] the conception of the neutron weapon dates back at least to the middle 1950's. Although successful tests of an ERW were conducted in the early 1960's, the principle has not yet found its way into the weapons stockpiles of NATO. One can piece together an idea of how the ERW functions based on freely available information. The physical principles of nuclear weaponry can then be used to deduce some numerical results, but one must use caution: these results are probably accurate to much better than a factor of two, but the dose/yield/slant-range relationship is surely uncertain to at least ten per cent.

In order to discuss the performance of an ERW, let us first review something of the physics of the weapon. It is well established that a neutron weapon must consist of a fusion explosive ignited by a relatively small fission bomb. Since fission produces most of the radioactive products which become fallout, it is clearly desirable to reduce the fission yield. If the goal is to shift the balance of weapons effects in favor of killing with neutrons, reduction of the fission yield is essential because most of the energy released in fission is available for blast and thermal effects.

A thermonuclear bomb contains a fission device, detonated by the implosion of conventional high explosives, and a fusion core compressed and ignited by the fission trigger. First, the conventional high explosives surrounding the fissile core detonate and compress the core. The fission trigger chain-reacts, releasing both neutron and high energy electromagnetic radiation, chiefly x-rays. The x-rays press against the fusion material which is a mixture of ^{6}Li chemically combined with deuterium and tritium, compressing and heating it at the same time as the neutrons convert some of the lithium-6 into helium and additional tritium. Some extra tritium is probably necessary in the original mixture since the T-T reaction produces surplus neutrons needed to convert the lithium-6. When the temperature and density rise enough, the D-T mix ignites, releasing energy, mostly as the kinetic energy of the neutrons. The hot gas of charged particles radiates x-rays which then interact with the weapon casing and the atmosphere to produce a fireball, and, finally, a blast wave.[2]

In a "classical" hydrogen bomb the neutrons produced by the fusion process are intercepted by a uranium casing which fissions, releasing even more energy. This increases greatly the blast and heat produced. A large fraction of the neutrons produced is trapped, and still more fallout is made.

The casing or "tamper" of an ERW is, however, constructed of materials which allow neutrons to pass almost freely. The neutrons escape with high energy, and can reach greater distances in lethal quantities than does the blast wave or heat pulse if the yield is

less than about 10 kT. Thus, at some distance from the explosion
the shock wave from an ERW might not damage buildings severely,
while the neutrons could reach inside to cause death.

Some of the nuclear ractions which are useful for our
understanding of an ERW are shown below. Only particles which
carry an electric charge can produce the electromagnetic radiation
which initiates the fireball. When plutonium fissions, the charged
fission fragments get most of the energy which then emerges as
(initially) electromagnetic radiation. In contrast, when deuterium
and tritium fuse to make helium and one neutron, the neutron gets
80% of the energy. Only if the neutron is absorbed in a nucleus
near to the very center of the explosion will the neutron energy
contribute very much to the destructive power of the weapon. This
is the principle by which an increased fraction of the energy of a
thermonuclear bomb can be directed to the neutrons and away from
the blast.

1. $n + {}^{239}Pu \rightarrow f.f. + 2n$

2. ${}_{6}Li + n \rightarrow {}_{3}H + {}_{3}He$

3. ${}_{2}H + {}_{2}H \rightarrow {}_{3}He + n \; ; \; {}_{3}H + p$

4. ${}_{2}H + {}_{3}H \rightarrow {}_{4}He + n$

5. ${}_{3}H + {}_{3}H \rightarrow {}_{4}He + n + n$

Instead of using pure liquid or solid tritium or deuterium in
a fusion weapon, a well known trick is used: the stable isotope
lithium-six is chemically combined with deuterium or tritium. When
a neutron from the fission bomb strikes the lithium-6, the lithium
decomposes providing a tritium nucleus in situ, very close to a
deuterium or another tritium nucleus with which it can fuse
(reaction 2).

The D-D fusion process (reaction 3) is not useful for an ERW
because only about one third of the reaction energy goes into the
neutrons. The rest goes into charged particles (proton, helium-3
and tritium). By holding the thermonuclear burning temperature
below the ignition point for the D-D reaction, the deuterium fusion
contribution can be suppressed.

Reactions four and five are the desirable ones. In the D-T
reaction the neutron carries off 80% of the energy. In the T-T
case, the neutrons take about 2/3 of the energy, leaving more for
other weapons effects than in reaction 4, but two neutrons are
produced. The D-T reaction may, nevertheless, be the better choice
for the majority of the energy release, because the more energetic
14 MeV/neutrons will travel farther.

One fission reaction releases around 180 MeV; one D-T fusion
reaction produces only 17 MeV. Thus, 10 fusion reactions are
needed to release as much energy as one fission. This permits us

74

to calculate the ratio of the numbers of neutrons produced by a pure fission weapon to those produced by an ERW.

What is "the" Enhanced Radiation Weapon? It is impossible to answer this question with precision because of the secrecy which necessarily surrounds the details of nuclear weapons engineering. Since there is no unclassified description of "the" ERW, let me propose a "candidate radiation enhanced weapon" (CREW, with apologies). This is not a design, but a system specification which will allow the estimation of the lethality of the weapon.

Assume a one-kiloton total energy release ERW of which 25% of the energy is derived from the fission trigger and 75% from the D-T fusion reaction. Since 20% of the fusion energy release is available for normal weapons effects blast and heat, and since about 70% of the energy of a fission weapon goes into those same effects, the fusion part of the weapon contributes an effective 0.21 kT and the fission trigger 0.25 kT, when the CREW is compared to a pure fission device. The "effective yield" of the CREW is thus at least 0.46 kT for the purposes of computing its normal military effects. The destruction radii for given overpressure from the CREW will be 75% $[0.46^{1/3} = 0.77]$ of those from a 1 kT fission weapon, and the thermal destruction radii of the CREW will be about 68% those from the same 1 kT device $[0.46^{1/2} = 0.68]$. A 1 kT fission weapon can cause severe damage to most buildings to a distance of 690 meters (5 p.s.i. contour) and moderate damage to a distance of 1280 meters (2 p.s.i. contour). The CREW causes similar damage[3] at distances of 520 m and 960 m.

It is said that the ERW will cause only minimal "collateral damage;" the CREW does less damage to structures than a fission explosive of equal yield, but it remains

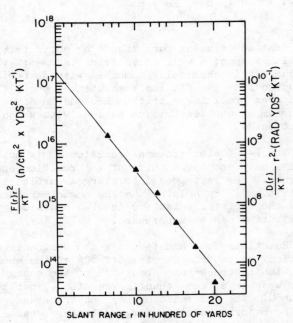

Fig. 1. Neutron dose per kiloton yield absorbed in actual tissue samples. From Ritchie and Hurst, ref. 4.

a considerable amount of damage. The calculation is somewhat sensitive to the ratio of fission trigger yield to fusion yield, but the blast destructive radius of a nuclear weapons varies only as the cube root of the equivalent yield. If the fission yield dropped to 0.1 kT, the equivalent yield would still be 0.36 kT, and the 5 psi contour would shrink to 490 m, 71% of that of a 1 kT fission weapon. The area destroyed by the CREW would be about half that destroyed by the fission weapon. Whether or not a tenth-kiloton fission explosion can ignite the fusion core cannot be answered from the open literature.

The neutron dose as a function of distance from a 1 kT fission bomb is well documented in the open literature.[4] Figure 1 gives neutron radiation absorbed in tissue as a function of distance for various energies. In order to assess the utility of neutron radiation from the CREW the ratio of the neutron production per kiloton of the CREW to that of a fission weapon must be calculated. The problem is not calculating the number of fusion neutrons, but rather estimating the number of fission neutrons leaving a fission bomb so that the published data can be used.[*] One finds that for equal energy releases, a pure D-T fusion explosion produces about 5 times as many neutrons as does a fission bomb.

The 1 kT CREW produces, however, only 4.1 times as many neutrons as a 1 kT fission weapon. This is because a fission trigger with 0.25 kT yield has been assumed. Even in the total absence of a fission trigger, the ratio of D-T neutrons to fission neutrons for the same energy release cannot exceed about 5:1 for the D-T reaction.

Much has been said about the additional range of 14 MeV fusion neutrons. In fact, after one collision with a nitrogen nucleus 14 MeV neutrons are reduced in energy to 10 MeV, the end point of the fission weapon neutron spectrum. Fig. 8.116b of Glasstone and Dolan[2] indicates that very fast neutrons constitute only about 15% of the total neutron fluence from a thermonuclear weapon. This fraction will be higher in a CREW, but not enormously so, because of the presence of a finite fission yield and because the neutrons are released in a quite dense plasma of hydrogen and helium and may suffer collisions while escaping. In Fig. 8.117b Glasstone and Dolan show that the fraction of very fast neutrons in the fusion

[*]Most of the energy release from a fission explosive comes from the last neutron generation. It is the last generation precisely because most of the neutrons produced do not cause additional fissions. Hence the neutron release is approximately given by $(2.3/e) \times [\text{yield}]/[180 \text{ MeV/fission}] + (1-1/e) \times [\text{yield}/180 \text{ MeV/fission}]$ since, roughly, one neutron/fission escapes in the earlier generations of the reaction. Yield must be translated into MeV units.

Fig. 1 includes the exponential dependence of dose on distance, but not the $1/r^2$ dependence; thus Fig. 1 is proportional to the total number of neutrons at a distance r. Knowing yield, distance, and the ratio of CREW to fission weapon neutron production, neutron dose rates from the CREW at given slant ranges can be estimated (Fig. 2). In order to calculate distances from

Fig. 2. Neutron dose absorbed in tissue sample for one kiloton weapons. The dose from the CREW has been estimated by calculating the number of D-T fusion events needed to release 0.75 kT of energy and assuming that all fusion neutrons escape the weapon casing.

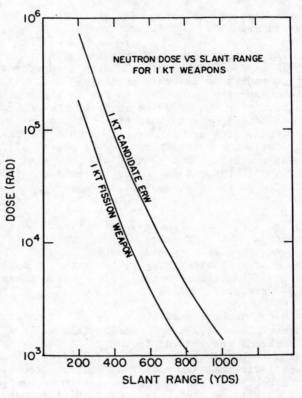

ground zero, a value for the height of burst (HOB) must be assumed. We have used 700 feet, (213 m). This value maximizes the area exposed to 10 psi overpressure, a value which can cause blast damage to military targets. The 5 psi overpressure region, within which civilian structures are usually destroyed, is reduced for this choice compared to what it would be if the HOB were chosen to maximize destruction to civilain structures. Cohen assumes a 500' (152 m) HOB which slightly reduces the area exposed to blast damage. It also expands slightly the distance at which the neutrons are effective. Dose as a function of distance from ground zero is shown in Fig. 3 for a 213 m HOB.

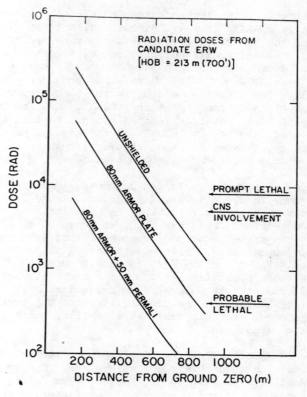

RADIATION DOSES FROM
CANDIDATE ERW
[HOB = 213 m (700')]

Fig. 3. Neutron dose
rates produced by a
1 kT CREW as a
function of radius
from ground zero.
Estimates of the
effectiveness of
shielding are
shown, as are the
dose levels at
which lethality
occurs or prompt
central nervous
system involvement
begins.

It is evident from Fig. 3 that unshielded victims above ground at the 5 psi distance (520 m), will receive a lethal dose of (10,000 rad) neutron radiation, which is enough to disable a victim within 5 minutes and to cause almost immediate death.[5] Only in earth covered basement shelters would civilians have a chance to survive. If 8000 rad is taken as the "prompt lethal dose," the CREW is instantly effective against unprotected personnel out to distances of 620 m from ground zero. Lethal doses of neutron radiation will be found at all distances less than 1100 m. However, unprotected victims at 1100 m will probably suffer death or serious injury from other weapons effects as well. The ERW would be highly effective against troop concentrations in barracks and staging areas.

The ERW is intended, however, as a way to stop tank attacks by killing tank crews, and the armor of a tank provides a certain measure of shielding against neutrons. Steel is not, however, a very good neutron absorber, particularly for fast neutrons. It is estimated that for every 178 mm (7") of armor the neutron flux is reduced by a factor of 10; the turret side and rear armor on a Soviet T-72 tank is believed to be 80 mm thick.[6] Because of air scattering, the neutrons from an ERW actually approach a tank very nearly equally from all directions, and not just on a radial line from the blast. Therefore, the average neutron will traverse $\sqrt{2}$ x 80 mm, or 113 mm, of steel as it enters the turret of a T-72. Armor elsewhere on the tank can be considerably thicker, up to at least 170 mm [6] The 80 mm armor provides a shielding factor of 4.13. The actual shielding to the tank crew will be greater since much of the armor is thicker than 80 mm. For 80 mm armor, however, the 8000 rad level is reached 415 meters from ground zero, and the 400 rad level at 845 m. Since the average armor on the tank is thicker, this calculation is overly optimistic from the point of view of a defender who employs an ERW.

Experiments performed on macaque monkeys (Macaca mulatta) by Curran, Young, and Davis of the U.S. Armed Forces Radiobiology Research Institute, confirm the anticipated effects of a dose of 8000 rad. On the other hand, at doses up to 4900 rads the same authors state: "the performance of most subjects generally approached base-line levels following the initial decrement period."[7] Since the West intends to use the ERW defensively, one must make pessimistic assumptions about its lethality. At least 8000 rad must be delivered to crews inside their tanks. Lower values leave open the probability that the tank crews will continue in control of their vehicles in a final kamikaze action inflicting heavy casualties.

Any borated hydrogenous plastic material will provide more protection than steel, and weigh less. One such material is Permali (by weight: H-6%; B-3%; O-49%; C-38%; Na-2%). It attenuates a beam of neutrons by a factor of 10 in 75 mm (3").[5] Assuming that the 80 mm tank armor received a further 50 mm of Permali (which has only 17% the weight of the same thickness of armor) cladding, an additional shielding factor of 9 is achieved. The weight of the tank would increase by much less than 10%. Doses inside shielded tanks are also shown in Fig. 3. If enemy tanks were equipped with a 50 mm Permali shield, 8000 rad dose levels would only be achieved out to 160 meters from ground zero. Since Soviet tank spacing is typically about 150 meters, one CREW might immediatedly stop one tank. Personnel in more distant tanks would be fatally injured, but not promptly incapacitated.

Military Review, a U.S. Army publication, reported in October, 1980 that the newer T-72S version had had its turret "encased in a box structure of welded plates, the cavity being filled with plastic materials capable of limiting the effects produced by hollow charges."[8] The defense must assume that many Soviet tanks will be so equipped and that the plastic may be borated as well. In any event, modern composite armor probably includes some plastic components and so is a better neutron shield than steel armor plate. This is an interesting case where Western emphasis on nuclear arms led Soviet designers to add light stand-off armor, which as a side effect reduces the effectiveness of conventional small diameter shaped charge ammunition.

Figure 4 shows, in a scale drawing, a Soviet tank force. The circles superimposd on the tanks indicate significant distances from ground zero. It is evident that one CREW might immobilize a

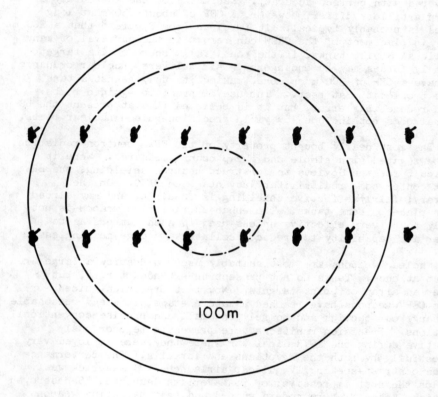

100m

Fig. 4. Scale drawing of Soviet tank force deployed in echelon and approaching contact with defending forces. The outer solid circle shows the area where crews inside permali shielded tanks receive a 400 rad dose. The dashed circle shows the range at which an 8000 rad (prompt lethal) dose is received inside tanks with no permali cladding. The dash-dot circle shows the 8000 rad radius inside a permali-shielded tank.

significant number of tanks with <u>no more than 80 mm</u> of armor. Any
additional steel shielding, particularly on the side facing the
burst, would add increased protection and so decrease the number of
tanks stopped. Thin hydrogeneous shielding on the tanks will
enormously reduce the military effect of the weapon. Simply
increasing the distance between tanks by 25% will also greatly
reduce the number of casualties.

If the ERW is fired from an artillery tube, it cannot be
delivered with perfect accuracy. Estimates of the inaccuracy of
modern artillery differ, however, a CEP of about 250 m at long
ranges is probably typical. It is clear from Figure 8 that a 200 m
error in placement of the CREW can result in the survival of many
additional Soviet tanks; if the round falls short of its target,
friendly forces may be endangered. Air delivered nuclear ordnance
may have a CEP of 300 m or more, making its use against a tank
force problematic at best.[9] The use of precision-guided bombs
would reduce this error, but it is believed that at present the US
arsenal does not include low yield precision-guided nuclear bombs.

The most dearly bought property of the ERW, neutron radiation,
can be negated with simple and cheap countermeasures. Were it
difficult for the Soviets to construct neutron shielding, the ERW
might confer some real military advantage on NATO. On the
contrary; fairly effective shielding is feasible, and may already
be in place on some tanks. The enhanced radiation weapon is not
likely to be very effective unless used in such numbers as to
render its capability to "reduce" collateral damage meaningless.

Nuclear weapons are most suitably used to destroy a large area
target at once. Tanks do not present an extended target, but
rather an array of point targets. Now that precision guided, or
"zero CEP" weapons have reached a mature stage, it seems reasonable
to change our tactics and to rely on small and precise conventional
munitions. Precision munitions were proven to be enormously
effective during the 1973 Sinai war where they were employed very
successfully by both the Egyptians and Israelis. The performance
of the contrast-seeking TV guided, airlaunched US Maverick was
perhaps the most impressive; of 58 Mavericks launched, 55 destroyed
Egyptian tanks.[10] More modern guided and self targeting weapons
from <u>Milan</u> and <u>HOT</u> to <u>SADARM</u> and <u>Hellfire</u> should be even more
effective. Modern long wavelength infra-red (10 micron) guidance
systems retain a large fraction of their effectiveness even in fog
or rain. When precision guided weapons are supported by powerful
unguided missiles, such as the German Panzerfaust III and the
French APILAS, a robust and affordable conventional defense is
clearly seen to be possible.[11] (These are all large diameter
shaped-charge weapons, useful against the T-72 with standoff
plates.)

The U.S. and Soviet nuclear arsenals include many other
tactical nuclear weapons. These weapons range in yield from ones

with very nearly the same power as the largest conventional bombs dropped in World War II up to ones with greater destructive force than the Hiroshima bomb. The specifically battlefield weapons are primarily those fired from artillery guns and have yields which are in the same general range as the ERW. The same kind of analysis which was applied to determine the utility of the ERW can be applied to these nuclear shells. The lethal blast range of a 1 kT fission weapon used against tanks is comparable to or less than the prompt-lethal radiation radius from the CREW. The lethal range from radiation effects is less. It is, therefore, clear that nuclear artillery shells would also not be particularly effective weapons in a European conflict, and would inflict many more civilian casualties even than the ERW. Recognition of this fact probably led to the development of enhanced radiation warheads in an effort to gain a more useful and useable atomic weapon.

Another proposed use for nuclear explosives is as land mines. The American arsenal lists two types of atomic demolition munitions: the MADM or Medium Atomic Demolition Munition with a yield of a few kilotons, and the SADM or Small ADM. The SADM has a yield estimated in the open literature at about 50 tons. Even if all of an ADM's yield goes into cratering, earth-moving, effects, 50 tons of nuclear energy release is about equal to 50 metric tonnes, of TNT. Since modern military explosives can be two to twenty times more powerful than TNT, perhaps as little as ten metric tonnes of chemical explosives might be equivalent to one SADM. Furthermore, the high explosive could be distributed into a shaped array which would be much more effective than a nuclear blast coming from a single point. Since a shallow underground nuclear burst is inherently radiologically dirty, the high explosive is vastly to be perferred.[12]

The armed forces of NATO are aware of the lack of military effect obtained from the use of single nuclear weapons, and for that reason it is planned to use such weapons in masses. Hypothetical scenarios routinely call for tens of nuclear weapons within only a few hours. One such scenario, taken from a US Army Field Manual, is shown in Fig. 5.[13] The number of weapons to be used in one small area and in one short time span is truly astonishing.

Perhaps one of the few genuinely important properties of tactical nuclear weapons is that the potential for their use forces the invader to disperse his tanks, which denies him the advantage of concentration. A dispersed force is inherently more vulnerable to a conventional defense. It is clear, then, that a non-nuclear strategy should employ multiple warhead indirect fire weapons which can attack several tanks at one time and at long range. The Assualt Breaker missile will provide a way to attack ten or more tanks in a circle at least a kilometer across. If Assault Breaker is deployed it will force tank dispersal in the same way that a TNW does. The Assault Breaker is not yet operational, and as is the

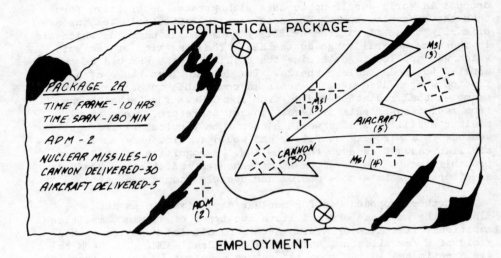

Fig. 5. Hypothetical nuclear package as shown in a U.S. Army field manual (ref. 13).

case with most high technology systems, it will probably not be as effective as its proponents claim in its first deployed form. Nevertheless, from its initial deployment it will pose a conventional threat which will force Soviet tanks to operate in dispersed formations.

It is often suggested that neutron activation of the soil by ERWs will pose a serious, long-term, problem. An article by C. Sharp Cook[14] gives some idea of the soil activity which might be expected from an airburst or ground burst. The airburst had a yield of 10 kT and a burst height of 150 meters. The scaling of the air burst is such that its effects can be considered similar to those a 1 kT ERW might produce. Thus, the dose rates as given in Cook's article should be correct for the CREW to a factor of two. The essential point is that most of the region surrounding ground zero has activity at the levels of a few rem per hour, not a few kilorems per hour. One hour after the burst the one rem per hour contour is about one mile across its smaller diameter. Three days later it is only about one half mile wide. After one week, it was negligible in extent. Thus, people occupying the region near ground zero a few hours after an ERW explosion and remaining there unprotected for a week might expect to receive a significant gamma dose with serious long term, but few short term complications. Since only about 3 moles of neutrons are produced per kiloton of fusion energy release, this result is not surprising.

According to the slide rule calculator included in Glasstone and Dolan[2], the fireball radius of a 1 kT fission weapon is around .05 mile or 80 meters. The fireball from a burst at an altitude of greater than 100 m should not touch the ground, and there should be little prompt fallout. Most ofthe residual radiation will come from neutron induced activity in sodium, manganese and aluminum. The residual activity from a surface burst is much higher, but results from fall-out and not neutron activation.

Seen purely from the physicist's vantage point, neutron weapons are an inelegant way to halt a <u>Blitzkrieg</u>. Nuclear weapons can destroy large "soft" targets, such as cities, with ease. If nuclear weapons have any military role which cannot be fulfilled by other weapons, it is just that: attacking extended targets. Armored forces present an array of point targets which now can be attacked with zero CEP weapons. That is <u>their</u> proper use. Even if the use of tactical nuclear weapons posed no political questions and would not lead to further escalation, countermeasures are simple, cheap and available. Shielding and dispersal negate the purely nuclear effects on an aggressor's forces, particularly if Federal Germany is not "to be destroyed in order to save it" by the use of more, and more powerful, nuclear weapons.

Nuclear arms are required as a deterrent, but their use must be tailored to suit their character. If nuclear deterrence fails, the enemy, to quote Alton Frye must "pay the entry price on his own territory."[15] This would call for a posture which emphasizes a robust conventional defense, made believable by an renunciation of the apparent NATO reliance on tactical nuclear weapons. At the same time, it entails the formulation of a strategy in which the distinction betwen nuclear and conventional warfare is made clearer and sharper. The Soviet Union must be made aware that NATO can and will resist a conventional attack with conventional weapons. And it must be made clear that NATO will respond to any use of nuclear weapons by using nuclear weapons against targets on Soviet and Warsaw Pact, not Western, territory.

Acknowledgements

Thanks are due to Dr. G. Allen Greb, Dr. James Digby, Dr. C. Ivan Hudson and Prof. Herbert F. York for stimulating conversations. Many active and retired officers of the United States Army and of the armed forces of other nations made significant contributions at crucial moments; in particular, Col. Carl Bernard (USA-ret) deserves special acknowledgement.

References

1. S. T. Cohen, "The Neutron Bomb," Institute for Foreign Policy
   Analyses, Special Report, November, 1978. (Cambridge, MA).
2. S. Glasstone and P. Dolan, The Effects of Nuclear Weapons,
   Third Edition, 1977. U. S. Departments of Defense and Energy.
3. Information extracted from the slide rule calculator supplied
   with Ref. 2 and from unclassified parts of the (U.S.) Defense
   Intelligence Agency publication, Physical Vulnerability
   Handbook - Nuclear Weapons (u), AP-550-1-2-69-INT.
4. R. S. Ritchie and G. S. Hurst, Health Physics 1, 390 (1959).
5. D. Gans, Military Review LXII, 19, (January, 1982).
6. Modern Tanks and Armored Fighting Vehicles, (London: Jones
   Publishing, 198).
7. C. R. Curran, R. W. Young and W. F. Davis, "The Performance of
   Primates Following Exposure to Pulsed Whole-Body-Gamma-Neutron
   Radiation," Armed Forces Radiobiology Research Institute,
   Defense Nuclear Agency, Bethesda, MD.   Report AFRRI SR73-1,
   January, 1973.
8. "First Deliveries of T72S with Cavity Armor," Military Review,
   October, 1980, p. 89., cited in Gans, Ref. 6.
9. Taken from a sample problem in DIA handbook, Ref. 3, Section IV
   B.
10. J. Digby, private communication.
    J. K. Miettinen, "Can Conventional New Technologies and New
    Tactics Replace Nuclear Weapons in Europe?" in Arms Control and
    Technological Innovation, Carlton and Schaerf, eds.,(London:
    Croom and Helm, 1977).
11. Col. Carl Bernard, U. S. Army, Retired. Private communication.
12. Data on nuclear weapons stockpiles are taken from publications
    of the International Institute for Strategic Studies, London
    and of the Center for Defense Information, Washington, DC.
13. U.S. Army Field Manual FM-100-5, Operations. Figure 5 is taken
    directly from FM-100-5 but has been re-drawn to permit better
    reproduction.
14. C. Sharp Cook, "Initial and Residual Ionizing Radiations from
    Nuclear Weapons," in P. H. Attix and E. Tochilin, eds.,
    Radiation Dosimetry, Second Edition (New York: Academic Press,
    1969) Vol. III, Chapter 24; and C. S. Cook, private
    communication.
15. Alton Frye, "Nuclear Weapons in Europe: No Exit from
    Ambivalence," Survival, 22 (May/June, 1980) pp. 98-106.

CHAPTER V

SEISMIC METHODS FOR VERIFYING NUCLEAR TEST BANS

Lynn R. Sykes
Lamont-Doherty Geological Observatory and Department of Geological
Sciences of Columbia University
Palisades, New York 10964

Jack F. Evernden
United States Geological Survey
Menlo Park, California 94025

and Inés Cifuentes
Lamont-Doherty Geological Observatory and Department of Geological
Sciences of Columbia University
Palisades, New York 10964

ABSTRACT

Seismological research of the past 25 years related to verification of a Threshold Test Ban Treaty (TTBT) indicates that a treaty banning nuclear weapons tests in all environments, including underground explosions, can be monitored with high reliability down to explosions of very small size (about one kiloton).   There would be high probability of successful identification of explosions of that size even if elaborate measures were taken to evade detection. Seismology provides the principal means of detecting, locating and identifying underground explosions and of determining their yields. We discuss a number of methods for identifying detected seismic events as being either explosions or earthquakes including the event's location, depth and spectral character.   The seismic waves generated by these two types of sources differ in a number of fundamental ways that can be utilized for identification or discrimination.   All of the long-standing issues related to a comprehensive treaty were resolved in principle (and in many cases in detail) in negotiations between the U.S., the U.S.S.R. and Britain from 1977 to 1980. Those negotiations have not resumed since 1980.   Inadequate seismic means of verifying a CTBT, Soviet cheating on the 150-kt limit of the Treshold Test Ban Treaty of 1976, and the need to develop and test new nuclear weaspons were cited in 1982 by the U.S. government as reasons for not continuing negotiations for a CTBT. The first two reservations, which depend heavily on seismological information, are not supported scientifically.   A CTBT could help to put a lid on the seemingly endless testing of new generations of nuclear weapons by both superpowers.

0094-243X/83/1040085-49 $3.00 Copyright 1983 American Institute of Physics

INTRODUCTION

The issue of a comprehensive nuclear test ban treaty has been on the arms control agenda longer than almost any other item. Serious scientific and political interest in a comprehensive test ban is now almost 30 years old. A major program of scientific research related to the verification of a complete test ban is almost 25 years old. In the United States more than $600 million has been spent on scientific research and instrumentation related to a possible comprehensive ban on weapons testing. 1983 is the 20th anniversary of the Limited Test Ban, which put an end to testing in the atmosphere, space and underwater by the 125 nations that have signed that treaty. It did not, however, limit underground testing since seismological understanding was then inadequate to permit reliable monitoring of underground explosions.

Announced policy of the United States Government has been to enter a Comprehensive Test Ban Treaty (CTBT) provided such a treaty could be adequately verified. A CTBT, of course, does not exist as of 1983. Two other treaties related to atomic testing have been signed by U.S. Presidents but have not been ratified by the U.S. Senate. The parties to those treaties, the United States and the Soviet Union, however, have stated that they would abide by the terms of those two treaties. The Threshold Test Ban Treaty (TTBT), which was signed in 1974 and went into effect in 1976, places an upper limit of 150 kilotons (kt) on the maximum sizes of underground weapons tests by the two countries. It also restricts testing to specific and designated test sites. The protocol to the treaty requires each party to provide certain geologic and geophysical data on each testing area to the other party when ratification is complete. It also calls for exchange of information on yields of calibration explosions for each geophysically distinct testing area. The Peaceful Nuclear Explosions Treaty (PNET) of 1976 also prohibits individual explosions larger than 150 kt, provides for on-site inspection of any grouped peaceful explosions larger than 150 kt and limits the aggregate yield of such a group to 1500 kt. Neither the U.S. nor the U.S.S.R. has detonated peaceful explosions larger than 150 kt since 1976.

A halt to the testing of all nuclear weapons was the original goal of the negotiations that led to the 1963 Limited Test Ban. After years of desultory discussion, apparently serious negotiations for a comprehensive test ban were resumed in 1977 by the U.S., the U.S.S.R. and Britain, but the talks were suspended in 1980. In July 1982, the United States Government announced that the test-ban negotiations with the U.S.S.R. and Britain would not be resumed. Three reasons were cited for that decision: 1) the lack of adequate seismic methods for verifying a comprehensive test ban; 2) an assertion that the Soviet Union either had or may have cheated repeatedly on the Threshold Test Ban by detonating explosions above the 150 kt limit; and 3) the need to rapidly develop and test new atomic weapons.

In this paper we address each of the three reservations cited for not continuing the negotiations for a CTBT. We expand on several of the points mentioned by Sykes and Evernden[1] in the October 1982 issue of <u>Scientific American</u> related to the verification of a comprehensive treaty. We also examine the determination of yields of U.S.S.R. weapons tests. A more extensive paper on that subject is in preparation by Sykes and Cifuentes. Our main conclusions are that a comprehensive test ban can be verified with high reliability for explosions as small as 1 kiloton even if extreme measures were taken to evade or cheat on a comprehensive treaty. The seven largest Soviet explosions since the TTBT went into effect in 1976 are of nearly identical yields; their yields being very close to 150 kt, the limit set by the Threshold Treaty. Since the three of us are seismologists, most of this paper is devoted to seismological verification of the TTBT and a CTBT. We do, however, address the proposition that the security of the United States and other countries has decreased and will continue to do so in the face of continued atomic testing. Our sense is that the issues to be resolved related to nuclear test bans are political even though statements continue to be made claiming that seismic methods are insufficient to permit a CTBT to be adequately verified.

The last two years have seen renewed public interest in arms control, a comprehensive test ban, and the prevention of nuclear war. A comprehensive test ban would be an important measure. By prohibiting the testing of nuclear weapons, a comprehensive treaty could inhibit the development of new weapons by the major nuclear powers. From our present perspective, it is clear that a comprehensive test ban in 1963 or soon thereafter could well have prevented or seriously slowed the development of the anti-ballistic missile (ABM), missiles carrying multiple warheads, the neutron bomb, the MX, Trident and Cruise missiles, and comparable Soviet weapons. During the last year, press reports have discussed the testing and development of a third generation of atomic weapons. A comprehensive test ban could help to put a lid on yet another round of the arms race by both the U.S. and U.S.S.R. A major purpose of a CTBT always has been to help prevent the spread of nuclear-weapons technology to other countries. Again and again, other signatories of the Non-Proliferation Treaty (NPT) have stressed to the U.S.A. and the U.S.S.R. the importance of a CTBT to the continued viability of the NPT. The Limited Test Ban Treaty and the NPT both obligate the United States and the Soviet Union to negotiate and proceed toward a comprehensive test ban. A CTBT could also help to provide a more favorable climate for additional arms control agreements.

NUCLEAR TESTING FROM 1945 TO 1982

Figure 1 shows the yearly number of nuclear tests conducted by the six countries that have tested atomic weapons between 1945 and 1982. The Limited Test Ban Treaty (LTBT) of 1963 had a positive effect in that the U.S., U.S.S.R. and U.K., which signed the treaty, did not test in the atmosphere after 1963. The LTBT put an end to the testing of very large weapons, i.e., those of tens of megatons. It clearly did not put an end to testing itself, which merely went underground after 1963. The Limited Test Ban Treaty had another negative effect in that public interest in test bans diminished greatly after 1963. Tests in the atmosphere, with their attendant

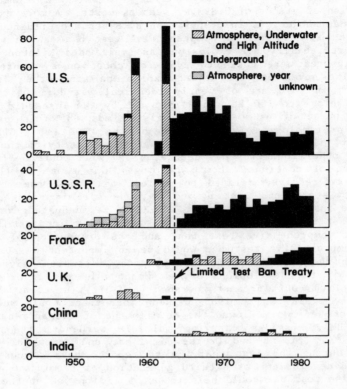

Fig. 1. Yearly number of nuclear explosions for the six countries that have tested nuclear weapons. Note that Limited Test Ban Treaty of 1963 (broken vertical line) did not reduce atomic testing nor halt the arms race. Testing merely went underground after 1963 by those countries that signed the Limited Test Ban. In 1982 the following numbers of nuclear tests (not shown in figure) were conducted: U.S. 17, U.S.S.R. 31, France 5, and U.K. 1.

fallout and mushroom-shaped clouds, were a much less abstract phenomenon to most of the public than tests underground. In our estimation, underground radioactive contamination is very much a fifth rate problem compared to the first order problem of the continuation of the nuclear arms race, which underground testing continues to fuel.

The number of weapons tests from 1963 to 1982 greatly exceeds the number of tests prior to 1963. China and India both became nuclear powers after 1963. Both France and China, which did not sign the LTBT, have tested in the atmosphere since 1963. France announced that as of 1974 she would no longer continue to test in the atmosphere. Hence, during the last ten years China is the only country that has tested in the atmosphere. About 10 to 20 percent of the tests conducted by the U.S.S.R. since 1974 have been peaceful nuclear explosions (PNE's). The United States has not conducted PNEs since 1973.

By examining the yields of U.S. and U.S.S.R. underground tests from 1963 to 1982, we find that both countries detonated a comparable number of tests of intermediate yield (20 to 200 kt) and a comparable number of larger tests. The largest underground test to date was the Cannikin explosion conducted by the U.S. in the Aleutian Islands in 1971. Our statements about the profile of testing as a function of yield rely upon a careful calibration of yields of Soviet tests, which is discussed later. In our judgment, claims that the U.S.S.R. tested underground explosions larger than 5,000 kt between 1963 and the time the TTBT went into effect in 1976 are based on a miscalibration of yields as estimated from short-period seismic waves. In contrast to the comparable profile of tests larger than 20 kt, the United States conducted many more smaller tests than the Soviet Union from 1963 to 1982. This asymmetry can be ascertained from published reports by the Stockholm International Peace Research Institute.

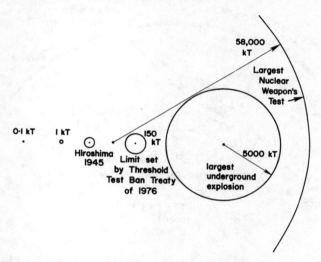

Fig. 2. Approximate sizes of nuclear weapons that have been tested, 1945 to 1982.

Figure 2 shows the great range in the size or yields of atomic weapons that have been tested thus far. One kiloton (kt) is equivalent in energy release to that of 1,000 tons of TNT. The sizes of weapons that opened the atomic age in 1945 were about 12.5 to 22 kt. With the advent of the hydrogen bomb, weapons of much larger yield were tested. The largest test thus far was the 58,000 kt Soviet explosion of 1961 in the atmosphere at their Arctic test site.

Some measure of the size of weapons tests can be realized from the fact that 5,000 kt, the approximate size of the largest underground explosion to date, is approximately equivalent in explosive power to all of the munitions expended in World War II. Announced yields of weapons carried by intercontinental ballistic missiles are in the range 40 to 9,000 kt. Both the U.S.S.R. and the U.S. have deployed thousands of weapons of that size range. While the Threshold Test Ban Treaty of 1976 put an end to the testing of larger weapons in that range, it did not limit the testing of all weapons carried by intercontinental missiles. As we discuss later, the smallest Soviet explosion that could be reliably identified under a comprehensive test ban is in the range 0.1 to 1-2 kt depending upon what assumptions are made about the deployment of seismic stations in the Soviet Union and about attempts to evade or cheat on a comprehensive treaty.

SEISMIC WAVES FROM EARTHQUAKES AND UNDERGROUND EXPLOSIONS

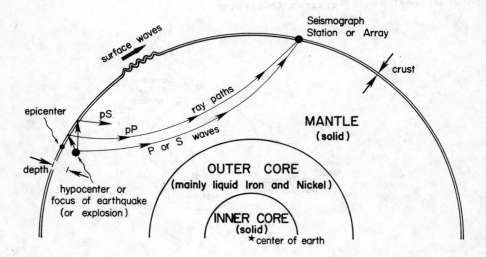

Fig. 3. Cross section of the earth showing ray paths of various types of seismic waves. The earth's crust is very thin beneath oceanic regions (about 6 km) and is about 30 to 50 km thick beneath most continental areas. Except beneath local volcanic centers the earth's mantle is solid for the propagation of seismic waves. A pP wave is a compressional wave that is produced by the reflection of a P wave from the surface of the earth just above the focus of an earthquake or an explosion. The hypocenter is the focal point of an earthquake or an underground explosion from which seismic waves radiate. The epicenter is a point on the earth's surface directly above the focus.

An underground explosion sets up elastic vibrations or seismic waves that propagate either through the earth or near its surface (Figure 3). Once the waves have been detected, the main task is to distinguish the seismic signals of explosions from those of earthquakes. This can be done with a network of several widely separated seismometers.

Two types of elastic vibrations can propagate through the solid body of the earth, that is, through the crust and mantle. The first waves to arrive at a seismograph station or array of instruments are compressional waves called P waves by seismologists. Slower body waves called shear waves (S waves) involve particle motion that is perpendicular to the direction of wave propagation (i.e., perpendicular to the wavenumber vector). Since the earth is nearly isotropic, only one type of shear wave propagates along each ray path. The ray paths of P or S waves in the earth's mantle generally follow paths that are concave upward since the appropriate seismic velocities generally increase with depth.

An underground explosion is a source of nearly pure P waves since a radially symmetric shock wave is generated by the explosion. An earthquake, on the other hand, derives its radiated energy from relaxation of a pre-stressed rock mass as rocks of the earth's crust or mantle slide rapidly past one another along the plane of a fault. This shearing motion at the earthquake source leads to the radiation of predominantly S waves. Thus, that source is fundamentally different from that of an underground explosion. A result of the spherical compressional symmetry of an explosion source is that all of the seismic waves it generates have a nearly radial symmetry about the focus (or hypocenter) and no shear waves develop other than by conversion upon reflection or refraction. In contrast, the highly directional character of an earthquake source gives rise to seismic waves with strongly asymmetric patterns (Figure 4). The asymmetry and amplitudes of the waves received at seismic instruments throughout the world provide the means by which seismologists can determine the faulting mechanism of a given earthquake.

In addition to the P and S body waves, there are also two types (again because of isotropy) of seismic waves that propagate only over the surface of the earth. They are called Rayleigh and Love waves after the scientists who first developed their theory. These waves result from complex reflections of P and S waves in the upper layers of the earth. (A non-dispersed Rayleigh wave does propagate on an elastic half-space.) A simple explosion can generate Rayleigh waves but not Love waves, whereas an earthquake generates waves of both types.

Earth noise (microseisms) dictates the pass bands over which instruments can be used to detect waves from small explosions and earthquakes. Microseisms are very large at most periods in the range 1.5 to 18 seconds. Seismic waves with frequencies higher than a few to about 10 hertz are usually attenuated such that they cannot be detected at great distances, i.e., thousands of kilometers. (Higher frequencies are sometimes observed for certain paths.) Thus, instruments for detecting P waves typically have pass bands of

92

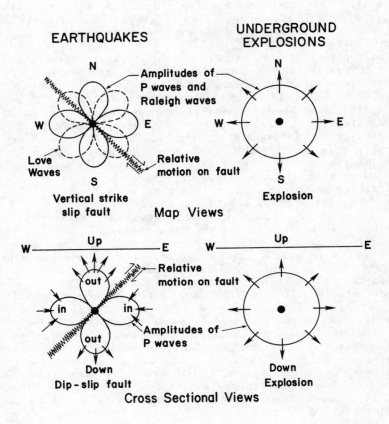

Fig. 4. Radiation patterns for various types of seismic waves for earthquakes and underground explosions. Underground explosions radiate P waves with spherical symmetry. Since earthquake sources involve shearing motion along a fault, the amplitudes and phases of waves radiated by earthquakes vary as a function of appropriate radiation angles. The relative movement of two blocks on either side of a fault can be described by three angles: the strike, dip and rake. These angles are akin to Euler angles. The strike of a fault is the azimuth as measured from north of a line formed by the intersection of the fault plane with a horizontal plane. The dip of a fault is the angle between a vector normal to the fault plane and a vector normal to a horizontal plane. The rake is the angle between a horizontal vector in the direction of the strike of the fault and a vector (the slip vector) in the direction of relative movement of one side of the fault with respect to the other.

about 0.7 to 5 Hz, whereas those designed to record surface waves have a pass band of about 20 to 100 seconds (0.01 to 0.05 Hz). Arrays or groups of nearby seismic instruments are often used to improve the signal to noise ratio of short-period P waves. Those arrays can be "steered" so as to increase the signal to noise ratio of waves arriving from certain azimuths and with certain apparent velocities (horizontal phase velocities). P waves of distant events typically have much higher apparent velocities across an array (since the ray paths are nearly vertical) than those of earth noise of comparable frequencies. It is possible to obtain useful peak magnifications of about 50,000 for long-period surface waves by installing instruments in boreholes about 200 m deep (to isolate them from atmospherically- induced tilts) and isolating the sensors from pressure fluctuations in the atmosphere. Digital recording has made the routine analysis of seismic data much easier and has enabled seismologists to more readily distinguish signals of small events in the presence of either earth noise or interfering signals of large earthquakes.

Seismologists characterize the size of a seismic event by means of magnitudes. A given event can be assigned several magnitudes, each one based on a different kind of seismic wave. A magnitude is the logarithm of the amplitude of a particular type of wave normalized for distance (and sometimes for depth of focus). We shall discuss only two magnitudes, which in seismological notation are designated M_S and m_b. The former is today generally based on the vertical component of Rayleigh waves with a period of 20 seconds. The skin depth of those surface waves is about 35 km. The m_b scale utilizes P waves with periods near 1 second. The magnitude of a seismic signal is related to the energy release at the site of the event. Magnitudes by themselves, however, do not have a physical meaning until they are calibrated against the yields of underground explosions or theoretically based estimates of the energy release in earthquakes. We will illustrate the use of these two magnitudes both for discriminating the signals of earthquakes and underground explosions and for determining the yields of underground tests.

METHODS TO DISCRIMINATE UNDERGROUND EXPLOSIONS FROM EARTHQUAKES

Each year there are thousands of earthquakes whose magnitudes are in the range corresponding to the yields of underground explosions, there being 15,000 or so earthquakes of m_b 4.0 or greater per year. The number of earthquakes in a given area generally increases exponentially as the size or magnitude of events decreases. For example, there are about ten times as many magnitude 4 earthquakes per unit time in most areas of high seismicity as there are magnitude 5 events. Hence, if we wish to identify explosions of smaller and smaller yield, we must deal with an increasingly greater number of signals from small earthquakes. We must be able to identify them successfully as earthquakes. The size of the smallest events that can be identified as either underground explosions or earthquakes is ultimately limited technically by earth noise in its several manifestations and operationally by the increasing number of small earthquakes that must be analyzed.

Seismologists have devised a number of schemes and methods for distinguishing an event as either an earthquake or underground explosion. We will describe a procedure or flow chart by which most of the world's seismic events can be easily and readily identified as earthquakes based on the location and depth of focus of those events. We will then discuss some other discriminants that can be used to identify the small percentage of remaining events as either earthquakes or underground explosions. The great majority of routinely detected events can be classified as earthquakes simply because they are either too deep or not at plausible sites for being underground explosions.

LOCATION AND DEPTH OF SEISMIC EVENTS

A location of an event in latitude and longitude is a powerful tool for classification. The position can be determined if the arrival time of short-period P waves is recorded at seismograph stations at several scattered locations in the world. The travel time of the P waves to each station is a function of distance and depth of focus. From the arrival times and travel-time curves developed over years of observation and analysis, it is possible to determine the location of the source with an absolute error of less than 10 to 20 km if the seismic data are of high quality.

The identification of seismic events occurring at sea is quite simple. It is assumed that the network monitoring a test-ban treaty would include a small number of simple hydroacoustic stations around the shores of the oceans and a few on oceanic islands to measure pressure waves in seawater. A hydroacoustic signal of an underwater explosion is so different from that of an earthquake and can be detected at such long range that the identification of a seismic event at sea as an explosion or an earthquake is simple and positive. We assume that, under a comprehensive test ban treaty, other national technical means (NTM) of surveillance will be employed. We assume that NTM will be available to readily identify the major logistics effort that would be needed to conduct a clandestine test in deep water. Hence, any event whose calculated position is at least 25 km at sea (a margin allowing for errors in seismic location) can be classified as an earthquake on the basis of its location, the character of its hydroacoustic and conventional seismic signals, and other intelligence data.

The accuracy with which the position of a seismic event can be determined using data of a world-wide network in an area offshore of an island arc has been tested with an array of ocean-bottom seismometers off the Kamchatka Peninsula and the Kurile Islands of the easternmost U.S.S.R. The tests indicate that the accuracy attainable with a high-quality world-wide seismic network in such localities is better than 25 km. Holding to that standard, well over half of the world's earthquakes are definitely at sea.

Another large group of detected seismic events have their epicenters on land but in regions where no nuclear explosions are to

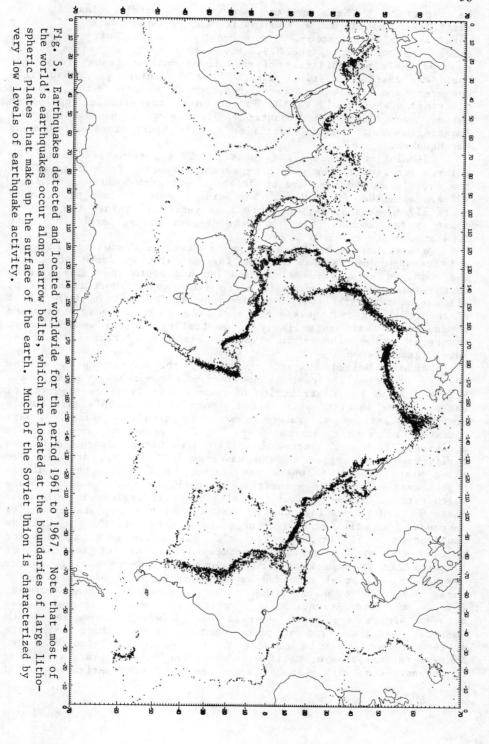

Fig. 5. Earthquakes detected and located worldwide for the period 1961 to 1967. Note that most of the world's earthquakes occur along narrow belts, which are located at the boundaries of large lithospheric plates that make up the surface of the earth. Much of the Soviet Union is characterized by very low levels of earthquake activity.

be expected. These events too can be safely classified as earthquakes based on a combination of seismic data and other intelligence information. For most countries of the world, (excepting, of course, the U.S.S.R.) either their societies are so open or their probabilities of conducting nuclear tests are so low that clandestine testing has a truly vanishing probability. Whenever and wherever there is doubt, however, the seismic discriminants we apply to the Soviet Union can also be applied to seismic events from other countries. In any case, the simple act of locating seismic events classifies a large percentage of them as earthquakes.

Calculating the depth of focus provides a means of identifying a large fraction of the remaining seismic events. From 55 to 60% of the world's earthquakes are at depths of more than 30 km; at least 90% are more than 10 km deep. Any seismic event as deep as 15 km is certainly an earthquake. No on has as yet drilled into the earth's crust as far down as 15 km, and the deepest nuclear explosions have been at a depth of about 2 km.

We will not repeat here the more extensive discussion in Sykes and Evernden[1] of various methods for calculating depth of focus. The combined effectiveness of location and depth in distinguishing earthquakes from explosions is quite impressive. More than 90% of all earthquakes are under oceans and/or at least 30 km deep. Most of the remaining earthquakes are of little interest because they are in countries that are unlikely to be testing nuclear weapons or in countries where clandestine testing would be detected by other intelligence methods.

For the United States, of course, the Soviet Union is the country of prime interest in monitoring a comprehensive test ban. Figure 5 shows the distribution of moderate to large earthquakes of the world for an eight-year period. Most of the world's earthquakes are not scattered at random over the earth's surface but are confined to rather narrow belts. The zones of most intense activity, such as those in the Aleutians, Japan and the Kurile-Kamchatka region of the easternmost U.S.S.R., are situated along plate boundaries where one of the large lithospheric plates that makes up the earth's surface is converging upon the other plate which remains at the surface. Geologists call regions in which one plate is underthrusting another and in which one plate gradually descends into the earth's mantle to depths of 200 to 700 km "subduction zones" or "island arcs". Subduction zones account for most of the world's greatest earthquakes. Zones of less intense activity can be seen in Figure 5 along the mid-ocean ridges where new volcanic material is added as the two plates on either sides of those ridges move apart. One zone of earthquakes, for example, follows the mid Atlantic ridge which is located nearly midway between Africa and South America and between Europe and North America. Another type of plate boundary occurs where one plate moves horizontally with respect to the other (as along the San Andreas fault system of California where the Pacific plate is moving to the northwest with respect to the rest of North America).

A lithospheric plate can be considered to consist of relatively cold material compared to that at depths of 75 to 200 km in the upper mantle where temperatures come closest to the melting point. That region can be considered to be the gliding layer of plate tectonics. Lithospheric plates have an analogy to cakes of ice moving around on a river in that most of the deformation (and earthquakes) are concentrated at the edges of the cakes of ice and little deformation occurs within each cake of ice. The advent of the plate tectonic model in the late 1960's has led to a great improvement in our understanding of why earthquakes, mountains and volcanoes occur where they do and why seismic waves are attenuated in different ways in different parts of the earth.

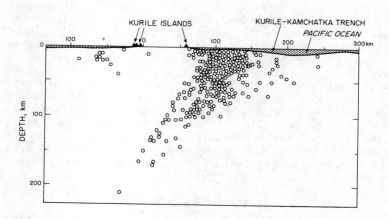

Fig. 6. Vertical cross section of the outer 200 km of the earth along a line running from continental Asia at the left to the Pacific Ocean at the right. Most of the earthquakes in that region, the most active seismic area in the Soviet Union, are related to the downward motion of the Pacific plate as it descends deep into the earth's mantle beneath Asia.

About 7% of the world's earthquakes occur in and near the Soviet Union. Of those, however, about 75% are situated near the Kamchatka Peninsula and the Kurile Islands. Figure 6 shows a vertical cross section through the outer 200 km of the region near the Kurile Islands. It extends from northwest at the left side of the figure, across the largely volcanic islands near zero distance, across the deepest part of the Kurile-Kamchatka trench to the Pacific Ocean. A section through the Kamchatka peninsula is similar to that through the Kurile Islands. In these regions the Pacific plate plunges beneath the mainland of Asia and in so doing produces a deep-sea trench where water depths are about twice those found in the normal oceans. A zone of earthquakes dips northwesterly. Earthquakes in those zones are located at shallow depths beneath the trench itself and extend to depths of 150 to 200 km beneath the volcanic chain. Most of the earthquakes in the upper 50 km occur along the boundary between the two plates. Deeper earthquakes occur within the Pacific plate as it plunges to great depths in the mantle. Sudden frictional sliding and brittle failure, which characterize earthquake motion, can only occur at temperatures less than a few hundred degrees C. Relatively low temperatures are found within the downgoing Pacific plate since the rate of plate motion (about 10 cm/yr) and the low thermal conductivity of earth materials require about 10 million years for temperatures to diffuse into the center of a downgoing plate which is about 75 to 100 km thick.

The distribution of earthquakes beneath the Kurile Islands and Kamchatka has important implications for seismic discrimination. Almost all of the shocks in these areas either have a focal depth greater than 50 km or are located well offshore. It can also be seen in Figure 6 that relatively few of the events within the upper 50 km are located within 25 km of the Kurile Islands. There are about 100 earthquakes per year (i.e., about 0.5% of the world's earthquakes) in the U.S.S.R. with depths of less than 50 km and location under land or less than 25 km at sea. These 100 earthquakes must be identified by other seismic discriminants. Magnitude m_b 3.8 corresponds to a well-coupled underground explosion (such as an explosion in hard rock in a continental crust of low attenuation (high Q)) with a yield of about 1 kt. We discuss later that very small seismic events in the very active Kamchatka-Kurile region can be monitored by hydrophones and ocean bottom seismometers placed just offshore. Hence, those areas do not lend themselves to clandestine testing.

SPECTRAL DISCRIMINANTS

Several powerful discriminants are based on the detailed character of the waves radiated by earthquakes and underground explosions, in particular on the relative amounts of energy in different period ranges. For example, a shallow earthquake generates 20-second Rayleigh waves with displacement amplitudes at least several times greater than those of an explosion that releases the same amount of short-period energy. In the somewhat careless jargon of seismology, the comparison of the two magnitudes is often referred to the "M_S:m_b ratio", although the analysis is virtually never made by a ratio comparison but by location on a two-dimensional plot of M_S vs. m_b. The point of relevance is that we compare levels of radiated waves of 1 sec and 20 sec periods.

Figure 7 shows the much larger surface waves from an earthquake of nearly the same short-period magnitude and at nearly the same distance as a nuclear explosion. The short-period magnitudes were determined as averages from many short-period recordings. The long-period Rayleigh waves for the earthquake are 10 to 100 times larger than those for the explosion. The Rayleigh waves can be seen, especially on the records of the explosion, to exhibit normal dispersion, i.e., waves of 80 sec. period arrive before those of 20 sec. period. In addition, the long-period P and S waves are much larger for the earthquake than for the explosion. The high-frequency character of the P wave of the underground explosion is distinct, and its first motion is clearly upward (i.e., away from the source).

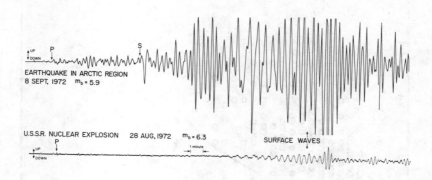

Fig. 7. Long-period seismograms recorded at Elat, Israel, for an earthquake in the Arctic region and a Soviet underground explosion. Both events occurred at nearly the same distance; both have nearly the same short-period magnitudes, m_b. Note that the surface waves and the long-period P and S waves are much larger for the earthquake than for the explosion. These differences in spectral excitation are reliable criteria for distinguishing the seismic waves of earthquakes from those of underground explosions.

Figure 8 shows an $M_S:m_b$ diagram for a large set of events for which the computed depths were less than or equal to 30 km. It can be seen that the events separate into two populations: earthquakes and underground explosions. Many Soviet events to date can also be readily identified as explosions because they occur at known testing areas and most were detonated exactly at an even hour or minute. Although this situation is unlikely to occur under a comprehensive test ban treaty, it gives us a good check on the power of the $M_S:m_b$ discriminant.

Only one earthquake in Figure 8 falls within the explosion population as defined by the line separating the two groups of events. This single event took place in the southwest Pacific Ocean, a region where the sensitivity of the network of seismic stations used in this study was poorer than it is in most of the Northern Hemisphere. Problem events of that type disappear as network coverage improves. The m_b values of explosions and earthquakes were adjusted to take into account regional variations in the amplitude of short-period waves. We will return later to this subject since it is central to the estimation of yields of Soviet explosions.

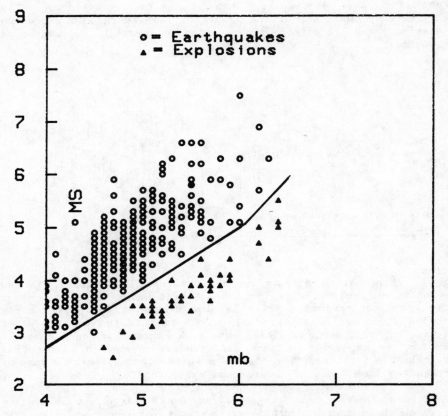

Fig. 8. Surface wave magnitudes of long-period Rayleigh waves (M_S) as a function of 1-second body wave magnitudes, (m_b). Note the clear separation of the earthquake and explosion populations. The earthquakes represented by the open circles have focal depths 30 km or less. Only one earthquake falls within the explosion population, as defined by the line separating the two populations. The bend in that line corresponds to predictions from theory of a change in the shape of the M_S:m_b relationship. The single problem event took place in the southwest Pacific, a region where the sensitivity of the network used was poorer than it is in most of the northern hemisphere. The m_b values for both earthquakes and underground explosions were adjusted to take into account regional variations in the amplitudes of short-period P waves.

In 1965 there was still a serious question about whether the M_S:m_b technique would be successful at m_b smaller than 5. While statements to the contrary are still occasionally made, abundant evidence was available by 1970 that the technique works as well for $3.5 < m_b < 5.0$ as it does for larger magnitudes[2,3,4]. In Figure 8 earthquake data have been added for 4.0

$\leq$ m_b $\leq$ 4.5 from a special study by Evernden et al.[2] in which m_b is an average for many stations and M_S is from a single station. While those data points based on a single station scatter more than values averaged over many stations, it is clear that all of the values for earthquakes fall above the decision line derived from data for larger magnitudes. Most of the other data points in Figure 8 represent averages for both M_S and m_b from many stations.

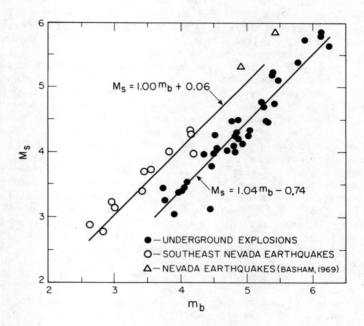

Fig. 9.   Adjusted M_S and m_b values from Blandford[4] for underground explosions and earthquakes in Nevada. Note that the two populations remain well separated down to magnitudes at least as small as m_b 3.7.   The individual slopes of either population are close to 1.0.

Evernden et al.[2] and Evernden[5] present several data sets that show that the M_S:m_b relationship for either earthquakes or underground explosions has a slope very close to 1.0 near m_b 4.0. In each case, however, the intercept value is such that the M_S curve for earthquakes is well above that for underground explosions.   Figure 9 shows a similar data set[4] for small earthquakes and underground explosions in Nevada.   The two populations remain well separated for events at least as small as m_b 3.7.   The M_S values in that figure are based on an average for four or more stations.   The separation of the two populations, of course, is not as good if a smaller number of stations is used for measurements of M_S. Thus, it can be seen that when data of good signal to noise ratio are available and the event is recorded by several stations, the separation of the earthquake and underground explosion populations is as good at m_b 3.7 to 4.0 as it is at m_b 5.0. An m_b of 3.7 is equivalent to a yield of 0.4 to 0.8 kt for well-coupled underground explosions.

A second spectral discriminant is based on the observation that long-period P and S waves are rarely or never seen in association with explosions, but one type or the other is routinely detected by simple seismometers for most earthquakes that have a 1 second P-wave magnitude of at least 4.5 (Figure 7). More sophisticated recording stations and digital analysis of the signals could lower the magnitude at which such waves can be detected.

Surface waves of the Love type are generated far more strongly by shallow earthquakes than they are by underground explosions, including even abnormal explosions. It was once thought that an explosion could not generate any Love waves at all. As we stated earlier, a simple (i.e., spherically symmetrical) explosion does not generate surface waves of the Love type. A phenomenon, which at that time was of unknown origin and was of great significance in thwarting President Kennedy's effort to achieve a comprehensive test ban in 1963 was the observation that nearly all underground explosions at the Nevada Test Site, particularly those in hard rock, generated unmistakable Love waves. The failure of the qualitative criterion "No Love Waves From Explosions" (at a time when such quantitative criteria as the comparison of the magnitudes of long-period and short-period waves were not adequately established) left seismologists unable to guarantee their ability to distinguish the seismic waves of underground explosions from those of earthquakes. Thus, the world's one and only real chance for a CTBT was lost.

The presence of Love waves in the Nevada tests has since been explained. What was not considered in the earlier analyses was the influence of the natural stress state of the earth on the waves generated by an explosion. The creation of a cavity and its surrounding shattered region by an underground explosion lead to the release of natural stresses, which in turn generate seismic waves, including Love waves, equivalent to those of a small earthquake consistent with the local stress field. The observed waves are a superposition of the waves from the explosion and from the release of the natural stress.

The release of natural stress also alters the amplitude of Rayleigh waves. The perturbation has never been large enough, however, to put in doubt the nature of an event identified by the $M_S : m_b$ ratio. Only rarely does the perturbation significantly affect the amplitude of P waves: it is not known ever to have changed the direction of their first motion. Moreover, if the magnitude M_S is determined from Love waves rather than Rayleigh waves, this $M_S : m_b$ ratio provides a third excellent discriminant.

A fourth characteristic feature of the seismic signals from explosions is that the first motion of the earth stimulated by P waves is always upward because the explosion itself is directed outward: the first P-wave motion of an earthquake can be either upward or downward (Figure 4). Note that the high-frequency P wave from the explosion in Figure 7 is clearly upward on the record. Clear first motions, however, are usually not evident at distant stations for events of m_b 5.0 or less.

The utility of yet a fifth discriminant based upon the spectral content of short-period P waves has been demonstrated in a study of 225 Eurasian earthquakes and 108 Soviet explosions, many being at PNE sites. The basic character of this discriminant involves the higher ratio of 0.4 to 1.0 Hz energy to 1.0 to 4.5 Hz energy for earthquakes than for exposions. All explosions at Soviet test sites were separable from all earthquakes when using data of one U.S. station (LASA). The explosions that could not be distinguished from all earthquakes were several of the PNE's. A multi-station analysis of similar data using a similar discriminant achieved total separation of all studied earthquakes and explosions. The explosion sets of the two studies, however, were quite different, there being many more PNE's in the first study than in the second. Whether the failure to achieve perfect separation in the first study derived from the different data set or single-station analysis is unknown at this time.

Several of the quite different properties of explosions and earthquakes account for these pronounced spectral differences. An important factor contributing to the separation of earthquakes and explosions on an $M_S:m_b$ diagram is the different radiation patterns of P waves. Explosions radiate short-period P waves equally in all directions while earthquakes have very asymmetrical patterns (Figure 4). The P waves that reach distant receivers leave the focus within 15° to 20° of vertically downwards. As illustrated in Figure 4, the P amplitudes for explosions are high in those directions while those for strike-slip earthquakes are small. Most earthquake sources show decreases of 0.4 to 1.0 magnitude units at large (teleseismic) distances from peak values when amplitudes are averaged over pertinent radiation angles. This factor contributes to the large spread in $M_S:m_b$ values for earthquakes. A simple explosion does not initially radiate any shear waves: earthquakes typically generate large shear waves. As a result Rayleigh waves generated by many types of earthquakes have a larger amplitude than the corresponding waves generated by underground explosions of the same m_b. This effect and the different radiation patterns provide discrimination for events of all sizes.

There is a characteristic time for the formation of the source of the seismic event: the time is equal to the maximum source dimension divided by the velocity of source formation. The source dimension for earthquakes is the length of the break where most of the energy is released; it is from 3 to 20 times greater, depending on the state of stress in the rocks, than the radius of the cavity and shatter zone of a comparable explosion. The velocity of source formation for earthquakes is from somewhat less to much less than the velocity of shear waves in the rocks surrounding the fault, whereas the relevant velocity for explosions is the velocity of shock waves in the rock which is essentially the velocity of P waves. As a result of these differences in the size of the source and velocity of source formation, the characteristic times for earthquakes and explosions differ considerably.

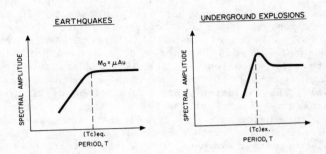

Fig. 10. Spectral amplitudes in the far field as a function of period for earthquake and explosion sources. At long periods, spectral levels are flat and proportional to a quantity called the seismic moment, M_O. Actual amplitudes are, of course, a function of radiation patterns and of attenuation of the waves from receiver to source. Spectral amplitude decreases near corner period, T_C for both earthquake (eq) and explosion (ex) sources.

Theory and observation indicate that at very long periods the amplitude spectra in the far field of both earthquakes and underground explosions become constant or flat (Figure 10). For earthquakes, the flat spectrum derives from the fact that the elastic energy radiated by earthquakes comes from relaxation of previously stressed rock. An explosion produces an outwardly propagating shock-wave with attendant non-linear processes. This shock-wave induces all subsequent seismic radiation phenomena. When the shock-wave finally reaches a distance where the rock media behave linearly, the still steep-fronted shock-wave induces a displacement field that has a time history similar to that of a delta function in time, thus generating a flat long-period spectrum in the far field. Even an explosion that vents will have a flat long-period spectrum.

Seismologists use the expression corner period, T_c, (or its reciprocal the corner frequency) to describe the approximate period at which there is a transition in the shape of the amplitude spectrum. For a given period longer than T_c, seismic waves radiated from various parts of the source arrive nearly in phase at a particular receiver or station. Thus, the source can be described as behaving like a point source for those periods. At shorter periods waves arriving at the receiver that are radiated from different portions of the source do not arrive nearly in phase. It is this interference pattern that leads to a reduction in amplitudes for periods shorter than T_c for both types of sources. The long-period spectral level for earthquakes is proportional to a quantity called the seismic moment, M_0. For a source like an earthquake involving shearing motion (i.e., relative motion parallel to the plane of the fault)

$$M_0 = \mu Au \qquad (1)$$

where μ is the rigidity modulus, A is the area of the fault over which rupture takes place (the dislocation surface), and u is the average relative displacement of the two sides of that portion of a fault that moves during an earthquake. A similar seismic moment can be defined for explosions where

$$M_0 = (\lambda + 2\mu) Au \qquad (2)$$

where λ and μ are the Lamé elastic constants, A is the area $4\pi r_e^2$, where r_e is the radius from the center of the explosion to points where deformation becomes purely elastic, and u is the radial displacement at r_e. The corner period for explosions scales approximately as yield to the one-third power. Hence, the larger the yield the longer the corner period. Similarly, larger earthquakes usually have longer corner periods.

These differences in spectral shape lead to varying amounts of discrimination between the signals from earthquakes and those from underground explosions for sources of different size (and hence corner period). Discrimination via the $M_S:m_b$ ratio is predicted to be greatest at larger magnitudes, which is in accord with the greater observed separation in the earthquake and explosion populations on an $M_S:m_b$ basis. For large earthquakes the lengths of fault breaks (many kilometers), the dimensions over which significant elastic strain is released, and the velocity with which faults rupture require more than one second for stored energy to be released. This leads to a reduction in the spectral amplitudes near 1 sec. for large earthquakes. Portions of faults that rupture in individual earthquakes are typically a mixture of areas of high strength and large areas of lower stress release. This irregular pattern as well as the time required for release of elastic energy lead to enhanced M_S/m_b values for earthquakes as small as m_b 4. For a more detailed discussion of the physical basis of discrimination, the reader is referred to a paper in press by Archambeau and Evernden.

In addition, observations of several U.S. explosions have demonstrated the existence of a phenomenon called overshoot. It is related to non-linear shock wave phenomena in strong rock, but it can be crudely thought of as arising from a linear elastic process in which cavity pressure increases rapidly to high values followed by a rapid decrease by a factor of 4 or so for hard rock. The lower pressure would then be maintained for many tens of seconds. The amount of overshoot is smaller or absent for explosions in less competent or incompetent rock. Overshoot may provide increased discrimination in two ways: 1) by increasing amplitudes in the m_b portion of the spectrum (Figure 10) and thus improving the $M_S:m_b$ discriminant; and 2) increasing the spectral contrast between earthquakes and explosions in the frequency range 1 to 10 Hz.

We have tried to show that several known properties of earthquake and explosion sources can be used to account for observed differences in the excitation of various seismic waves. To illustrate the utility of these discrimination methods we present data on all world-wide earthquakes of magnitude m_b 4.5 or greater that occurred during a 162-day period in 1972. There were 977 such events. At the nine stations for which we had data, 29 of the earthquakes either had their surface waves mixed in those of larger earthquakes or had uninterpretable long-period waves. Therefore the analysis is conducted on the data of the remaining 948 earthquakes. We return later to the problem of mixed surface waves. The discriminants used are: 1) location equal to or more than 25 km from shore; 2) calculated depth of 50 km or greater; 3) calculated depth of 30 km or greater; and 4) $M_S:m_b$ value on the earthquake side of the decision line of Figure 8. These criteria are applied to the data in various orders in Table 1 to illustrate the usefulness of each and the great overlap in applicability.

TABLE 1. Identification By Location, Depth and $M_S:m_b$

| Method | Criterion | Events Identified as Earthquakes | |
		Number	Percentage
A.	Depth $\geq$ 50 km	375	40
plus	Location > 25 km at sea	801	84
plus	Depth $\geq$ 30 km	856	90
plus	$M_S:m_b$	948	100%
B.	Depth $\geq$ 50 km	375	40
plus	$M_S:m_b$	944	99.6
plus	Location > 25 km at sea	948	100%
C.	Depth $\geq$ 30 km	565	60
plus	$M_S:m_b$	946	99.8
plus	Location > 25 km at sea	948	100%

Method A in Table 1 applied the discriminants in the order 2, 1, 3, 4. A depth of 50 km or greater or a location more than 25 km at sea identified 801 of the 944 earthquakes. An additional 55 events were classified as earthquakes when all events with calculated depths of 30 km or greater and not at sea were treated as identified. The $M_S:m_b$ criterion identified all of the remaining events. Method B used 2, 4 and 1 in that order. Of the total of 948 events, 944 were classified as earthquakes by depth of 50 km or greater or by $M_S:m_b$. Location at sea was required as a discriminant for only four events. Method C replaced depth of 50 km or greater in Method B with depth of 30 km or greater. The only change is that the location criterion is now required for only two of the 948 events. Both of these events were of the smallest magnitude recorded in the study and were located in the southwest Pacific, far from most of the nine stations. Thus, 100% of the 948 events were identified as earthquakes by either method. None of the 56 explosions in Figure 8 would be classified as earthquakes since all were on land, all had calculated depths of less than 30 km and all had $M_S:m_b$ values in the explosion population.

Long-period S waves were measured from the seismograms for the same earthquakes. Such waves were measured for nearly all of the events of m_b 5.0 or greater, most of those without measured S values resulting from off-scale amplitudes on the available analogue records. For earthquakes of m_b less than 5.0, a discriminant based on location more than 25 km at sea, depth of 30 km or greater, and reported long-period S waves successfully identified 239 of 277 earthquakes, 140 of these earthquakes having had detectable S waves on available analogue records. Digital analysis of the seismograms certainly would have detected long-period S waves for many more of these earthquakes.

The discussion above was limited to events of magnitude 4.5 and greater because of the limited number of stations providing data. Nonetheless, special studies of large numbers of earthquakes of smaller magnitude have been conducted as mentioned earlier[2,3,4]; they confirm the applicability of the depth and $M_S:m_b$ criteria to small events. Comparable data sets do not exist for any of the other spectral criteria discussed above. Nevertheless, adequate analysis has been done to confirm they would also be of great utility in monitoring a CTBT.

The methods discussed are not necessarily optimum ways of achieving discrimination. For example, the use of the largest Rayleigh waves in the period range 20-50 sec. tends to average out lows in the spectra of Rayleigh waves and to give greater separation between explosions and earthquakes than when M_S is determined solely from 20 second waves.

Reports that earthquakes occasionally have $M_S:m_b$ values like those of explosions have been cited as a factor reported to confound the effective monitoring of the CTBT[6]. Note that we found only one possible example of that type in Figure 8 (plus another event on the decision line). Those two events were located far from the area in which the network gives its best results. As networks have improved, the maximum magnitudes of such problem events have decreased. In 1972 the United States' Government submitted a report to the U.N.'s Conference of the Committee on Disarmament[6] describing 23 events that appeared to be earthquakes but which could not clearly be distinguished as such on an $M_S:m_b$ diagram. All of the events were located in Central and Southern Asia.

Tatham et al.[7] studied that suite of so-called "anomalous events". Upon reanalysis, about half had $M_S:m_b$ values in the earthquake population. Most of the original magnitudes had been determined from one or two stations: much existing data had not even been consulted. By substantially increasing the number of observations, poor sampling of the radiation patterns was avoided and the events ceased to be "anomalous". Many of the remaining problem events had depths of 25-50 km. It is known theoretically that for those depths earthquakes of certain focal mechanisms poorly excite Rayleigh waves. Theory predicts, however, that Love and higher mode Rayleigh waves often are well excited under those circumstances. Measurements on waves of those types identified several additional problem events as earthquakes.

When this had been done only a single sequence of events at one place in Tibet remained as a problem. Long-period seismographs in boreholes have been developed and operated since that sequence occurred in 1965. Those instruments suppress background noise and permit Love waves to be detected for small earthquakes. Love and higher-mode waves need to be used for those few understandable circumstances where fundamental mode Rayleigh waves are poorly excited. Data from a few close stations can provide better estimates of depth which could also identify events of that type as earthquakes. Several stations recorded first motions of P waves that were downward (dilatations) for the remaining problem sequence of Tibet. Thus, they must have been earthquakes. The first motion criterion, which was given great prominence in early test ban negotiations in 1958, usually works well only for larger events. One of the factors that probably contributed to a small M_S:m_b ratio for the Tibetan sequence was a focal mechanism that radiated P waves near the maximum of the radiation pattern but which was located near a node for 20 second Rayleigh waves. Hence, factors that helped to foil one technique allowed another to be applied.

Thus, it seems reasonable that for the networks we describe below, there should no longer be any problem events at m_b 4 or greater provided proper instrumentation is operated and care is taken to use several discriminants. We know of no Eurasian earthquake with a 1 second P-wave magnitude of 4 or more of the past 20 years whose waves are classified as those of an explosion. Of course, numerous such Eurasian earthquakes, especially during the early part of that period, went unidentified because of inadequate data. Furthermore to our knowledge not one out of several hundred underground nuclear explosions set off in the same period radiated seismic waves that could be mistaken for those of an earthquake. Our experience indicates an extremely low probability that a detected event will remain unidentified when all available techniques of discrimination are brought to bear.

SEISMIC NETWORKS FOR MONITORING A CTBT

We now consider some of the types of networks that could be used under a CTBT for monitoring the Soviet Union. We discuss several of the schemes that have been suggested as potential ways of evading surveillance and thus of conducting illegal tests. A CTBT would undoubtedly contain a rapidly executable escape clause, which a country wishing to test, if it felt its national security was endangered, would likely invoke rather than resort to clandestine testing. The Limited Test Ban Treaty contains such a clause. Nevertheless, in designing networks for verifying a CTBT we will make the worst assumptions that can be made about possible cheating.

No monitoring technology can offer an absolute assurance that even the smallest illicit explosion would be detected. We presume that an ability to detect and identify events whose seismic magnitude is equivalent to an explosive yield of about 1 kiloton would be adequate. It has been asserted that, for the United States to subscribe to a comprehensive test ban treaty, a 90% confidence of detecting and identifying each and every violation would be required. Developing new atomic weapons, however, generally requires a series of tests, and the probability that at least one underground explosion in a series will be detected rises sharply as the number of tests increases. In addition, a 90% level of confidence for even a single explosion probably does not appear realistic to us. For a country seeking to evade the treaty, the probability of detection and identification of a single event it would be willing to accept would certainly be much lower than .90, say as low or lower than .30. Unless the probability of identification were low, a country whose national interests seemed to demand a resumption of testing would presumably invoke the escape clause rather than risk being caught cheating. Use of a probability of .30 or less for a single test follows from the simple logic that a country not wishing to be caught cheating will not test weapons of a number and yield that will ensure at high probability that it will be caught cheating. It should be realized that there will always be an asymmetry in the probability considered acceptable to a potential evader and the probability considered desirable by a country monitoring compliance of a CTBT.

Given these standards of reliability (we use 90% and 30% confidence for purposes of discussion), it is possible to specify the surveillance capability of any seismic network to be used to verify compliance with a CTBT. Three kinds of networks will be considered for seismic surveillance of the Soviet Union. One consists of 15 stations located entirely outside the borders of the U.S.S.R., a second adds 15 internal stations to the 15 external ones, while the third network adds 10 more internal stations. We will first examine the monitoring capabilities of these networks assuming that the relevant seismic noise is the constant background of atmospherically-induced seismic energy called microseisms. For this type of seismic noise we will consider both explosions that are well coupled into the surrounding rocks, i.e., into seismic waves, and those for which some attempts have been made to reduce the amplitudes of seismic signals by detonating them in very large cavities.

The second scenario assumes it to be useful to the evader to try to hide the seismic signals of the explosion in the signals of large earthquakes. For the U.S.S.R., the only even conceptually credible scenarios require use of distant earthquakes since all of its seismic areas other than Kamchatka and the Kurile Islands have inadequate rates of seismicity to allow credible evasion scenarios based on nearby earthquakes. Very detailed monitoring of seismic activity in the Kurile-Kamchatka area is possible because of its location near Japan and the Aleutian Islands and because ocean-bottom seismometers and hydroacoustic sensors could be placed just offshore. Thus, the most seismically active areas of the U.S.S.R. are not regions in which clandestine testing is likely to go undetected, if serious attempts at monitoring are undertaken.

Multiple-shot evasion (setting off a series of explosions such that some of their signals look similar to those of an earthquake) is useless if it is executed against normal earth noise. Multiple shots in well coupled media would, of course, produce very large seismic signals. While the shooting sequence may be conducted in an attempt to confuse one of the seismic discriminants like M_S-m_b, it would not work against a combination of several discriminants that utilize data from several spectral bands and wave types. A sequencing of shots over several seconds which would be designed to reduce the maximum size of a given part of the P wave signal while allowing their 20-second surface waves to add constructively, could be identified as a series of explosions by various spectral techniques, i.e., by the short-period discriminant using amplitudes near 5 and 1 Hz or by merely changing the method of estimating the short-period magnitude. Thus, multiple-shot evasion does not appear to be an effective means of cheating on a CTBT.

TABLE 2. Evasion Possibilities in Kilotons for External Seismic Network Assuming Normal Earth Noise

Source Type U.S.S.R.	Type of Coupling	External Net Alone Maximum Yield for Probability of Detection	
		>90%	>30%
Seismic Areas, Hard Rock or Below Water Table	Fully Coupled	1 kt	0.5 k
Granite	10 times decoupling	10***	5
Salt Domes	75 times decoupling	10**	10**
Dry, Porous Alluvium	Full coupled but m_b reduced 0.9 units compared to hard rock	1-2*	1-2*

* Limited by maximum thickness of unsaturated alluvium.
** Limited by maximum sizes of cavities that can be constructed clandestinely and maintained without collapse.
***Assumes small hole packed with energy absorbent such as fine-granite carbon. Experimental results set the factor of ten.

Table 2 shows the evasion possibilities against the external network for various evasion scenarios, assuming that the pertinent noise is normal levels of microseisms. For most areas of the U.S.S.R., a capability to detect events of m_b 3.8 will achieve detection at 90% confidence for explosions of 1 kt or larger in well-coupled strong rock, salt or saturated alluvium. This is, of course, a significant capability without resorting to internal stations. The latter stations are primarily used to address the problems of significant decoupling in salt and hard rock. The assumptions about signal-to-noise-ratio, minimum number of stations detecting an event (four in the above case) and the program for estimating network capabilities are described in a series of three papers by Evernden[8].

Magnitudes larger than m_b 3.8 probably would be generated by a 1 kt explosion in hard rock in the Kurile-Kamchatka area and by explosions in saturated alluvium in most of the U.S.S.R. Seismic waves from U.S. explosions in the Aleutians at Amchitka, an area similar in geological and geophysical properties to the Kurile-Kamchatka area, are larger than those from many other test sites for explosions of the same yields. Nearly all seismic regions of the U.S.S.R. are situated along its international borders so that mere detection of a seismic event at m_b 3.8 in most other parts of that country would immediately flag it as a suspicious event for which other discrimination techniques and NTM could be fully utilized.

If we make what seems to us the logical assumption that anomalously low M_S/m_b ratios for earthquakes would be removed with adequate instrumentation and data analysis, a hypothetical network must be able to detect surface waves of M_S 2.6 to properly complement an m_b capability of 3.8, assuming relevant noise to be only microseisms. (M_S 2.6 is the surface wave magnitude of the decision line in Figure 8 for an m_b of 3.8.) Those requirements can be met for a network of 15 stations located external to the U.S.S.R. In fact, the capabilities of that network are somewhat better for the seismically active areas of the U.S.S.R.[1]. While the hypothetical network used for these calculations does not exist, stations of better quality have been operated and their data studied extensively for many years. Evernden et al.[2], for example, found that the M_S detection limit for a single long-period station of high quality is about M_S 3.0 at distances up to 4,000 km from seismic events. Digital analysis of the data, which was not used in the cited study, as well as the operation of several stations of high quality within that distance range could reduce the detection limits for surface waves appreciably. The 90% detection threshold[6] for Rayleigh waves for sources about 3500 km from an array of long-period instruments in Alaska (called ALPA) is M_S 2.6. Thus, there do not appear to be serious impediments to achieving a surface wave capability of M_S 2.6 or better for the Soviet Union with a network of external stations.

It must be stressed, however, that the noise condition here assumed is sometimes inappropriate. Several earthquakes of less than m_b 4.5 occurring in the U.S.S.R. each year have unmeasurable surface waves due to their being mixed with signals from large earthquakes elsewhere in the world. Assuming the percentage of mixed signals is about 3.5%, this translates into a few earthquakes per year of m_b 3.8 or greater either in or within 25 km of the U.S.S.R. with depths of less than 50 km. Of greater importance is that there are many opportunities per year to hide surface waves of underground explosions in the persistent and highly dispersed codas of large world-wide earthquakes. Both of these types of masked events require internal stations for their identification.

If in addition to the 15 external stations, 15 simple stations are also operated within the U.S.S.R., greater identification capabilities can be obtained. Fully coupled explosions of 0.1 kt could then be identified at 0.90 probability in much of the U.S.S.R. A greater number of events in the Soviet Union would also be identified as earthquakes based on improved capabilities to estimate depth of focus and on more accurate analysis of short-period spectra of both P and S waves. This network would thus reduce the number of still present events for which their long-period surface waves are masked by signals from larger distant earthquakes.

The coupling of energy into seismic waves must be known if yields or energy release are to be calculated from seismic waves, i.e., from magnitudes. Yields have been published for a number of U.S. and for a few other underground explosions. The coupling of explosive energy into seismic waves is not very dependent upon rock type providing underground explosions are set off in either hard rock or water-saturated rock (i.e., below the water table). For a large test (say one near 150 kt) to be contained in the earth and not to blow out at the surface, it must of necessity be conducted below the water table. Coupling is only reduced significantly when either a small explosion is set off in dry porous media like dry alluvium or an attempt is made purposely to muffle its signals by detonating it in a large cavity. The possibilities of achieving low coupling are negligible for large underground tests and are limited but real for small explosions. We return later to a more detailed discussion of estimates of yield.

The maximum thicknesses of dry porous alluvium in the Soviet Union[8] limits non-cratering explosions to a maximum yield of 1 to 2 kt, and then only to a few extremely restricted areas, which are also subject to easy monitoring by other NTM. It is those limits that control the capabilities in Table 2 for the external network alone. Cheating opportunities in dry alluvium are reduced to a maximum of 1 kt for 0.90 probability of detection with the internal networks to be described. Attempts to cheat on a test ban by conducting explosions in dry alluvium are also a particularly risky endeavor. That material is unstable enough that an explosion might well produce enough deformation to form a surface depression which could easily be detected by satellite photography or other NTM. Some non-seismologists[9] discuss tests of 5 to 10 kt in alluvium and their possible non-detection as if that were a reasonable evasion possibility for the Soviet Union. Neither we, the Military Geology

Branch of the U.S. Geological Survey nor the Defense Advanced Research Projects Agency consider this to be a possibility.

Low coupling in dry alluvium is achieved because a great deal of energy is expended in compaction of that material because it contains a large amount of air-filled pore space. The nature of dry alluvium is such that it cannot be used for "big-hole" decoupling. Any significant cavity in alluvium would collapse during attempts to excavate it. This is one of several cases in which two or more evasion possibilities are not additive.

In the estimation of nearly everyone, detonation of explosions in large cavities so as to reduce amplitudes by a factor of 75 or so is the most serious evasion possibility. The largest size of cavity that can be both reasonably constructed and maintained is unknown, the upper limit built to date being holes adequate in size for decoupling of about 1 kiloton[8]. Holes adequate in size for decoupling of 10 kt (about 0.5 million cubic meters, i.e., a cube about 75 meters on a side) are often assumed possible. Highly irregular holes of such volume do exist. Their utility for decoupling scenarios was totally rejected 10 years ago, but their possible use is now being discussed.

A point of significance is that the theory of cavity decoupling requires elastic continuity of the cavity walls. Fractures in the walls, even for stable cavities, will result in unattenuated or less severely attenuated signals at sites along ray paths passing through the fractured portion of the walls. This effect was observed for the decoupled STERLING explosion in the salt cavity created by the SALMON explosion in Mississippi. The prevalence of such fractures in granitic and metamorphic rocks may negate use of cavities in such materials even if cavity stability is achieved. Also, remember that such cavities must be excavated in a totally clandestine manner or without suspicion being aroused. National technical means (NTM) other than seismology would place severe constraints on the feasibility of such decoupled explosions. Although large cavities in salt may be easier to build or at least conceptually to build, there are only a few such areas in the U.S.S.R. where this might be done. Our internal networks, particularly that containing 25 internal stations, are designed specifically to address this problem.

The vast areas of bedded salt to the northwest of Irkutsk (Lake Baikal) in Figure 11 are almost certainly improbable sites for multikiloton, fully decoupled explosions. It was assumed that other NTM would do what was necessary in most of this area when designing the network of 15 internal stations[8]. For the 25 station internal network, special emphasis was placed on this area. Certain seismic waves of the P and S type that travel through the crust are quite large up to distances of a few hundred kilometers. The crust in most of the Soviet Union is such as to transmit those waves very efficiently. If we assume normal background noise, evasion is not likely to be attempted at yields larger than 1 kt for areas of thick salt deposits if stations are deployed in their vicinity.

Detection by one station from the major salt dome area north of the Caspian Sea (Figure 11) would also virtually constitute

114

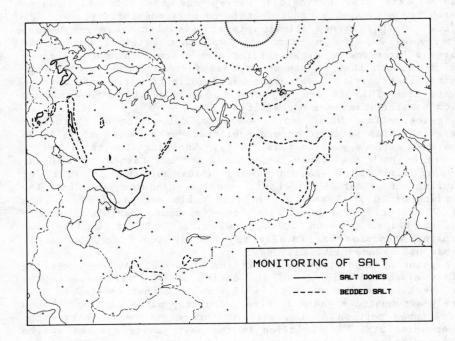

Fig. 11. Areas of U.S.S.R. containing salt deposits of significant thickness.

identification at the 1 kt level since that area is. non-seismic and since a single station is assumed to have a vertical and two horizontal components from which the azimuth and distance of a nearby event could be determined. Thus, detection of an event by a single station in that salt dome area would immediately flag it as an explosion.

It has been suggested by Evernden[8] and by others that a useful element in an evasion scenario would be hiding even the short-period waves of an explosion in those of a distant large earthquake. In the cited article, it was pointed out that even under the erroneous assumption of comparable spectral content for the P waves of explosions and earthquakes, an evader would experience great operational complexities in exploiting any calculatable increase in evasion posibilities. Considering only 15 internal stations, his evasion opportunities per year would be very limited and an effective test program would be limited to small yields and to a very few events a year unless the evader maintained several clandestine test sites. He would have to maintain his weapons in constant readiness for firing and would have to execute the explosions within 100 seconds of the time of arrival of the short period waves of the earthquake. He must estimate the amplitudes of the waves of the earthquake very accurately (both maximum amplitude and decay rate); he must be certain of the amplitudes of P waves generated by his explosion to within 0.1 m_b units; he must be content to allow monitoring networks to detect the event at one or two stations with high probability and at three stations with 10% probability; and he must install and operate both his test site (including development of the required large cavity) and his own extensive seismological network in a totally clandestine manner over a few to many years.

Thus, the requirements placed on a successful evader are staggering while those of the monitoring nation are simply to operate networks of high quality and to pursue data processing in a vigorous and determined manner.

Rather than repeat previous discussions[1,8] of the 15 internal network, we will give results of new analyses based on a set of 25 internal stations, these analyses having been completed since our Scientific American article was published. As pointed out earlier, the network of 15 internal stations depended implicitly on other NTM and the patterns of Soviet population, climate and vegetation to achieve high monitoring capability in many of the remote areas of the U.S.S.R. The rationale behind study of a 25 internal network is to strive to achieve a significantly greater seismological capability and thus to require less dependence on other NTM. When the time hopefully and finally comes for developing U.S. concepts of the total surveillance system to be used in monitoring of a CTBT, the relative roles and capabilities of the several NTM will require thorough discussion.

An as yet unresolved aspect of monitoring of a CTBT is the required magnitude or yield threshold of the monitoring network. Most discussions speak in terms of 1 kt yields as if there is some kind of tacit if not explicit agreement that this is indeed the agreed-upon magic number. Such is not the case. In 1973 in response to testimony that a feasible network with a one kiloton threshold could be deployed within the U.S.S.R., John Foster, then Director of Defense Research and Engineering, told the U.S. Senate that monitoring capability to one-quarter of a kiloton would be required under a CTBT to insure national security interests. No feasible newtork can be deployed to monitor such a threshold, particularly in the face of the potential evasion scheme of cavity-decoupling. As long as that testimony of Foster's stands as the position of the U.S. Government, any discussion of seismic monitoring of a CTBT is bound to be barren.

However, more recent testimony[13] to the House Armed Services Committee, suggests a much more tractable reality. On pages 192 and 193 of that document is a letter from Harold Agnew of Los Alamos

Laboratories to Congressman Jack Kemp. Two quotes from that letter are of present interest. When discussing strategic-related testing, Agnew wrote "I don't believe testing below say five to ten kilotons can do much to improve (as compared to maintaining) strategic posture." As regards both strategic and theatre significance, he said "The military significance to either the U.S.S.R. or the U.S.A. of conducting clandestine tests below five to ten kilotons is per se of relatively little importance today." If these statements can be considered as representative of present reality, and if one can thus replace Foster's 0.25 kiloton with Agnew's 5 kiloton, a nearly intractable monitoring problem is converted into one of simplicity itself.

One further topic should be discussed before proceeding to analytical results for the 25 internal network. A question always asked is the negotiability as part of a CTBT of such a network within the U.S.S.R. Two major points: 1) under a cooperative program of seismic research pertinent to earthquake prediction, for several years U.S. scientists have been permitted to operate both long and short period seismometers within 12° or so of the Soviet test site at Semipalatinsk (eastern Kazakh); 2) in early discussions of a CTBT (1958-1960), the Soviets agreed to manned monitoring stations within the U.S.S.R, agreeing to an initial set of 15 internal stations for the first four years of the treaty. The U.S. was requesting 21; other aspects of our still ongoing confrontation, however, terminated those discussions. Nevertheless, the acceptability to the Soviets of seismic monitoring stations within their borders as an aspect of a CTBT is an established fact. They have always distinguished between fixed-site monitoring stations and U.S.-selected sites for on-site inspections of specific seismic events. The failure of the U.S.A. to appreciate and understand the logic of this subtlety has long confounded discussions. The varying intricacies related to all treaty discussions are such as to lead to uncertainty about how many stations will be agreed upon in any future negotiations. It does seem to us to be a credible reading of history that the number of monitoring stations would never be a critical stumbling block in negotiations between the U.S.S.R. and U.S. if all really vital political aspects of the treaty were resolved.

Now for a few critical parameters of the network analysis. The nearly fully decoupled 0.38 kt explosion STERLING was clearly recorded at a distance of 111 km but not at 240 km. There were no intermediate stations. We assume that, in similar terrane, this explosion could have been clearly recorded at 150 km. This leads to a predicted range of observation of a 1 kt decoupled explosion of 250 km. Hence, we model detection to 2.25 degrees (250 km) for a 1 kt decoupled explosion. This is done, of course, assuming normal noise levels.

What about times of abnormal noise? The typical coda of a distant large earthquake (the only earthquakes statistically useful in a hiding-in-earthquake scenario) is of even longer period than the normal microseismic noise. Thus, separation of the explosion signal from interfering "noise" will be even easier for earthquakes than for microseismic noise. Therefore, we consider that earthquake codas are of no importance for hiding the much higher-frequency signals of explosions, particularly when considering detections at a few hundred kilometers. It is useful to point out that, on both observational and theoretical grounds, it is possible that seismic signals of "decoupled" explosions decouple less at high frequencies than at 1 Hertz. If this is indeed true, the network capability calculated here is poorer than that actually attainable.

The other possible noise problem is storm-associated, high-frequency noise. The particular significance of this problem is that an evader would know when noise levels were high and might try and execute a clandestine explosion during that time. There are inadequate empirical data to give quantitative answers to this problem. Dr. Eugene Herrin (personal communication) has assured us that, based on his experience in the central U.S., this factor will be a problem at certain times. Extensive experience in California in the operation of high-sensitivity, high-frequency seismic systems distributed from the coast to the mountains provides some specifics that suggest the problem is probably a minor one on a network basis. Stations near the coast certainly are often swamped by microseisms induced by near-shore phenomena, these microseisms being at frequencies that seriously degrade detection capability. As distance from the coast increases, however, the amplitudes of these microseisms markedly decrease until at distances of 50 to 200 km (depending on the region), they do not influence station detection capability at frequencies of a few Hertz or higher. At distances of several hundred to several thousand kilometers from shorelines (as are most of the 25 internal stations of the modeled network in the U.S.S.R.), these microseisms should be of no significance.

The only other problem is higher-frequency noise (generally 10 Hz or higher) induced locally by high winds. In many areas, such noise never reaches amplitudes adequate to confound detection at somewhat lower frequencies though it might confuse some spectral discriminants at individual stations. Such noise is induced locally and is sensed locally at high amplitude. Therefore, blackout of even a dense regional network essentially never occurs. In an investigation of noise levels at the several subarrays of the Large Aperature Seismic Array (LASA) in Montana, it was found that there is no increase in mean noise levels in the winter and that there was always extreme variability in noise levels at the several subarrays, some having very low noise levels when others had very high levels. An even much more scattered network such as that being investigated for deployment in the U.S.S.R. should be even less affected on a network basis. Thus, if we find adequate probability of detection at five or so stations, wind-induced noise will probably not cause this to go below 3 to 4 station detection. For the ranges of detection relevant in our analyses and since all of these stations have three-component systems, three station detection should be more than adequate for detection, location, and identification.

Therefore, we conclude that it is legitimate to conduct our analysis based on typical (i.e., microseismic) noise conditions. We do stress that all identification at the low magnitudes of events to be considered must depend upon spectral composition, wave-type content, and phase rleationships on the short-period seismograms.

Figure 11 shows locations of Soviet salt deposits that might be sites for at least 1 kt decoupled explosions. The 25 internal stations of the network we studied are indicated by solid triangles on the five parts of Figure 12. We assume that a 1 kt decoupled explosion generates what we refer to as an m_b 2.0 signal (3.9 minus $\log_{10} 75$) and is detectable at high probability at a distance of 2.25 degrees. The rate of signal decay is assumed to be that of the eastern U.S. A 3 kt explosion will generate nearly an m_b 2.5 signal while a 10 kt explosion will generate an m_b 3.0 signal. Figure 12a presents calculated capabilities in terms of number of detecting stations at 0.90 probability for a 3 kiloton explosion (m_b 2.5); Figure 12b for 0.30 probability. Figures 12c and 12d are for a 10 kt explosion (m_b 3.0).

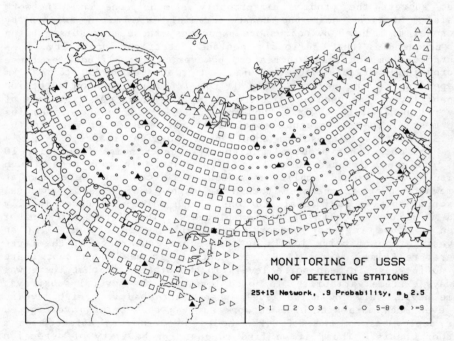

Fig. 12. (a-e) Detection thresholds throughout U.S.S.R. for network consisting of 25 stations internal to Soviet Union (solid triangles) and 15 external stations. Various symbols denote number of stations detecting an event of a given magnitude, m_b, at probabilities of either 0.30 or 0.90.

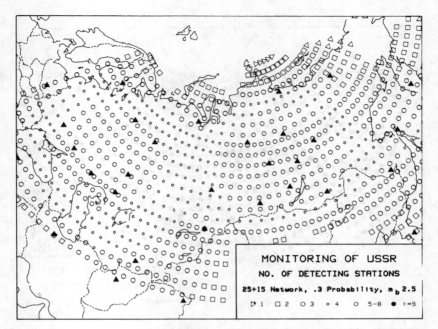

Figure 12b

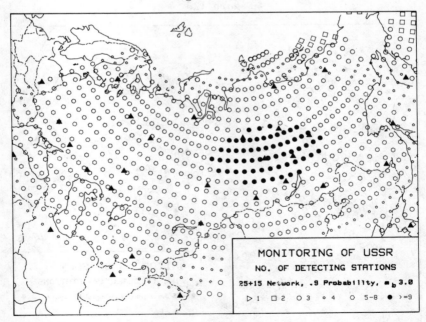

Figure 12c

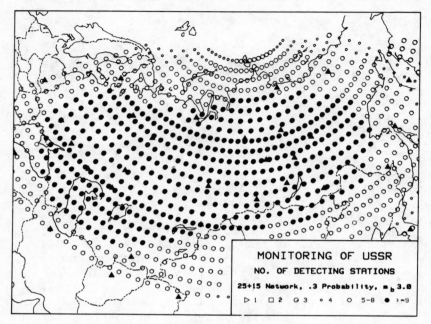

Figure 12d

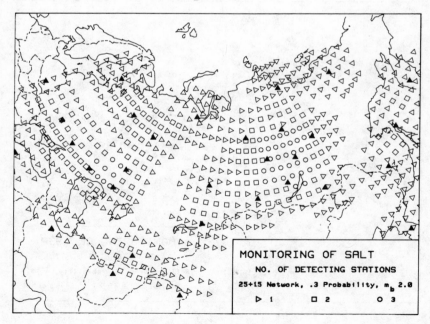

Figure 12e

At 0.30 probability, a fully decoupled 3 kt explosion in all large salt areas of the U.S.S.R. would be detected at 5 to 9 seismic stations. For some of the smaller salt masses, fewer stations would detect but their signal-to-noise ratio would be very high (because of network configuration). The levels of detection are adequate for location and identification and are probably adequate to thwart evasion by use of high-wind noise associated with the infrequent storms in most of the areas of significance. In most hard rock areas, four or more stations would detect a 3 kt fully decoupled explosion. If a greater capability were to be required in the far northeast, redesign of the network or more stations would be required.

The 10 kiloton/0.30 probability map (Figure 12d) indicates detection at 9 or greater stations, thus implying detection of 5 kilotons at 0.30 probability at 7 or more stations. These capabilities are so high that it is pointless for an evader to consider decoupling with large cavities of even 2 or 3 kt, much less 5 to 10. Therefore, the entire issue of large cavity decoupling is moot if such a network as that studied is installed as part of a CTBT.

Figure 12e is relevant to 1 kt explosions and the same network at 0.30 probability. It indicates a high capability throughout much of the U.S.S.R. It would seem that augmentation of this network with other NTM might well give adequate monitoring capability even at 1 kiloton throughout the U.S.S.R.

SOVIET COMPLIANCE WITH THRESHOLD TEST BAN TREATY AND ESTIMATES OF YIELD

The concept of magnitude has no direct relationship to the size or energy release in an event unless some type of data are used to calibrate particular magnitude scales. The U.S.S.R. has not announced or published yields for any of its weapons tests. It has announced yields for a small number of its peaceful explosions. The magnitude-yield relationships used to calculate the sizes of Soviet explosions from the amplitudes of their seismic waves must come mainly from U.S. explosions. The few yields provided by the Soviets, however, agree with the estimates derived below for U.S. experience and procedures.

The wave types that can be used to estimate yield are once again short-period P waves and long-period surface waves for which m_b and M_S are the respective magnitudes. Each of these wave types has its particular advantages and disadvantages for estimation of yield. Accurate determinations of yield have, of course, become of vital importance since 1976, the date that the Threshold Test Ban Treaty went into effect. The limit of 150 kt is well above the size of underground explosions that can be detected and discriminated by both M_S and m_b with very high levels of confidence by networks external to either the U.S. or the U.S.S.R.

Short-period P waves can be detected from smaller explosions than can surface waves. With the advent of digital recording and

122

quiet long-period stations, however, this is no longer a significant factor for yields near the 150 kt limit of the TTBT. The standard deviation of the mean magnitude for an event near 150 kt as determined from P waves is about the same as that determined with a much smaller number of surface wave observations. A Soviet event near 150 kt will have published data available from 50 to 100 world-wide stations recording P waves and about 5 to 10 such stations recording surface waves. The standard deviation of the mean magnitude in each case is about .02 to .03 magnitude units. This corresponds to about a 5% variation in yield if the testing medium and testing conditions remain the same. Almost all of the arguments about the yields of Soviet explosions come from disagreements about the nature of systematic errors rather than from uncertainties about the magnitudes themselves.

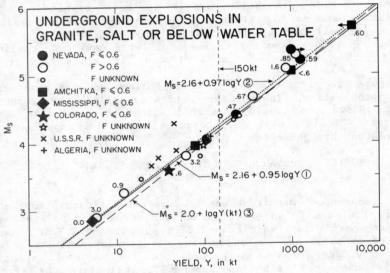

Fig. 13. Surface wave magnitude (M_S) as a function of yield for underground explosions in granite, salt or below the water table. Data from Marshall et al.[10]. F is the amplitude ratio of components of natural stress (tectonic) release/pure explosion. Tectonic release is derived from natural stress energy in the earth that is relieved by the sudden formation of the explosion cavity and a large crushed region around it.

Figure 13 shows surface wave magnitudes, M_S, as a function of yield for explosions at various test sites for which yields have been published. The data for underground explosions in granite, salt or those below the water table show very little scatter about a linear relationship between M_S and logY over a factor of 1,000 variation in yield and independent of location of the explosion. The slope of that relationship is very close to 1.0, as predicted theoretically. If exploded in high-porosity materials, explosions near the 150-kt limit of the TTBT must of necessity be conducted below the water table so that the explosions are contained. Contained explosions in low coupling media such as dry alluvium or in large cavities are not possible near 150 kt. The small amount of scatter in Figure 13 reflects statements made by many seismologists over the last 10 years that there is a nearly universal M_S-logY curve. It is for this reason that we prefer to use surface waves (i.e., M_S) to calibrate yields for various Soviet test sites.

Equation 2 of Figure 13 is the best fit[10] to all of the published data. As mentioned earlier underground explosions can trigger or release various amounts of natural strain energy. This strain energy or tectonic release affects long-period waves but it has not been shown to significantly affect short-period waves. Equation 1 in Figure 13 represents a best fit to those data points for which the tectonic release was small (i.e., less than 0.6, where F is the ratio of the amplitudes of tectonic and pure explosive components). The predominant type of focal mechanism for tectonic release at the Nevada and Amchitka test sites is of the strike-slip type. Tectonic release at Amchitka has also been small. A strike-slip mechanism of faulting does not significantly bias M_S values of the explosion providing an average is obtained over a network of stations well distributed in azimuth about the source. As can be seen, equations 1 and 2 in Figure 13 differ very little near 150 kt.

In estimating M_S for Soviet explosions, Sykes and Cifuentes (in preparation) used the same distance correction factors as Marshall et al.[10] did for the data in Figure 13. For that reason we use equations 1 and 2, which were derived from those data, in estimating yields of Soviet explosions.

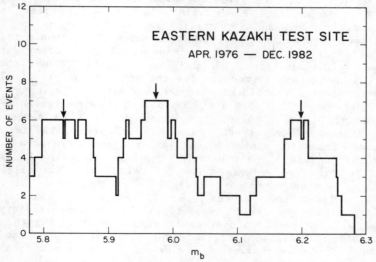

Fig. 14. Number of underground explosions at eastern Kazakh Test Site of U.S.S.R. as a function of magnitude, m_b. Data were stacked (added together) assuming a box-car function of ±0.4 m_b units about the mean magnitude for each explosion. Underground tests are concentrated near magnitudes indicated by three arrows.

Figure 14 shows the frequency of occurrence of Soviet explosions at their eastern Kazakh test site of $m_b \geq 5.8$ as a function of m_b from the date the Threshold Test Ban went into effect, 31 March 1976, through 1 December 1982. The largest Soviet explosions since 1976 have been detonated at that test site. Each event was assigned a width in m_b (box-car function) extending from its minus to its plus 90% confidence limits (about ±0.04 m_b units about its mean magnitude). Seven explosions, all of which have occurred since June 1979, define the peak centered at m_b 6.20. The individual magnitudes and their standard deviations are consistent with all of the seven explosions having the same yield. Thus, the largest Soviet explosions since the TTBT went into effect appear to be of the same or nearly the same yield. We are certain that Alewine and Bache's[14] assertation that two of these seven events were decidedly larger than the others is erroneous and results from the use of limited and uncalibrated data (without station corrections) by the U.S. Geological Survey in its program of preliminary estimates of magnitudes. Hence, we are not dealing with a situation in which one or two explosions may have exceeded the 150 kt limit for some reason. Whatever the yield, it must be the same or nearly the same for all seven explosions.

Sykes and Cifuentes have attempted to calibrate what m_b 6.20 is in terms of yield using Soviet explosions for which the tectonic release was small ($F \leq 0.4$). For those explosions the error in M_S resulting from the interference of signals from the explosion itself and tectonic release must be small. When no correction is made for tectonic release, all of the values fall below M_S 4.227 and 4.271, the size of the surface waves for explosions of 150 kt by equations 1 and 2 of Figure 13. For F values less than 1.0, the correction for tectonic release is negligible for strike-slip faulting mechanisms. The maximum positive correction for tectonic release is that for a purely thrust fault mechanism. In the latter case, some of the M_S values for $F \leq 0.4$ are slightly above those two numbers. The actual correction for tectonic release is likely to be somewhere between that for thrust and strike-slip faulting. Hence, we use M_S values for U.S.S.R. explosions of low tectonic-release to ascertain that m_b 6.20, the magnitudes of the seven largest Soviet explosions corresponds to a yield close to 150 kt. We then estimated yields for explosions of larger tectonic release from this calibrated m_b-yield relationship for the eastern Kazakh test site. In this way we were able to obtain yields for all U.S.S.R. explosions at that test site regardless of the amount of tectonic release. Thus, the largest U.S.S.R. explosions since the TTBT went into effect had yields close to 150 kt. The uncertainty in the data as well as possible systematic errors in applying the M_S-logY relationships of Figure 13 to Eurasian paths lead to an uncertainty of about 25% in estimates of yields of Soviet explosions at the eastern Kazakh test site.

In estimating the m_b values in Figure 14 station corrections were computed for each of the 119 stations used in the analysis of Soviet explosions. Figure 15 shows the frequency distribution of

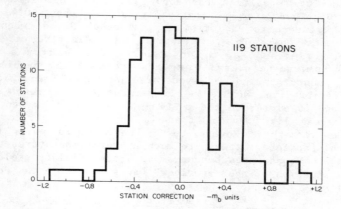

Fig. 15. Frequency distribution of station corrections for 119 seismic stations used to determine magnitudes of Soviet explosions. These corrections were made for each explosion since certain stations consistently report either high (or low) magnitudes for a given event.

those corrections. Individual stations systematically record waves whose magnitudes differ from the mean for a given explosion. For example, stations in Scandinavia, a region which has not been affected by volcanic or tectonic activity for several hundred million years, consistently report magnitude values well above the mean. Stations in regions that have been affected more recently by earth deformation and volcanic activity consistently register magnitudes below the mean. It is necessary to take these systematic variations into account since the same stations or groups of stations do not always report each explosion. Station corrections for m_b are usually much larger than those for M_S. It can be seen in Figure 15 that if two stations were picked at random their station corrections would likely differ by several tenths of an m_b unit. The same is likely to be true by reciprocity if we pick two testing areas at random. Hence, it is not too surprising that m_b values for explosions at Nevada and eastern Kazakh differ systematically by about 0.4 m_b units.

Much of the U.S. testing experience comes from Nevada, a region of young earth deformation and volcanic activity. The three main Soviet testing areas in eastern and western Kazakh and the Arctic, however, are all situated in areas that have not been significantly affected by earth deformation or volcanic activity for hundreds of millions of years. Hence, for an explosion of a given yield, the waves arriving at various seismograph stations will be larger from Soviet explosions than from explosions in Nevada. If the Nevada m_b-logY relationship is applied to P-wave data from Soviet tests without correction, the yields of Soviet explosions will be overestimated by a factor of about 3 or more.

It has been realized for about 15 years that M_S/m_b data for Soviet explosions are systematically offset about 0.4 magnitude units with respect to data from explosions in Nevada when recorded at common stations. This, of course, is indicative of systematic variations in either M_S, m_b or both. Evidence from a number of sources is now quite overwhelming that most of the problem resides with m_b. Short-period P waves are attenuated more in passing through regions in which the temperature at a given depth is elevated and is closer to the melting point (as it is beneath young geologic areas like Nevada) than is the case for older geologic areas like the three Soviet test sites. Data on a number of geophysical parameters attest to much higher temperatures at a given depth beneath Nevada than beneath most areas of older geology. Most of the attenuation of short-period P waves occurs in the depth range 50 to 200 km.

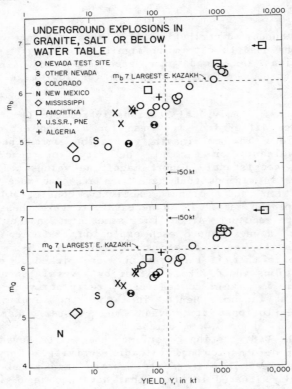

Fig. 16. Short-period magnitudes, m_b and m_Q, as a function of yield for underground explosions in granite, salt or below water table. Data from Marshall et al.[10]. Note that for a given yield, m_b values scatter more than do m_Q (below) or M_S (Figure 13) values. Arrows on two data points indicate that published yield was only indicated as being either greater than or less than a certain value.

Figure 16 (upper half) shows m_b data for explosions of published yields[10] at various test sites. It can be seen that the m_b data scatter much more than do the M_S data of Figure 13. For explosions of a given yield, m_b is highest for tests at Amchitka in the Aleutian Islands where the ray paths to distant stations pass through mantle materials with a high Q; it is lower for explosions at the Nevada Test Site, and is lower yet for explosions on the Colorado Plateau, a region of great seismic attenuation in the upper mantle[11]. If the three explosions at Amchitka are used to calibrate the Soviet test sites (under the hypothesis that the upper mantle has a high Q for both areas), it can be seen that m_b 6.20, the mean magnitude of the seven largest eastern Kazakh explosions since 1976, corresponds to a yield close to 150 kt. As noted above similar yield values are determined using surface waves.

Marshall et al.[10] derived a new magnitude scale, m_Q, by making two corrections to values of m_b. One correction attempts to account for differences in attenuation in the upper mantle beneath various test sites. That correction is estimated indirectly from the P velocity of the uppermost mantle beneath various testing areas. A decrease in that velocity was found long ago to correlate with an increase in attenuation. The second correction that they attempt to model is for the supposed interference of the direct P wave with a P wave that has undergone a reflection at the earth's surface near the source (pP of Figure 3). It can be seen in the lower half of Figure 16 that m_Q data scatter less as a function of yield than do the raw m_b data. The m_Q data define two linear relationships, a lower one for explosions in tuff (a fairly soft volcanic rock) and sedimentary rocks and an upper curve that includes explosions in granite, salt and andesite from various test sites. The upper curve is the appropriate one to use for the larger explosions at the eastern Kazakh test site[14]. Data for four Soviet peaceful explosions also fall along that curve. This also indicates that the yields provided by the U.S.S.R. have not been altered or manipulated. The best fitting linear relationship between m_Q and logY gives an m_Q of about 6.35 for Y = 140 kt, which is very close to the m_Q estimated for the seven largest explosions at the eastern Kazakh test site. A similar result is obtained if the data for explosions in granite, salt and andesite are only corrected for attenuation beneath each test site.

Several other lines of evidence argue that the m_b-logY curve for explosions in Nevada cannot be used without correction in estimating yields of Soviet explosions at their three main test sites. Nuttli[11] used a third type of seismic wave called L_g, a short-period surface wave that travels mainly in the continental crust, to estimate yields of Soviet explosions. His work indicates a bias in m_b (i.e., the difference in m_b for explosions of the same size and in the same medium at the eastern Kazakh and Nevada test sites) of about +0.35 magnitude units. Der[12] calculates a somewhat larger bias from studies of the differential attenuation of short-period P waves from the two test sites that are recorded by common stations. He also finds that at many stations P waves from the eastern Kazakh test site are much richer in high frequencies than those from Nevada. The spectra of the U.S.S.R. explosions are similar to those for P waves that pass through some of the oldest geologic regions of the earth. This argues strongly for a large, positve m_b bias. An average bias of about +0.33 units between eastern Kazakh and three Nevada explosions in granite is obtained from Marshall et al.'s[10] correlation of m_b with seismic velocities in the uppermost mantle.

A strong correlation has also been found between m_b values measured at individual stations and P-wave travel times to those stations. The U.S.S.R. routinely publishes seismological bulletins that include P-wave arrival times of earthquakes; it is a straightforward procedure to interpret the times for stations in Central Asia in terms of the expected pattern of m_b values in the general vicinity of the eastern Kazakh test site. That approach indicates a bias of about +0.4 magnitude units.

Several of these results were reported at a symposium of the American Geophysical Union in June 1983 devoted to estimates of yields of Soviet explosions. In all of the presentations (except one for which the data were from an area about 1000 km from eastern Kazakh) the bias in m_b between the eastern Kazakh and Nevada test sites was estimated between +0.35 and +0.45. Our sense is that the correction is about +0.35 to +0.40 m_b units. Thus, it is clear that in estimating yields using data largely from the western United States, m_b values for the eastern Kazakh test site must be corrected by several tenths of an m_b unit. When this is done, the estimated yields of the largest seven Soviet explosions are all close to 150 kt. If that correction is not made, yields will be overestimated by a factor of 3.

Several statements have been published either in the popular press or in scientific journals (Table 3) about the yields of Soviet explosions under the TTBT. Those estimates place the sizes of the largest Soviet explosions from 260 to larger than 800 kt. All of those estimates involve in our estimation an incorrect calibration of yields using m_b data. There appears to be no scientific justification for any of the estimates in Table 3.

It should be noted that the United States consistently overestimated the yields of U.S.S.R. underground explosions as well for the period prior to the TTBT by not calibrating m_b data correctly. These overestimations have undoubtedly affected U.S. estimates of Soviet nuclear capability and military strength in many ways over the past 20 years.

In a syndicated column for 10 August 1982, Jack Anderson alleged that in 1978 U.S. scientists secretly changed their methods of calibrating the kiloton, in effect doubling the size of permitted test explosions by the Soviet Union. He goes on to say "When the Soviets subsequently doubled the size of their blasts, intelligence experts were alarmed; they suspected a leak to the Kremlin at the highest levels." The silliness of Anderson's intelligence experts is evidenced by the fact that M_S magnitude values also rose and only after the rise reached values equivalent to 150 kt at all other sites in the world.

TABLE 3.   Published Statements About Yields of Largest U.S.S.R.
Explosions Under Threshold Test Ban Treaty

Jack Anderson - column in Washington Post, August 1982

"...the Soviets appear to have exceeded the 150-kiloton limit
at least 11 times since 1978. One test in September 1980 was
clocked at a likely size of 350 kilotons, according to my
sources."

"As recently as July 4, the Soviets set off a huge nuclear
blast. It was estimated at a likely 260 kilotons,..."

Harold M. Agnew - letter to Science, April 8, 1983

"...subsequent tests appeared to us to range as high as 400
kilotons..."

Alewine and Bache[14]

"In U.S. experience an m_b greater than 6.2 (as measured for
the largest Soviet events) has only been seen for yields of
600-800 kt and larger."

Judith Miller - N. Y. Times, July 26, 1982

"One official said that there had been several Soviet tests,
many at one particular site, that had been estimated at 300
kilotons."

All of the largest Soviet explosions since the TTBT went into
effect have been detonated at a portion of the eastern Kazakh Test
Site near the Shagan River. From Figure 17 it can be seen that the
maximum magnitudes of tests gradually increased from about 5.8 to
6.2 from 1976 to 1979. The largest underground test at the eastern
Kazakh site prior to 1976 was of m_b 6.25 in July 1973. All of the
largest Soviet tests prior to the TTBT were detonated at their two
Arctic test sites in Novaya Zemlya. Thus, the largest Soviet test
at eastern Kazakh prior to 1976 had a yield of 150 to 200 kt. The
test of July 1973 was followed by a six year period during which the
magnitudes of the largest tests at eastern Kazakh were in the range
m_b 5.8 to 6.0. Prior to 1976 the Soviet Union chose, for whatever
reasons, to conduct tests larger than 150 to 200 kt at their two
remote test sites in the Arctic whereas after 1976 their largest
tests were detonated at the eastern Kazakh test site.

130

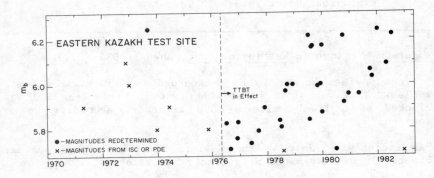

Fig. 17. Short-period magnitude, m_b, for underground explosions from 1970 to 1982 at the eastern Kazakh Test Site in the U.S.S.R. Circles denote magnitudes redetermined using station corrections. Those determinations are more precise than those indicated by X's.

CONCLUSIONS

The issues related to the monitoring of a comprehensive test ban can be summarized as follows. Our understanding of the subject of seismology and the testing of seismograph networks are sufficiently complete to ensure that compliance with a comprehensive test ban treaty could be verified with a high level of confidence. The only explosions with a significant likelihood of continually escaping detection would be those of very small yield, less than 1 kiloton provided the monitoring system includes stations in the Soviet Union. The yields of underground explosions that might go undetected or unidentified under a CTBT are therefore less than 5 to 10% of the yields of the nuclear weapons of 1945 and less than 1% of the 150 kt limitation specified by the Threshold Test Ban Treaty of 1976.

From a verification viewpoint a CTBT would actually establish the equivalent of a very low threshold, since weapons of an extremely low yield could be tested underground without the certainty of being detected and identified. A treaty that imposed a threshold near the limit of seismological monitoring capability might therefore be considered an alternative to a CTBT. Such a treaty might be preferable to the present high threshold.

In this regard, it is pertinent to return to Harold Agnew's letter[13] to Congressman Kemp. In addition to the themes already discussed, Agnew presented the oft-heard argument against a CTBT that depends upon the fact that such a treaty is indeed a low threshold treaty as regards surveillance. The argument goes that, under a CTBT, a country such as the U.S. would be bound not to test at any yield while another country might choose to test clandestinely at yields below the monitoring capability of the U.S. surveillance systems. Though such tests may be of no particular significance as regards weapon design, they might be enough to maintain viability of that nation's weapon laboratories while the U.S. laboratories disintegrated. Thus, at some future time, the U.S.S.R. would have developed a great and favorable asymmetry in national capabilities and could legally withdraw from the treaty and return to a testing program while the U.S. would not be able to do so for many years. This is an old argument and has been one of the main lines of defense against a CTBT. The fact that today we can conceive of a possibly deployable network that has a greater capability than required for U.S. or U.S.S.R. national security interests, however, finally provides the opportunity to reach a rational solution to a low threshold treaty. The argument goes as follows:

Given: a) monitoring to 5 kt is adequate for national security; b) a more than adequate capability to monitor fully decoupled 5 kt explosions can be achieved by the 25 stations of the network internal to the U.S.S.R.; c) all uncertainties of yield estimate would never confuse 1 and 5 kt explosions; and d) the 25 station internal network also has a high deterrance capability for 1 kt decoupled explosions, particularly when augmented by other national technical means.

Consider a Yield Threshold Treaty set at 1 kt: a) all explosions of significance to national security would be easily identified if they ever occurred; b) the capability to legally test below 1 kt would allow the U.S. and U.S.S.R. to do equal levels of limited testing in order to maintain a minimum cadre of competence at their weapon laboratories; and c) no significant advantage would accrue to anyone who tried to test say a 2 kt explosion. The uncertainty of yield estimates for small explosions would not be significantly greater than for 150 kt events because the slope of the m_b vs. Yield curve is markedly greater at low yields, meaning that twice the uncertainty in proper m_b leads to no greater uncertainty in relative yield.

A comprehensive test ban treaty certainly would have an important advantage over a low threshold treaty in that all technological uncertainties would work against the potential evader. A country planning a clandestine test could not know the exact seismic- detection capability of other nations or the exact magnitude of the seismic waves that would be generated by a clandestine test nor would he know capabilities of all other NTM. A ban on nuclear explosions of all sizes would also have the important conceptual value that nuclear weapons, no matter what their size, would be recognized as being inherently different from conventional weapons. Hopefully with time this could produce a wider "fire break" between conventional and nuclear weapons such that the boundary between them would not be as likely to be crossed during a time of international tension. A complete test ban may also provide greater incentives for additional countries not to develop, test or acquire nuclear weapons.

It is sobering to consider how the state of the world might be different if a complete test ban had been achieved in 1963 or soon thereafter. The number and accuracy of nuclear weapons that can be delivered intercontinental distances has grown tremendously since 1963. Those weapons and the emplacement of intermediate-range missiles on land and sea have shortened the response time of the superpowers to warnings of possible attack to a matter of a few minutes, a length of time that is too short for rational decision making. The loss of life and social damage that would be caused by a major nuclear exchange today are vastly greater than they were 20 years ago. Both the United States and the Soviet Union are less secure now than ever before. This arises not from a failure to develop and test nuclear arms as rapidly as possible but from the growing stockpiles of weapons, the inability of any nation to defend itself in a meaningful way against atomic attack, and the shorter response times that now exist. While the Limited Test Ban Treaty greatly reduced the radioactive pollution of the atmosphere from nuclear testing, it did not significantly alter the pace of atomic testing nor the development of major new weapons systems. Testing merely went underground after 1963.

Of the three reasons that were cited in July 1982 for the U.S. not continuing the negotiations toward a CTBT, the two reservations that are rooted in seismological observations -- the reliability of seismic methods for verification and alleged Soviet cheating on the Threshold Test Ban Treaty -- are not well founded scientifically. Focusing on those two erroneous issues diverts attention from several larger issues that deserve more extensive discussion: 1) do the advantages of a CTBT outweigh the possible military disadvantages arising from possible clandestine testing at yields of a fraction of a kiloton; 2) is the security of either the United States or the Soviet Union, in fact, decreased by a continuation of atomic testing; and 3) does an adequate infrastructure exist in the U.S. Government to conduct a vigorous and thorough monitoring effort as will be needed under a CTBT?

A comprehensive test ban agreement should not be regarded as a substitute for disarmament. Meaningful reductions in the nuclear threat must include a continuing and serious process of arms control. In this process, however, a CTBT could have an important part. It would certainly be a necessary ingredient of any meaningful nuclear freeze. It is a portion of freeze proposals for which extensive scientific research and negotiations have already been conducted. Hence, it is a part of nuclear freeze proposals that could be implemented rather rapidly.

Finally, it should be remembered that the problems of a CTBT are overwhelmingly political rather than technical or scientific. The subject of seismological verification of a CTBT has, in fact, become mature in the last decade. Major discriminants for explosions and earthquakes were developed in the 1960's. No new surprises in seismic methods or instruments have occurred in relationship to the verification of a CTBT and none appear on the horizon. Likewise, new methods of evasion have not been developed. Verification of a CTBT will require a determined monitoring effort.

Under a CTBT there will undoubtedly continue to be a need for seismological research. These research needs, however, should not be construed as technical impediments to a CTBT.

ACKNOWLEDGMENTS

We thank Drs. Paul Richards and William McCann for critically reading the manuscript. The views of this paper are not necessarily those of either the U.S. Geological Survey or the United States Government. Lamont-Doherty Geological Observatory Contribution Number 3495.

REFERENCES

[1] L.R. Sykes, and J.F. Evernden, Sci. Am. 247, 47-55 (1982).

[2] J.F. Evernden, W.J. Best, P.W. Pomeroy, T.V. McEvilly, J.M. Savino, and L.R. Sykes, J. Geophys. Res. 76, 8042-8055 (1971).

[3] T.V. McEvilly, and W.A. Peppin, Geophys. J. R. astr. Soc. 31, 67-82 (1972).

[4] R.R. Blandford, Seismic Data Laboratory Report 245, Teledyne Geotech, Alexandria, Va. (1972).

[5] J.F. Evernden, Bull. Seis. Soc. Am. 65, 359-391 (1975).

[6] Conference on the Committee on Disarmament, CCD/388/Corr. 1, 31 August (1972).

[7] R.H. Tatham, D.W. Forsyth, L.R. Sykes, Geophys. J. R. astr. Soc. 45, 451-481 (1976).

[8] J.F. Evernden, Bull. Seis. Soc. Am. 66, Pts. 1-2, 245-324; Pt. 3, 549-592 (1976).

[9] P.C. Hughes, and W. Schneider, Jr., in Arms Control and Defense Postures in the 1980's (Westview Press, 1982), pp. 21-37.

[10] P.D. Marshall, D.L. Springer, and H.C. Rodean, Geophys. J. R. astr. Soc. 57, 609-638 (1979).

[11] O.W. Nuttli, EOS Trans. Am. Geophys. Un. 64, 193 (1983).

[12] Z.A. Der, Ref. 11, p. 193.

[13] U.S. House of Representatives, Effects of a Comprehensive Test Ban Treaty on United States National Security Interests, House Armed Services Committee Document 95-89, August 14-15 (1978).

[14] R.W. Alewine, and T.C. Bache, Ref. 11, p. 193; letter submitted to Science, June (1983).

CHAPTER 6

IMAGE ENHANCEMENT BY DIGITAL COMPUTER

B. R. Hunt
Science Applications Inc., Tucson, Arizona 85711
University of Arizona, Tucson, Arizona 85721

ABSTRACT

In the past 15 years digital computers have been used
to greatly improve the quality of images, and to thereby
increase the extraction of information from images. In
this paper we review and discuss a number of techniques
for the enhancement of imagery by digital computer. The
implications of this technology to verification of arms
control treaties is left to the reader.

INTRODUCTION

During the past 15 years it has been difficult (perhaps impossible) to have been a regular reader of news-
papers and magazines and not seen remarkable pictures of
the Moon, Mars, Saturn, Jupiter, and a score of their
satellites. The pictures released by NASA were obtained by
remote satellite spacecraft, and are remarkable in their
quality and depiction of detail in our planetary neighbors.
What the casual viewer of these images does not know, however, is that these pictures are the product of substantial
amounts of computing. A variety of digital image processing techniques are often applied by NASA to the images
received from the spacecraft. The finished or "enhanced"
images are often a dramatic improvement over the original
or "raw" images. It is these improved images of which
the public is aware, rather than the original unprocessed
images.
    Technology is ubiquitous. What can be done to im-
prove the imagery from Mars can be also done to the image
acquired by a military reconnaissance sensor. Thus, the
question arises: what is digital image processing, and
how is it to be applied to imagery? Answering these questions is the purpose of this article.
    It is an element of open public discussion in the
news media that the United States possesses various reconnaissance systems that are capable of obtaining images
of regions to which free access is denied. The exact
nature of the reconnaissance systems is also widely specu-
lated upon in the news media. However, the Government of
the United States does not formally admit to any of the
various speculations, nor offer information at any open
official level concerning what the nature of the sensor

0094-243X/83/1040134-16 $3.00 Copyright 1983 American Institute of Physi

systems might be, nor make any official admission concerning the quality and nature of the images obtained by
any possible sensor system. Therefore, there is no openly
available hard data concerning the utility or applicability
of any such sensors, in general, or the utility and applicability of such sensors in verification of arms control,
in particular.

The applicability of digital image enhancement in
arms control is in the verification issue. If NASA can
improve the quality of spacecraft images from Jupiter,
then one can speculate about applications to improve the
quality of images produced by sensor systems employed
for the purposes of arms control. Because the U. S.
Government does not officially acknowledge any such
sensors or their operating characteristics, it is important to note that any discussion of this general area can
be only speculation - albeit, intelligent speculation
guided by basic principles of optics, engineering, and
computing. Thus, I issue the following disclaimer: all
the material in this article is written on the basis of
publicly available basic research in methods of image
processing, and can not be interpreted in any way as being
representative of, or confirming of, any specific technology employed by the U. S. Government. The discussion
that follows is, therefore, "generic", and discusses the
general nature of enhancing images produced by any sensor
system, rather than making any unverifiable speculations
or any methodology employed for purposes of Arms Control
(or any other purposes of the U. S. Government).

SOURCES OF DEGRADATION IN IMAGERY

We begin by more precisely defining what we mean by
image enhancement and image restoration. By image enhancement we mean any process which leads to an improvement in visual quality of an image. The keywords in the
previous sentence are "visual quality", since they imply
a subjective judgement that the image has been improved.
On the other hand, image restoration is more specific.
In image restoration we are concerned, first, with identifying a specific class (or classes) of degradation effects in the image formation process which created the
image and, second, with removing those degradation effects.
Thus, we "restore" the image from effects of degradation.
Image enhancement often reduces to a "bag of tricks",
i.e., the development of a repertoire of processes which
can be applied to an image and which are known to generally or usually have a beneficial effect on image quality.
However, image restoration is much more precise. To
carry-out image restoration, a mathematical model or

description of image degradation effects is created. The creation of an algorithm for image restoration can then be posed as the solution of the equations of the degradation model for the values of the restored image.

Image restoration is a much more scientific process than image enhancement. The usage of digital image processing to improve imagery is usually carried out by implementation of various schemes of image restoration [1]. However, the term "image enhancement" is more widespread in both technical and general literature, and is often applied to situations that are more accurately described as image restoration. Certainly, the highest quality processing, to produce an image with greatest fidelity to the original objects being imaged, represents tasks in image restoration. We will be more lax in the remainder of this article and follow the popular (but erroneous) practice of using "image enhancement" as a terminology to describe even image restoration problems.

Before attempting to remove sources of degradation in images obtained by a sensor system, it is first necessary to describe the processes that occur in image formation that can degrade an image and, thereby, make desirable the process of image enhancement (restoration). There are three broad classes of image degradation effects. The three classes are:

   (1)   Object and Scene Radiometry;
   (2)   Optical Image Formation Mechanisms in the Sensor;
   (3)   Image Detector Mechanisms in the Sensor Focal Plane.
We discuss each of these in the following paragraphs.

OBJECT AND SCENE RADIOMETRY

Figure 1 is a crude schematic of object and scene radiometry. The objects are described as a two-dimensional pattern of reflectivity $r(x,y)$. This pattern is illuminated by a solar flux, I_0, which we assume to be constant for simplicity but which can also be a spatial distribution in the case of complex haze or cloud cover. The product $I_0 r(x,y)$ is attenuated by scattering in suspended aerosols in the atmosphere (water vapor, dust). In addition, aerosols directly scatter some fraction of the solar flux back into the optical axis of the sensor. Thus, the total flux observed by a sensor system is:

$$s(x,y) = \alpha_1 I_0 + \alpha_2 \alpha_3 I_0 r(x,y) \tag{1}$$

where α_1 is the scattering fraction of solar flux into the sensor, α_2 is the coefficient which accounts for attenuation of the reflected flux, and α_3 represents attenuation of incident solar flux.

Detection of objects is strongly dependent upon the

<u>visible contrast</u>, i.e., the relative change in shades of gray at the boundaries between two objects. When the change between two shades of gray at an object boundary approaches 1% to 2% of the maximum image brightness, the human visual system loses the ability to distinguish the boundary and, hence, to detect or perceive the object.

The importance of contrast in radiometry can be understood in terms of equation (1). Suppose that the magnitude of the $r(x,y)$ term is small compared to the term $\alpha_1 I_0$. The control of exposure* for image formation in the sensor must be based upon the maximum visible flux, which is controlled by the term $\alpha_1 I_0$ under our above supposition. If the exposure is set to accomodate $\alpha_1 I_0$ the result will be to capture an image which has little variation in contrast. The term $\alpha_2 \alpha_3 I_0 r(x,y)$ being small compared to $\alpha_1 I_0$, its dynamic range of contrast is compressed relative to the $\alpha_1 I_0$ term. The result is a <u>low-contrast</u> image. If the contrast compression is severe enough, the object details in $r(x,y)$ may become invisible to the unaided eye.

Inherent photon noise also becomes a problem in low contrast imaging. Since photons obey a Poisson law, then the standard deviation of photons at the sensor is:

$$\sigma_{sensor} = \overline{(\alpha_1 I_0 + \alpha_2 \alpha_3 I_0 r(x,y))}^{1/2} \qquad (2)$$

where the overbar represents ensemble flux average and the square-root is usual for Poisson statistics. Obviously if σ_{sensor} is of the same order of magnitude as the $r(x,y)$ term, then the low-contrast image is further corrupted by noise. Image noise is visible as "salt and pepper" flecks overlaying the visible scene. The common "snow" attributed to TV reception in a fringe broadcast area is a fine example of image noise.

Low contrast imaging is a degradation that is associated with three major phenomena. First, the inherent reflectivity variations may be small. Second, the incident flux I_0 may be small, e.g., objects within the shadow of a cloud or in the shadow of another larger object. Third, the transmission coefficients α_2 and α_3 may be small and/or the aerosol scattering coefficient α_1 may be large (relative to α_2 and α_3).

---

* "exposure" is a term that is valid for both film-based imaging or imaging by electro-optical sensors such as Charge-Coupled Devices. [2].

138

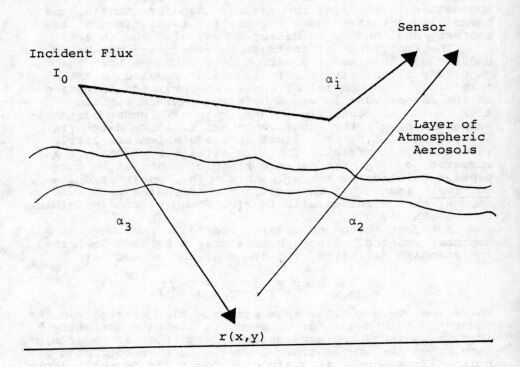

Incident Flux

I_0

Sensor

α_1

Layer of
Atmospheric
Aerosols

α_3

α_2

$r(x,y)$

Figure 1: Simple Scene Radiometry

IMAGE FORMATION MECHANISMS IN THE SENSOR

The flux which propagates from the scene to the sensor is collected by an optical system and brought to concentration in a focal plane. The image within the focal plane can be described by a specific two-dimensional convolution, which we do not prove here. The resulting equation is:

$$g(x,y) = \int_{-\infty}^{\infty} \int_{-\infty}^{\infty} h(x-x_1,y-y_1)s(x_1,y_1)dx_1dy_1 \qquad (3)$$

where $s(x,y)$ is as described in equation (1), $g(x,y)$ is the image observed in the focal plane, and $h(x,y)$ is the image formation point-spread-function of the optical system. (See [3] for details in the derivation of equation (3)). It is possible to prove that equation (3) has an equivalent description in the Fourier frequency domain,

$$G(u,v) = H(u,v)S(u,v), \qquad (4)$$

where G, H, S are the two-dimensional Fourier transforms of the corresponding quantities [3].

The important effects in image formation are obviously embodied by the point-spread-function $h(x,y)$ or its Fourier transform, $H(u,v)$. If $H(u,v)$ approaches zero for any values of (u,v), then for those values of (u,v) no information from the objects of the scene, $S(u,v)$, are transferred to the image $G(u,v)$. Even when $H(u,v)$ does not approach zero, if it becomes small relative to other values, the loss of contrast at those spatial frequencies represents an inability to detect object structure, as discussed above. Image degradations which can be present, either singly or in combinations, in the image formation process are:

(1) Diffraction optical effects. Every optical aperture has a cut-off beyond which no finer object resolution is possible. At spatial frequencies below that cut-off, the magnitude of $H(u,v)$ can be so small as to cause an effective loss of information [1,3).

(2) Optical aberrations, such as an out-of-focus lens, can cause attenuation of spatial frequency structure and effective loss of information.

(3) Relative motion during the period of exposure between objects being imaged and the focal plane can cause a loss of object structure in the image.

IMAGE DETECTOR MECHANISMS IN THE IMAGE FOCAL PLANE

A detector is a mechanism which intercepts the flux

concentrated in the focal plane by the optics. The detector is itself a source of degradations which are similar to the types of degradations previously discussed. The degradations are:

(1) Detector Transfer Function. The detection mechanism, whether film or semiconductor, integrates a finite area of the image to produce a response. This integration can be described by an equation similar to equation (3) above, but with a different function h. The Fourier transform of the detector function which replaces h in equation(3) is referred to as the Detector Transfer Function.

(2) Detector Sensitivity. Detectors are not devices with perfect sensitivity to photons. Most imaging detectors require more than one photon to produce a unit of detector response, e.g., a photo electron or a developed grain of silver in a film. The detector quantum efficiency (DQE) measures this sensitivity, and if the DQE is small, then loss of contrast occurs, particularly in the presence of detector noise.

(3) Detector Noise. No detector is perfect. The capturing of image information is always accompanied by random uncertainty in the image information. In detectors that capture image information the noise is usually due to electronic thermal noise (for electronic or semiconductor image detectors) or the random size and shape of silver grains in a photographic emulsion.

For general reading and references of the modelling of image degradation effects the reader should consult [1].

The basic equation of image formation can be summarized as:

$$g(x,y) = \int_{-\infty}^{\infty} \int_{-\infty}^{\infty} h_c(x-x_1,y-y_1) s(x_1,y_1) dx_1 dy_1 + n(x,y) \quad (5)$$

where h_c is a composite spread function of a number of effects (optics, motion, detector), $s(x,y)$ is as given in equation (1), and $n(x,y)$ encompasses all noise sources. Equation (4) assumes the image detector is linear in response. If the detector is nonlinear (photographic film is the most prominent example of a nonlinear detector), then the equation for the image recorded on the detector becomes:

$$g(x,y) = \phi\left[\int_{-\infty}^{\infty} \int_{-\infty}^{\infty} h_c(x-x_1,y-y_1) s(x_1,y_1) dx_1 dy_1\right] + n(x,y), \quad (6)$$

where $\phi[\cdot]$ is a function that describes the nonlinear detector response.

Equations (1), (5), and/or (6) describe mathematically

the image restoration (enhancement) problem. Given the
image $g(x,y)$ we must infer $s(x,y)$, which in turn must be
used to infer $r(x,y)$. This is <u>not</u> an easy problem.
Students in mathematics recognize equation (5) as a
Fredholm integral equation of the first kind, and such
equations are notoriously difficult to solve. It is this
problem, as well as the even more difficult problem in
equation (6), which represents the challenge in image
restoration.

TECHNIQUES OF DIGITAL IMAGE RESTORATION AND ENHANCEMENT

The employment of a computer for image restoration/
enhancement must begin with conversion of the image onto
a form suitable for digital computation*. This conversion
consists of two steps: scanning the image and extracting
samples from the image, each sample representing the image
intensity at a particular point; and quantizing the extracted samples into a digital number which can be entered
directly into a computer or recorded on a computer storage
medium such as magnetic tape.
The extraction and conversion of image information
into digital form is governed by some basic constraints.
There is an upperbound or maximum spacing between samples
which must be adhered to, or the sampling process will
lose information [4]. Likewise, each sample must be
quantized to a minimum number of bits in a binary computer
representation or the quantization process will introduce
an unacceptable amount of error [4].
The computer operations which can be applied for
image enhancement can be divided into two broad classes:
<u>point operations</u> and <u>neighborhood operations</u>. (The
dividing line is not precise. It is possible to modify a
point operator with neighborhood statistics.) A single
sample from an image is usually referred to as a picture
element, or <u>pixel</u> in the common technical vernacular.
Point operations can be characterized as operating upon a
single pixel to produce an altered value of the pixel, i.e.,
a single spatial picture sample has its value changed without reference to any other picture sample. By contrast,
a neighborhood operation will alter the value of each
pixel as a function of the pixels which surround each
pixel within some neighborhood or spatial region.
To understand some simple bases for point operations,
consider the radiometry in equation (1). Since $r(x,y)$

---

* Conversion to digital form is required for film imagery.
Modern electro-optical sensors can be engineered so that
their output is intrinsically digital, and no digital
conversion is necessary.

is the desired object reflectivity that may be obscured by
atmospheric conditions, then solving equation (1) for r(x,y)
yields:

$$r(x,y) = \frac{s(x,y) - \alpha_1 I_0}{\alpha_2 \alpha_3 I_0} \qquad (7)$$

This equation states that to obtain an estimate of r(x,y)
we would subtract an estimate of the reflected flux and
then divide by the incident flux times the attenuation
coefficients. What makes this a point operation is obvious
in this equation. The value of the image in pixel s(x,y)
is the input to a calculation which does not require any
other pixels neighboring to s(x,y). In fact, since the
denominator is fixed by atmospheric conditions, the operation is obviously <u>linear</u>.

Point operations need not be linear, of course.
Given a pixel at location (x,y), it is legitimate to
consider an arbitrary transformation of the pixel value.
Thus mathematically, we express the enhancement as:

$$g_e(x,y) = p[g(x,y)], \qquad (8)$$

where p[·] is an arbitrary linear or non-linear operator,
and g_e is the pixel value after enhancement. What matters
is that the operator p[·] accomplish some desired objective
with respect to making scene detail more useful visually.
The choice of p[·], therefore, may be closely associated
with the nature of the degradation which afflicts the
imagery. Point operators are ofter referred to as <u>contrast
Operators</u> or <u>contrast alterations</u>, because they function,
in a generalized mathematical way, like the contrast adjustments that can be implemented by darkroom photographic
technique.

Important categories of point operator processes, and
the type of degradation they correct, are listed in the
following:

(1) Linear Scalings. Operators, similar in form to
equation (7), can be realized by combinations of additions/
subtractions and multiplications by a constant. The
correction of degradations such as radiometry errors is
obvious, but is also applicable to any type of degradation involving an equation of the type of (1), i.e.,
attenuation of the information, plus an added bias term.
The actual enhancement need not be re-computed for every
pixel. Since pixel intensities are usually quantized at
a small number of bits, say 8 bits, the actual results of
equation (7) can be precomputed, stored in a table, and
looked-up from the table for a given pixel intensity.

(2) Non-linear Scalings. The more general point
operator p[·] in equation (8) has to be chosen to match

the type of degradation which must be corrected. For example, in an over-exposed picture, a statistical histogram of pixel intensities would be dominated by the largest pixel intensities. Any operation which increases the magnitude of small numbers more in proportion than large numbers would be useful to correct for the over-exposure. A logarithm function for p[·], followed by a linear rescaling to preserve dynamic range, would be valid. For an under-exposed image, the opposite operation, such as an exponential, would be useful. The actual utility of any non-linear operation is determined by the extent to which the functional form of p[·] matches a particular form of degradation. For example, if the non-linear function ϕ in equation (6) is known, then it may be removed from the imagery by choosing a function ϕ-1 for the operator p[·].

(3) Histogram-Derived Transformations. The discussion in the previous paragraph accentuates the usage of statistics for determining the point operator. The final refinement of this is the histogram equalization operation. A histogram of pixel intensities is accumulated, from which it is always possible to determine a transformation that distributes pixel intensities over the entire dynamic range of intensities in such a way that the transformed intensities occur with equal probability, i.e., the histogram of the transformed image is uniform or "flat". The result is an image with very strong accentuation of minor contrast differences [5].

(4) Human Interaction. Recent developments in computer hardware make it possible to implement any point-operation with a look-up table. The table can be referenced so fast that imagery can be enhanced at the video refresh rate of a conventional color CRT. This in turn means that any operator interaction device can be linked - for example via microprocessor - to mathematical models of enhancement, and the parameters of the enhancement adjusted in real-time. The resultant enhanced image is instantly visible, since the operator adjustment can be used, via the enhancement model, to alter the contents of the look-up table. Thus, the human operator can adjust "knobs" and see the enhanced picture instantly. The knobs may be physically real or only a linkage to a mathematical enhancement model of substantial complexity. However, the ability to instantly view the results of adjusting knobs allows the human operator to optimize visible image quality. A completely mathematical optimization of visible image quality is not possible, of course, because no complete model of human visual response is known. Using the interaction by human and computer enhancement model, no human visual system model is needed. Instead the operator directly optimizes visual quality for the

particular enhancement model he is employing.

As stated above, the second broad class of enhancement is by <u>neighborhood operations</u>. The broadest analytical model of a neighborhood operator is given as:

$$g_e(x,y) = \rho[g(x,y)] \qquad \text{for } (x,y) \, \varepsilon R \qquad (9)$$

where $\rho[\cdot]$ is the functional form defining the neighborhood operator and R is the actual dimension of the neighborhood itself.

The most common neighborhood operation is that of convolution, or weighted averaging. For a set of pixels we have:

$$g_e(x,y) = \sum_{s=1}^{M} \sum_{r=1}^{N} h(r,s) g(x-r,y-s) \qquad (10)$$

where h is a set of MN weights in the average. For the simplest case, $h(r,s) = 1/MN$ for all s, and we see that equation (10) is a moving average. In general, there is a well-developed theory corresponding to equation (10), and one can show that it is the sampled-image version of equation (3), and that a discrete Fourier transform of equation (10) exists which is the discrete (sampled) analogy of the continuous transform seen in equation (4) [1,5].

Given the simplicity of the operator in equation (10), it is not surprising that what makes the difference is the specification and design of the weights $h(r,s)$. Equivalently, from the discrete analogy mentioned above we have that:

$$G_e(m,n) = H(m,n) G(m,n) \qquad (11)$$

where m,n are discrete frequency indices, and in discrete Fourier space the problem is the design of the Fourier transfer function $H(m,n)$. If the transfer function is appropriately chosen, then a variety of degradations in the imagery may be corrected. The processes which are invoked to design either $h(r,s)$ or $H(m,n)$ belong to the study of the theory of optimal filtering, which we will not attempt to explicate here. (See, for example, the books by Andrews & Hunt [1] or Pratt [5] or Oppenheim and Schafer [4].) However, for an appropriate optimality criterion in the design of the filter, there exists the implementation of a linear neighborhood operator, in the form of either equation (10) or (11).

It is important to note that, since certain image degradations are expressible in the form of equation (3) (or in equation (4), which is the Fourier analogy to equation (3)), the correction for degradations of the form

of equation (3)˙is to apply a digital version of a neighborhood operator which is identical in form to the original
degradation. The only difference between the h in equation (3) and the h in equation (10) is the following.
Since equation (3) describes a degradation process, then
the enhancement is achieved by choosing h values for equation (10) which are the inverse, <u>with respect to the moving
average operator</u>, of the h values in equation (3). In
terms of Fourier descriptions, the H(m,n) values in equation (10) must be 1/H(u,v) from equation (4).

The following types of degradations have been removed
from imagery by the simple linear operator of the form of
equation (10):

(1) Diffraction limit. Although information cannot be
obtained beyond the optical diffraction limit of an
aperture, substantial loss of image quality occurs for
objects that approach in scale the smallest spatial frequency details visible through the aperture. Substantial
improvement in image quality is achievable by using a
filter which attempts restoration, out to the optical
cutoff, of the diffraction limited optical transfer function of an optical system.

(2) Optical aberrations. Many optical aberrations can
be described as an equation such as (3). A common example
is when the optical system is out of focus. Again, if the
optical transfer function of the aberration process is
known, then it is possible to find a processing by
neighborhood operation, in the form of equation (10), which
will correct for the aberration.

(3) Motion. If the optical focal plane and the object
are in relative motion, and if the motion of the image
in the focal plane is rapid during the time the image is
being exposed, then the focal plane image will be blurred
by the motion. Depending on the complexity of the motion,
a transfer function can be written to describe the motion
blur. This transfer function can then be used, in the
form of either equation (10) or (11), to restore the image
from motion degradation.

(4) Noise filtering. As seen in equation (5), noise is
present in any real image. The magnitude of the noise
is the issue. The noise magnitude is usually quantified
by the signal-to-noise ratio (SNR). A typical measure of
SNR is the variation in image signal divided by image
noise. Using statistical variance as our measure of
variation, then:

$$\text{SNR} = \frac{\sigma^2_{\text{image}}}{\sigma^2_{\text{noise}}} \tag{12}$$

When the SNR is large enough, no visible noise effects
are perceived. As the SNR drops, the noise becomes visible

and objectionable. To reduce the noise effects, a filter
of the form of equation (10) can be used to average the
noise in a spatial neighborhood and reduce its visible
effects.
      Neighborhood operators need not be linear. Again,
it is the type of degradation which is most important in
choosing and specifying the type of nonlinear enhancement
operator. One very prominent nonlinear neighborhood
operator is the moving median filter. That is, instead
of computing an average of pixels in a moving neighborhood, the median of the pixels in the moving neighborhood
is computed. The resulting filter has been shown to be
very powerful for removal of certain types of noise. Noise
due to isolated defects, such as a scratch in the emulsion
of a film, can be removed by a median filter. The median
filter has little effect on edge sharpness, if properly
designed, so that isolated streaks, scratches, or spot
flaws can be almost magically removed without affecting
other visual properties of the image. This is quite important, because a more conventional noise filter using
a moving average will tend to reduce visual sharpness of
objects in the image as it smooths out the visible noise.
      Other nonlinear filters are based on the nonlinear
model of equation (6). If the degree of image blur caused
by h in equation (6) is too great, and if the response
characteristic ϕ in equation (6) is too nonlinear, then
the optimum solution of (6) for the undergraded image s
must be made by a nonlinear operator. The actual rationale for solution involves Bayesian probability analysis,
but the actual computation can be shown to be equivalent
to linear neighborhood operators combined with nonlinear
point operators [1].
      This latter point indicates a more general aspect of
image enhancement. Frequently, the best improvement of
the imagery can not be obtained by a single step. Instead,
to produce the optimum results it is necessary to combine
several linear and nonlinear operators of both the point
and neighborhood type. Again, modern computing technology
is quite important, because it is possible to build ecomoni-
cal, but quite powerful, special purpose computers in which
a human can interact with the image enhancement processes,
choosing and applying the appropriate one at each step.
Again, in lieu of a general theory of visual image quality
on which to base comprehensive enhancement models, the
availability of interactive image computers is essential
for optimum improvement of image quality.

THE NECESSITY FOR MORE IMAGE ENHANCEMENT RESEARCH

      When the visible improvements in image quality from
image enhancement are seen, it is sometimes natural for

the non-expert to believe that image enhancement is magic, and that there are no limits to the degree of improvement that can be achieved. There are, however, very distinct and basic limits to the degree of improvement in image quality which can be achieved. The fundamental limit lies in our ability to deal with an unfavorable SNR. In equation (5), as noted above, we are faced with solving a Fredholm integral equation of the first kind in order to restore the image from the degradations. Such equations are ill-conditioned, which means that small errors or uncertainties in the data can be translated into very large errors in the solution. Noise, as in equation (5), represents such a source of errors. It is possible to show that many enhancement processes have the effect of amplifying whatever noise is present in the degraded image. If noise amplification is too great, then the enhanced image quality is suppressed by the visually objectionable aspects of the amplified noise.

Noise is a fundamental limit, like that of entropy in thermodynamics. (Indeed, there is legitimate justification for considering any kind of noise in a measurement process as a manifestation in various ways of the Second Law.) Reduction of the uncertainty which noise represents can only be done by the injection of other information into the enhancement paradigm. For example, if one knows the image being enhanced is that of an airplane, then the knowledge about the object represents considerable a-priori information which can be added into the enhancement process. But the exact mechanism by which to do this is uncertain. The research necessary to achieve this a-priori constraining of restoration options has not been carried out.

Image enhancement/restoration has not been adequately discussed in this short article. Indeed, entire books have been written on the topic [1]. The discussion has been complete enough to make clear the following points, however:

(1) Image enhancement/restoration is based upon the physics and mathematics of image formation processes. The understanding of these models is the most complete aspect of our theory of image enhancement.

(2) Much of image enhancement must be characterized as a "bag of tricks", that is, a selection of techniques which must be applied by a human analyst. Hopefully, the analyst may have available an interactive image processing computer which will make it possible to optimize the quality of the enhanced image by allowing the analyst to instantly see the results of his enhancement processes. However, there is one great deficiency to the "bag of tricks" approach. Not all human operators are equally clever or resourceful in using the tricks at their disposal. Consequently, an image which causes one analyst to conclude

no enhancement is possible may be treated with great success
by another analyst. For images with great significance,
such as those which might be used in weapons verification
monitoring, it is disturbing to think of the consequences
of an analyst failing to produce the optimum visual
quality from a given image.
(3) The frontiers of image enhancement lie in trying
to make better use of marginal imagery, i.e., images in
which the SNR is so low that current enhancement technology
will fail. It is most unfortunate that currently little
research is going on in this area. The remarkable successes of image enhancement/restoration have led many to
believe that image enhancement is a "solved problem",
when only the easiest parts of the problem have been
solved as of yet. It is ironic that research on image
enhancement/restoration is dwindling at a time when the
discussion about weapons control and verification are
emphasizing the virtues of obtaining maximum information
from an image.

The dwindling of image enhancement research is a
natural consequence of the successes of this technology,
as well as the perception that perhaps enhancement is not
needed. The engineers who specify, design, and build
sensor systems always have plans to create a bigger and
better sensor, one which produces images of even higher
quality than anything seen so far. In this environment
image enhancement seems of secondary importance. Nonetheless, image enhancement remains of importance. True,
each new and improved sensor creates higher quality imagery
than its predecessor. But no complex sensor system always
operates exactly as designed and programmed. Occasionally,
it fails in some major or minor way and the resulting
image, be it underexposed or out-of-focus or whatever, will
be a candidate for image restoration or enhancement. If
that particular failure is associated with an image of
critical importance to a specific weapons verification
problem, then do we not want available the best enhancement
technology?

SUGGESTIONS FOR FURTHER READING

Since image enhancement is broad, and we have only
surveyed the basic outlines within this article, the reader
may find the following suggestions useful if further reading is desired.

The only booklength treatment of image restoration
is by Andrews and Hunt [1]. Several books on digital
image processing are available, which cover enhancement
and restoration, as well as other topics. See, for example,
the books by Pratt [5], Gonzalez and Wintz [6], Rosenfeld
and Kak [7], Hall [8].

Several journals carry research articles on image processing, particularly Applied Optics, Journal of the Optical Society of America, IEEE Transactions on Acoustics, Speech and Signal Processing, and Computer Graphics and Image Processing. Two special issues of the IEEE Proceedings, from July 1972 and May 1981 contain a number of excellent articles on digital image processing.

REFERENCES

[1] H. C. Andrews, B. R. Hunt, Digital Image Restoration, Prentice-Hall, Englewood Cliffs, N. J., 1977.

[2] J. E. Hall, et al, "A CCD imager for image processing applications," Proceedings 1978 IEEE Electron Device Meeting, Washington, D. C., Dec. 1978, pp. 415 - 418.

[3] J. W. Goodman, Introduction to Fourier Optics, McGraw Hill, New York, 1968.

[4] A. V. Oppenheim, R. Schafer, Digital Signal Processing, Prentice-Hall, Englewood Cliffs, N. J., 1972.

[5] W. K. Pratt, Digital Image Processing, Wiley, New York, 1978.

[6] R. Gonzalez and P. Wintz, Digital Image Processing, Addison Wesley, 1977.

[7] A. Rosenfeld and A. Kak, Digital Picture Processing, Academic Press, 1979.

[8] E. Hall, Digital Image Processing and Recognition, Academic Press, 1982.

CHAPTER 7

TEACHING ABOUT PHYSICS AND THE NUCLEAR ARMS RACE

Dietrich Schroeer
Department of Physics and Astronomy
University of North Carolina
Chapel Hill, NC 27514

The arms race has an important scientific and technological component; but it inevitably also involves political problems. In thinking about the arms race, physicists tend to emphasize the quantitative aspects, while policy makers and society as a whole focus on the political choices. A personal experience of teaching a course on the nuclear arms race to non-science students is described to show how teaching quantitative aspects in an order-of-magnitude fashion can help non-scientists decide when numbers are important, and when they may be irrelevant or misleading to policy considerations.

I. INTRODUCTION

In response to social pressures some physicists have expanded their teaching to include public-policy issues that combine science and social concerns. One such social issue is the nuclear arms race. Not only is it of rising political interest at the moment, but it has special appeal to physicists, both because they have been historically involved and because it contains physics. The nuclear arms race can be a source of illustrations applying basic physics principles to actual technologies. Nuclear physics specifically is the basis for the nuclear-weapons technology; thermodynamics accounts for diffusion processes and energy releases; electricity and magnetism helps understand isotopic separation processes, the electromagnetic pulse, and computers; etc. Further, the nuclear arms race is susceptible to an order-of-magnitude approach; many of its major issues can be examined in a quantitative way without understanding the details of the technologies involved. For these technical reasons, and because of the social importance of the nuclear arms race, a substantial number of physicists have taught a course on this subject at some time in their teaching career, and more are doing so right now.[1,2]

I have been teaching a course on "Science, Technology and the Nuclear Arms Race" at the University of North Carolina since 1976. This article is a description of this course to indicate how teaching on the arms race might be done. I would like to go beyond a pure course description, to consider some of the implications of a physicist teaching such a course. The problems faced in teaching about the arms race are characteristic of those faced by all teachers of "relevant" physics courses, whether they be about the energy situation, the environment, or science policy. Teachers of such courses face several problems. They are aware of one of these: they must talk about issues that in part go far beyond their

0094-243X/83/1040150-18 $3.00 Copyright 1983 American Institute of Physics

expertise. The second problem they are less likely to recognize: their professional biases as technically trained persons inject themselves into their class presentations. These two problems come to a head when decisions have to be made about how far to go in quantifying material in such a course. I want to describe one way of resolving the conflict of wanting to be quantitatively professional without obscuring the important political issues.

That physicists lack some of the expertise needed for discussing the nuclear arms race is obvious. The important aspects of that topic often go far beyond traditional physics. Other sciences play a major role in the nuclear arms race, such as the computer science of digital image processing, the chemistry of nuclear-waste reprocessing, the geology of seismic detection, the physiology of radiation effects on human cells, the psychology of catastrophic events, and the political science of treaties and world stability. Very complex technologies are involved, including the most advanced computers, metallurgy, and nuclear power reactors. Public policy considerations abound; deterrence demands certain philosophical assumptions, and mutual assured destruction may be deadening some ethical nerves in mankind. A recent bibliography on physics and the nuclear arms race shows the wide scope of the questions to be considered, even when the orientation is technological.[3] How is a "mere" physicist to understand these other sciences, be knowledgable in the sophisticated technologies, and discuss these moral questions?

A second problem presented by an arms-race course is more subtle. Physicists bring certain attitudes toward their teaching, attitudes instillled in a good physicist (good in the sense of being able to do good physics) through the norms of science.[4] These norms encourage loyalty to professional peers, and they promote quantitative work. The performance of scientists is judged by how well they choose scientific problems to tackle. Only if a scientific problem promises to be soluble in a quantifiable way with good precision, only then is it accepted as a good problem worthy to be worked on. The quantifiability insures that theories can be tested and experiments repeated so that all peers know when the problem has been successfully solved. Numbers are the lifeblood of physics. We all have the greatest admiration for Enrico Fermi for his back-of-the-envelope estimates for everything, even the number of piano tuners working in Chicago. Not only is this estimating facility highly prized among physicists, but it is also admired by non-physicists. Conversely the norms also instill an insensitivity toward non-quantifiable parameters. The search for quantitification may influence not only what aspects of the arms race are seen as important, but also may control how they are analyzed. Unfortunately this bias may be very much unrecognized as being so important and as overlooking the uncertainties inherent in the issue. One purpose of this paper is to help physicists not only to recognize this bias, but also to put it to constructive use.

My own physics training has not particularly helped me in deciding how to teach an arms-race course. If I am able to teach on this topic, it is in good part simply a result of practicing for some years. I also bring to the teaching of the nuclear arms race some expertise that I have acquired through my work on the interaction between science and society. I am involved in a public-policy-analysis program here at the University of North Carolina at Chapel Hill. As part of this involvement I have taught a course on "Science and Policy" within our Department of Political Science. This course considers the nature of technical expertise and the structure of the scientific community; teaching it has made me aware of my own biases as a "technical expert." My conclusion is that the problems of technical expertise are unavoidable, that an emphasis on quantification is inevitable since both sides do use numbers. The only way to minimize this problem is to beconstantly aware of it. Not only must we acknowledge our limits in knowledge, but we must also be sensitive to the limits in a technical approach. However, once the limits are recognized, they can be exploited. I want to point out our limits as physicists, and how to exploit them in teaching nuclear arms-race courses.

I will begin by briefly indicating how one might inject arms-race material into regular physics courses, and what points one might illustrate with such examples. I will then outline my own course on the nuclear arms race as a comprehensive listing of what material really is needed in a nuclear arms-race course that is technically oriented. I shall give two examples to show how to do such teaching. The course outline and the examples will illustrate the problems and opportunities of teaching about the arms race. They will show the advantages contained in quantifying the nuclear arms race, and what limits should be placed on such quantification.

II. PHYSICS IN THE NUCLEAR ARMS RACE

One major attraction of the nuclear arms race to physicists is its contact with basic physics, and how some of its issues can be analyzed in an order-of-magnitude way. I here want to give physics teachers a feeling for the physics contained in the nuclear arms race by listing some examples in Table 1, and by describing two illustrative applications of physics to the arms race.

Table 1. Some physics in the nuclear arms race. Reference 3 lists a selection of relevant technical sources.

A. Nuclear Weapons

1. Chemical Explosives: Atomic physics - chemical energies defined by the Bohr atom.
2. Fission Bomb: Nuclear binding-energy curve; $E=mc^2$ compared to potential energy, kinetic energy, thermal energy; radioactive isotopes - nuclear decay schemes; exponential decay and growth; theory of fission; chain reaction - neutron multiplier; critical mass - mean free path of neutrons; preignition - spontaneous fission; rate of chain reaction.
3. Fusion Bomb: Fusion reactions and their energy releases, like $D + T \longrightarrow \alpha + n + 17.6$ MeV, quantum-mechanical tunneling rates;

production of tritium by Li + n $\longrightarrow$ T + α + energy; U-238 fission - neutron capture crossection.

4. Effects from Nuclear Explosions: Shock waves - velocity of sound and overpressures; temperature of fireball - blackbody radiation and Wien's law; double light flash - atomic ionization; electromagnetic pulse - electricity and magnetism; prompt radiation - nuclear mass curve; α, β, γ, n absorption; fallout - radioisotope halflives, radiation doses, and atmospheric physics; energy releases and scaling laws - e.g. inverse-square; ozone depletion - atomic spectra and reaction rates; medical effects - radiation damage.

B. Delivery Systems

1. Bombers: Aerodynamics - Bernoulli effect; engines - thermodynamics; ram jets - fluid flow; strength of materials - solid state physics; radar - microwaves; electronic countermeasures - electronics.

2. Missiles: Rocket acceleration and flight mechanics - Newton's laws and classical mechanics; gravity - mass concentrations; escape velocity - Kepler's laws; strength of materials - solid state physics; guidance systems - gyroscopes; computers - solid-state devices; satellite guidance - Doppler shifts; missile stellar guidance - astronomy.

3. Submarines: Buoyancy - fluid dynamics; accoustics - e.g. water thermal refraction; echo triangulation; 100% U-235 nuclear reactors; anti-submarine warfare - accoustics; ultra-low-frequency waves; neutrino communications.

4. Deterrence: Nuclear balance - effects of nuclear weapons, $1/r^2$, $1/r^3$; reliability - systems analysis; zero-sum games - game theory; strategies - operations research.

C. Strategic Alternatives to MAD

1. Civil Defense: CO removal - partial pressures; cooling of shelters - thermodynamics; absorption crossections for radiations; fallout - its distribution and flight times - atmospheric physics; evacuation and industrial recovery - systems analysis.

2. Antiballistic Missile Defenses: Rocket accelerations and flight times - Newton's laws; radar - wave phenomena; damage to missiles - x-ray radiation damage; ablation of missile skin - thermodynamics; neutron damage - preignition; chaff - metal dipole reflectors; beta blackout - ionization of air by electrons.

3. Chemical and Biological Warfare: Toxicity - chemistry and biochemistry.

4. Tactical Nuclear War: "Pure" fusion neutron bombs - D + T reactions; dirty bombs - fast neutrons and U-238; weapons effects - scaling laws, e.g. $1/r^3$ for blast; absorption of neutrons in matter - e.g. elastic collisions; fallout - halflives, atmospheric physics; infrared guidance systems - black-body radiation.

D. Arms Control

1. Nuclear Proliferation: Nuclear reactor technology - moderators and elastic collisions; fission products - radioactivity; isotope separation - thermal diffusion and centrifuges; absorption

versus fission crossections; reactor power output – thermodynamics; denaturing – the breeding of Pu-239 and Pu-240; breeder reactors – reprocessing.

2. The Nuclear Test-Ban Treaty: Earthquakes – their wave aspects; seismic arrays – interference phenomena; decoupling theories – impedance matching; plate tectonics – geophysics.

3. Strategic Arms Limitations: Monitoring test telemetry – electricity and magnetism; satellite monitors – Kepler's laws; verification photography – f stops, focal length, film resolution; optical and ground resolution – air effects on light, e.g. Brownian motion; improved photography – Fourier analyses; multispectra photography – infrared and ultraviolet radiation; sideband radar – microwaves; satellite power – solar cells.

4. Disarmament: International relations – energy needs and supplies; guns or butter – economics; technological progress – the growth of science; the technological imperative – computer capabilities.

To demonstrate the relevance of physics to the nuclear arms race as cited in Table 1, I will here give two examples in which basic aspects of physics are applied to arms-race technologies.

1. Thermodynamics

Consider the problem of storing a nuclear weapon. Must its core be refrigerated to remove the heat emitted by the radioactive decays within the critical mass? The critical mass for a bomb made out of Pu-239 is on the order of 10 kilogram, or a sphere with a radius of 10 cm, including a proper tamper of 5cm thickness that reflects neutrons back into it. How many atoms is this, how many atoms decay each second, how much energy is released by these decays at 7.7 MeV per decay, considering the halflife of ^{239}Pu is 24,400 years? If contamination by Pu-240 of this critical mass is small, we can neglect the heat produced by that shorter-lived isotope. The number of Pu-239 atoms is

$$N = \frac{10 \text{ kg}}{239 \text{ kg/kmole}} \times \frac{6 \times 10^{26} \text{ atoms}}{\text{kmole}} = 2.5 \times 10^{25} \text{ atoms}$$

In the critical mass of 10 kilogram each second how many atoms of Pu-239 are decaying?

$$\frac{dN}{dt} = -\frac{\ln 2}{T_{1/2}} \times N = -\frac{0.693}{24,400 \text{ yrs}} \times 2.5 \times 10^{25} = 2.3 \times 10^{13} \text{ sec}^{-1}$$

How much energy is released in these decays at 7.7 MeV per decay?

$$P = \frac{dE}{dt} = 2.3 \times 10^{13} \text{ sec}^{-1} \times 7.7 \times 10^{6} \text{ eV} = 28 \text{ watts}$$

So this critical mass emits 28 watts of heat. Using the black-body radiation equation at a temperature T=300 K, and an $\sigma=6.7\times10^{-12}$ J/sec/cm^2 $^\circ$K^4, we can estimate how warm that mass might get. For a small temperature difference between the Pu and the surroundings, the radiative energy ΔR will be

$$R = \sigma \, 4T^3 \, \Delta T \, 4\pi r^2$$

$$\Delta T = \frac{\Delta R}{4\pi r^2 \, \sigma \, 4T^3} = \frac{28 \text{ watts}}{4\pi(10 \text{ cm})^2 (5.67 \times 10^{-12} \text{ Jsec}^{-1} \text{cm}^{-2} \text{K}^{-4})(300 \text{ K})^3}$$
$$= 38 \text{ K}$$

That temperature rise might be of some concern, but since the critical mass plus tamper is surrounded by other materials which do not generate heat, the equivalent surface area will be much larger. Radiation of the heat most likely will cool the Pu-239 core sufficiently to avoid the need for cooling.

2. Wave Properties

Wave interference phenomena, like multiple-slit interference patterns in optical physics, are involved in the nuclear arms race. In the Safeguard ABM system a Parameter Acquisition Radar unit is to detect incoming missile reentry vehicles at long ranges of perhaps about 200 miles. The radar consists of a grid of detectors, 80 detectors on a side separated by a=1 meter. The wavelength of the radar is also 1 meter. The angular resolution of this unit can be found by analogy to the angular width of the interference pattern of N=80 slits. Thus the angular resolution is about:

$$\Delta \sim \frac{\lambda}{aN} \text{ radians} = \frac{57.3^\circ}{80} = 0.7^\circ = 0.0125 \text{ radians}$$

At a range of 200 miles this corresponds to

$$r = D\Delta\theta = 200 \text{ mi} \times 0.0125 = 2.5 \text{ miles}$$

Clearly the resolution is not very good, at least not if the radar is to distinguish the reentry vehicle from surrounding decoys in order to guide the ABM to come into direct contact with the reentry vehicle. To destroy the RV under such radar guidance the explosion of the ABM warhead must have a very large destructive radius on the order of miles. Hence a nuclear explosion is required for radar-guided ABMs.

Both of these examples show how deeply physics is embedded in the nuclear arms race, and how examples from that race can illustrate very basic physics principles. Relevant examples have always been useful in teaching physics courses. Arms-race examples can serve two functions. They can enliven physics courses in a dramatic fashion. And they can introduce students to the topic in a relatively unbiased way, whetting their appetite for further explorations of the nuclear arms race. These narrow science questions can lead to broader discussions.

III. OUTLINE OF A NUCLEAR ARMS-RACE COURSE

The course on "Science, Technology and the Nuclear Arms Race" was developed within the Department of Physics and Astronomy under the sponsorship of our Curriculum on Peace, War and Defense (which supervises both the ROTC programs and students interested in international security problems). It requires no science background, and is taken by students from many academic disciplines; Some of them feel comfortable with a somewhat quantitative approach, while others fear numbers. This dual sponsorship and diverse student audience has made the course goals broader than might normally have been the case. My own long-standing interest in science-and-society issues led to the incorporation of policy implications into the course content. But in spite of this broad approach, the course content and style still often reveal that the teacher is a physicist. Table 2 lists the topics covered in this course; for each topic is listed a range of questions that ought legitimately be considered: the science and technology, the drive behind the technical developments, and the policy implications.

Table 2.   Topics covered in a course on "Science, Technology and the Nuclear Arms Race." After each topic is listed a gradation of questions that might reasonably be addressed, ranging from very technical matters to problems of public policy. Some of the scientific background is listed in more detail in Table 1.

A. NUCLEAR WEAPONS

1. Pre-Nuclear War:
    a. The Scientific-Technical Background: Chemical energy energy releases.
    b. The Drive behind the Developments: As chemistry and physics developed, the explosive yield (e.g. of TNT) and the accuracy of delivery systems (e.g. airplanes, rockets and submarines) were greatly improved; the ability to cause large-scale destruction by conventional became very advanced.
    c. The Policy Implications: Mass destruction through conventional chemical explosives is quite feasible, particularly through strategic aerial bombardment of civilians in cities, as developed in World War II.

2. The Fission Bomb:
    a. The Scientific-Technical Background: The explosive energy release from nuclear fission chain reactions, nuclear reactors.
    b. The Drive Behind the Developments: The scientific drive toward the use of nuclear energy started at the turn of the century; the technological application of nuclear fission was obvious in 1939; the war effort made the development of nuclear weapons inevitable.
    c. The Policy Implications: The shock of the A-bomb helped end the war, air warfare seemed superior to other military operations; scientists became heroes, the post-war Baruch plan and the cold war, and the push for nuclear reactor applications.

3. The Hydrogen Bomb:

a. The Scientific-Technical Background: Explosive energy release from nuclear fusion reactions; the fission-fusion-fission bomb and limits to bomb size.

b. The Drive Behind the Developments: The Russian A-bomb; the McCarthy era; the Air Force and Teller versus the AEC and Oppenheimer.

c. The Policy Implications: Strategic bombing required big bombs; worries about an inferiority of the U.S. fueled the nuclear arms race; first the strategy of massive retaliation, then the concept of deterrence.

4. Full-Scale Nuclear War:

a. The Scientific-Technical Background: Nuclear weapons effects; psychology of catastrophes; population and industrial concentrations; social structures and civilization.

b. The Drive Behind the Developments: Large-scale nuclear war became possible; the inaccuracy of aiming plus the huge size of the weapons made tactical nuclear war an infeasible strategy; conventional warfighting capabilities were much more expensive.

c. The Policy Implications: Through the 1950s to the 1980s civilians have been held hostage by nuclear weapons as a guarantee that no one used nuclear weapons anywhere. Massive retaliation, deterrence and MAD.

B. Delivery Systems

5. Intercontinental Bombers:

a. The Scientific-Technical Background: Improving bomber and electronics technology.

b. The Drive Behind the Developments: The sucess of strategic bombing during World War II produced the Strategic Air Command; jet engines improved with better materials.

c. The Policy Implications: With the model of strategic bombing of World War II fresh in mind, the bomber forces grew, emphasizing strategic over tactical bombing, big H-bombs over tactical nuclear bombing, and emphasis on deterrence rather than warfighting. The Oppenheimer controversy reflected this desire for hydrogen bombs useful in strategic bombing.

6. Intercontinental Ballistic Missiles:

a. The Scientific-Technical Background: Chemistry and alloys provided the fuels and engines capable for intercontinental ranges; improved inertial missile guidance; miniature A-bombs and H-bombs; MIRVing of payload; improved accuracy through in-flight navigation by stellar guidance; communications from satellites, or inflight mapreading.

b. The Drive Behind the Developments: Transistor technology became available after about 1950, making missile guidance control feasible; improving H-bomb technology provided small warheads; Sputnik and the apparent missile gap of 1960 spurred rapid deployment of ICBMs; continuing improvements in computer capabilities have driven improving accuracies of ICBMs.

c. The Policy Implications: The initial low accuracy of missiles made them unsuitable for anything but retaliation on "soft" civilian target, producing mutual assured destruction (MAD); now the increasing accuracy of missiles, and MIRVing, has made the ICBM not only a potential instrument for nuclear warfighting, but also has made it vulnerable to attack, and hence less suitable as a component of MAD.

7. Submarine-Launched Ballistic Missiles:
a. The Scientific-Technical Background: Compact and long-lived nuclear power reactors; solid-fuel rockets; inertial guidance systems for missiles and submarines; extremely-low-frequency communications systems; anti-submarine warfare; knowledge of ocean currents and temperature layers.
b. The Drive Behind Developments: Hyman Rickover single-handedly pushed this very logical development - nuclear-powered submarines were the only way to achieve indefinite submersion times; Rickover's victory over the aircraft-carrier admirals is still being contested; so far ASW has not kept up with submarine performance.
c. The Policy Implications: Missile-launching nuclear submarines are essentially invulnerable to attack; their missiles are so far relatively inaccurate, making them a good nuclear deterrent; future satellite guidance may make SLBM suitable for nuclear warfighting.

8. Deterrence:
a. The Scientific-Technical Background: Operations research; zero-sum games; minimax problems; the Richardson model; accuracy of weapons' delivery; MIRV; ABM; accuracy versus yield in enhancing kill parameters.
b. The Drive Behind the Developments: Deterrence combines capabilities with perceptions of credibility; the lack of any alternative capabilities in the past made deterrence inevitable; unMAD strategists have long wanted to remove this policy of all-or-nothing with some more reasonable, and more credible alternatives, such as conventional or nuclear warfighting scenarios - this has encouraged the search for lower level technological fixes like neutron bombs.
c. The Policy Implications: Deterrence is the policy at present not just by choice, but also by necessity, as no other alternatives are so far reasonable; in an insane way, defense is bad, offense is offensive, and unethical suicide threats are the way to continuing peace; rational analyses, like operations researchers play, may lead to rational, but destructive actions.

C. ALTERNATIVES TO DETERRENCE

9. Passive Defense - Civil Defense:
a. The Scientific-Technical Background: Nuclear weapons effects and how to protect against them; social planning for post-war recovery.
b. The Background Behind Developments: Deterrence has always seemed slightly irrational; it seems one ought to develop defenses; as limited nuclear warfighting becomes more likely, civil defense seems more rational and feasible; there is the supposed Soviet civil defense effort to encourage a U.S. effort.
c. The Policy Implications: Civil defense of countervalue

targets destabilizes deterrence; civil defense adequate to protect large numbers of civilians against a strategic nuclear attack is hard to do and might require changes in American lifestyles.

10. Active Defense - Antiballistic Missiles:
   a.   The   Scientific-Technical   Background: Nuclear-tipped interceptor missiles; guidance by phased-array radar; penetration aids; MRV, MIRV, MaRV; lasers and particle beam weapons; space-based ABM systems; terminal guidance for interceptor missiles.
   b. The Drive Behind Developments: As missile silos become vulnerable to attack, defending them becomes more desirable; increasing computer capabilities make the ABM appear more feasible.
   c. The Policy Implications: Extensive ABM systems would violate SALT I agreements; how little would the offense have to spend to overcome the ABM?   Defending ICBM silos is stabilizing, but city-defending ABM is destabilizing; the ABM treaty tried to limit that technological imperative.

11. Chemical-Biological Warfare:
   a.   The   Scientific-Technical   Background: Organic   chemistry, biology, physiology, binaries; aerosols; vaccinations; gas masks and other protection; Agent Orange; VX weapons; myotoxcins; mustard gases; recombinant DNA techniques.
   b. The Drive Behind the Developments: Is CBW an inexpensive alternative to nuclear deterrence?   How does the user protect himself against it?   There is fear of technological surprises by the enemy; the r-DNA technique presents a long-term future technological imperative.
   c. The Policy Implications: So far there has been little use of CBW; the Geneva Convention of 1925; in the case of the U.S. it may be due to the failure of defoliation in Vietnam; hopefully the U.S.S.R. is discovering the same problem with CBW in Afghanistan; have humanitarian considerations played a role in suppressing CBW use?

12. Tactical Nuclear War

   a.   The   Scientific-Technical   Background: Enhanced-radiation bombs; prompt and permanent biological incapacitation; tank warfare; smart bombs - precision-guided missiles; AWACS; multiple-target-tracking radar; fallout patterns; civil defense; Pershing II, SS-20; cruise missiles; TERCOM.
   b. The Drive Behind Developments: In Europe the conventional balance has numerically favored the Warsaw Pact nations.   Tactical nuclear weapons have been seen as the answer for NATO.   As missile guidance has improved, smaller nuclear warheads, such as the neutron bomb have been developed, to make nuclear warfighting seem more feasible.   Would a limited tactical nuclear war escalate to a full-scale nuclear war?
   c.   The   Policy   Implications: Tactical   nuclear   war   seems attractive in Europe, as long as war seems conceivable and conventional war is much more expensive.   Would use of nuclear weapons damage the defender's civilians much more than the attacker's soldiers?   Is limited nuclear war likely to escalate into

a full-scale nuclear war?

D. ARMS CONTROL

13. Technological Imperatives

a. The Scientific-Technological Background: Solid-state physics; growth of computer capabilities; Josephson junctions; accuracy versus yield; MRV, MIRV, MaRV, TERCOM, NAVSTAR; cruise missile; the sociology of science and technology; science policy; bureaucratic imperatives; action-reaction phenomena; technical creativity; technological literacy; centralized planning versus competition – capability for political control; internal versus external competition; worst-case reasoning in international security.

b. The Drive Behind the Developments: Sometimes new technologies are so sweet that it is difficult, if not impossible, to resist developing them. Once they are available, then the technological imperative pushes them toward deployment.

c. The Policy Implications: The technological imperative insures that bureaucratic resistance to truly innovative technologies can be overcome. But this may lead to technologies that are undesirable and destabilizing within the political context.

14. Nuclear Proliferation

a. The Scientific-Technical Background: Nuclear reactors and fissile materials; national and global energy demands and supplies.

b. The Drive Behind the Developments: The technology of nuclear weapons requires only money and effort, not science; Great Britain went nuclear through reactor technology; France through deGaulle; China with Russian help; India through reactor technology. Nuclear power, breeder reactors make proliferation technologically feasible; political control and incentives are necessary to keep it from happening.

c. The Policy Implications. Uncontrollable nuclear proliferation threatens nuclear deterrence; non-proliferation treaties and nuclear umbrellas; can one have breeders with Pu reprocessing without encouraging proliferation?

15. Nuclear Test Ban Treaty

a. The Scientific-Technical Background: Problems of radioactive fallout; the detection of nuclear explosions and distinguishing them from earthquakes.

b. The Drive Behind the Developments: Guilt feelings about nuclear weapons produced ban-the-bomb movement; estimates of the effectsof fallout; equilizing of nuclear capabilities on both sides; final official support of informal scientific exchanges; controversy over earthquake data; Kennedy and Krushchev relationship; satellite verification and monitoring of above-ground tests; disagreement over on-site inspections.

c. The Policy Implications: Limited test ban treaty; end of fallout pollution; laser and particle-beam fusion test simulations; improvements of nuclear warheads through underground nuclear tests; scientists as political advisors.

16. Strategic Arms Limitations

a. The Scientific-Technical Background: Better computers lead to improved missile accuracy, ABM feasibility and antisubmarine warfare; fractionally orbiting missiles; optical and electronic verification.

b. The Drive Behind the Improvements: SALT control of numbers of weapons; ABM treaty to limit poor and expensive technology; but the technological imperative of MIRV was not controlled; megatonnage is now decreasing, but nuclear warfighting capability is increasing, destabilizing MAD; verification of weapons quality is difficult; U.S. would like to preserve its technological lead, but can do so only by continuing development work, contrary to hope of SALT; motivation for research is fear of surprise, imagined new threats, and the desire for bargaining chips.

c. The Policy Implications: How does one compare different weapons systems; does one need bargaining chips at SALT talks; do political pressures settle SALT agreements, rather than better monitoring techniques? Wohlstetter and the OSRA debate on experts and scientific advising; what goes on at SALT negotiations? Does SALT save money, does it stabilize, or is it a license to build? Detente between the superpowers.

17. Disarmament

a. The Scientific-Technical Background: Economic theories of guns versus butter; modelling the arms race; theories of deterrence and detente; verification.

b. The Drive Behind the Developments: The arms race costs social effort, more in the U.S.S.R. with its strained production capacity; arms are seen by some as distorting society and reducing control of government by citizens, fewer arms are seen by some as decreasing probability of war.

c. Policy Implications: Ban the bomb, better red-than-dead, and the nuclear freeze are popular expressions of the desire for disarmament; butter versus guns – i.e. the defense budget versus social services; detente; linkages between strategic arms control and political relationships (e.g. grain sales).

IV. SAMPLE EXERCISES

To illustrate this wide range of approaches toward issues in a technically oriented arms-race course, I present here two homework exercises done by the students in my own course. Each of the exercises involves a concrete activity, in one case looking up material in a reference source, in the other doing a rough measurement of the size of a building. Simple calculations have to be carried out to obtain order-of-magnitude estimates of parameters related to the nuclear arms race. Finally the implications of the numerical results have to be discussed. The measurements and calculations are to give the students a feeling for the certainty in numbers so they can better understand how far such calculations really determine policy.

1. Targeting Exercise:

In the targeting exercise the students are given a table listing how various nuclear weapons effects scale with the yield of the weapons.[5] Based on this table they must answer a series of questions.

(i) "Consider our table of nuclear weapons effects, in which we have listed the damage radii of prompt radiation, fallout, various blast overpressures, thermal radiation, and cratering, for nuclear weapons of various yields. Extrapolate the effects in this table from 10 Mt to 100 Mt." In answering this question the students explore and observe the factors of $10^{1/3} \simeq 2$ and $10^{1/2} \simeq 3$ increases in the radius of destruction for blast and fallout effects respectively. That gives them a feeling for the $(\text{Yield})^{1/3}$ and $(\text{Yield})^{1/2}$ dependencies of the destructive radii.

(ii) "Compare these listings to the reported effects of the Hiroshima and Nagasaki bombs, to judge which of these effects most closely describes the actual effects expected from a nuclear-bomb explosion." The H&N bomb calibrations show that 5 psi seems to be the most relevant bomb damage effect for civilian fatalities.

(iii) "Based on this table calculate the area of destruction of one 10-Mt bomb and of 1000 10-kt bombs. These have the same total yield, but which has the bigger destructive area?" The students here observe that small bombs are more effective per kiloton of yield in areal destructiveness.

(iv) "From the table the destructive area of an air-bursting 1-Mt nuclear weapon is about 30 mi^2. Within that area essentially 100% of the civilians would be killed – those surviving inside that area would be balanced by those killed outside that area. Assume you are a Soviet military planner and are requested to target 25 1-Mt bombs on U.S. cities to kill as many civilians as possible. Which cities would you select as targets, how many bombs would you drop on each, and how many casualties would you expect? If you had an additional 25 bombs, which cities would you choose and how would this increase the casualties?" Using a statistical listing of populations and population densities the students find about 14 million fatalities for the first 25 bombs, and about 7 million for the second 25 bombs, as shown in Table 3. This exercise gives them a feeling for the requirements for MAD, and for the saturation effects of nuclear bombing.

(v) "In the light of these calculations is the strategy of deterrence feasible?" This is a question about the political consequences of these calculations, with no one correct answer.

(vi) "Would the survivors of a nuclear war envy the dead?" This is obviously a philosophical question based on the order-of-magnitude calculation.

From this assignment the students learn the enormous destructiveness of nuclear weapons when used on cities. They also recognize the idea of saturation, that after some point additional nuclear weapons are relatively useless. Thus they begin to understand in what sense the term "overkill" is right, and in what sense it is wrong, when applied to the current nuclear balance. They may also discover how insensitive the results are to the details of the calculation as they try different targetting strategies. The hope is that after this assignment they understand MAD – and how little it requires expert calculations.

Table 3. A simple plan for targeting fifty 1-Mt hydrogen bombs onto U.S. metropolitan areas with the objective of maximizing civilian casualties. The destructive area of a 1-Mt bomb is assumed to be 30 mi^2. The population data is taken from the 1981 edition of the U.S. STATISTICAL ABSTRACT.

City	Pop. Density	Area	Bombs	Deaths
New York	26,000/mi^2	300 mi^2	10	7.8 M
San Francisco	15,700/mi^2	45 mi^2	1	0.5 M
Philadelphia	15,100/mi^2	128 mi^2	4	1.8 M
Chicago	15,100/mi^2	220 mi^2	7	3.2 M
Boston	14,000/mi^2	46 mi^2	1	0.4 M
Washington	12,300/mi^2	61 mi^2	2	0.7 M
First 25 1-Mt bombs			25	14.4 M
Newark	16,300/mi^2	23 mi^2	1	0.4 M
Baltimore	11,600/mi^2	78 mi^2	2	0.7 M
Detroit	11,000/mi^2	138 mi^2	4	1.3 M
St. Louis	10,200/mi^2	61 mi^2	2	0.6 M
Cleveland	9,900/mi^2	76 mi^2	2	0.6 M
Pittsburgh	9,400/mi^2	55 mi^2	2	0.5 M
Milwaukee	7,600/mi^2	95 mi^2	3	0.7 M
San Francisco	15,700/mi^2	(15 mi^2)	1	0.2 M
Minneapolis	7,900/mi^2	55 mi^2	2	0.4 M
Long Beach	7,400/mi^2	49 mi^2	1	0.2 M
Seattle	6,400/mi^2	83 mi^2	2	0.4 M
Los Angeles	6,100/mi^2	463 mi^2	3	0.5 M
Second 25 1-Mt bombs			25	6.6 M
TOTAL of 50 1-Mt bombs			50	21.0 M

2. Cavity for Hiding Underground Nuclear Tests:

   (i) "Assume that the U.S.S.R. want to cheat on the 1987 total-test-ban treaty which says that no nuclear tests are allowed at all, not even underground. Assume the U.S.S.R. want to test a 10-kt warhead for their SS-25 missile. They try to hide that test by decoupling the explosion of the warhead in a spherical cavity, 800 meters ($\frac{1}{2}$ mile) underground. The question is whether they can get away with such a test without the U.S. detecting it. From Figure 8.3 of Bolt's text,[6] what must be the diameter of such a cavity to successfully muffle that explosion? Compare that diameter to the length of a football field." The required 100 meters is equal to the length of the field, this should impress them somewhat.
   (ii) "What is the volume V of that cavity in cubic meters (m^3), where $V = 4\pi r^3/3$, and r is the radius of the sphere?" They will find the volume to be about 500,000 m^3; a number that won't mean much to

them.

(iii) "Go to the main library, and estimate its volume in m^3. Walk along the front and the side, remembering that a decent pace is about one meter. The height may be estimated to be the number of floors times the height of each floor, or the number of stairs times the height of each step. Compare that volume to that of the dirt pile from the secret cavity." UNC has a quite large library, which still is only about 1/10 the size of the dirt pile. They are impressed by the size of the dirt pile.

(iv) "At a density of 2 tons per m^3 (two times that of water), what is the mass of the dirt from this secret cavity? A large dirt truck can haul 20 tons of dirt. How many truck loads does that hole represent?" They find that it represents about 50,000 truck loads, which they realize requires quite a number of trips.

(v) "What do your results say about a complete nuclear test ban treaty?" The students realize that this much activity cannot be hidden from satellite reconnaisance. Hence they understand that test violations cannot be made for fairly large explosions by decoupling.

(vi) "How many violations of a nuclear test ban can we let the U.S.S.R. get away with because we might not detect them?" Answers here usually deal with technological improvements obtainable from a few tests, and with matters of mutual confidence between adversaries.

This exercise forces the students to do an order-of-magnitude estimate of a physical parameter, realizing then that the resulting numbers are not very certain. In this case they recognize that even uncertain numbers can in some cases be useful, that decoupling an explosion of this magnitude is not possible to do in secret. Then discussions of limiting large underground nuclear test by treaty make sense to them. They appreciate what a reasonable cutoff for a nuclear test ban treaty might be. The question of cheating brings them back to the point that these numbers ultimately are to be seen through a political filter, that they may not dictate a policy when perceptions are involved.

V. QUANTITY VERSUS QUALITY

I cite the course content and the two illustrations not just to help potential teachers of such courses, but also to show the wide range of possible approaches to the material in a nuclear arms-race course. A nuclear arms-race course requires the discussion not only of technical but also of social issues. This is the most important point to be made in such a course.

Then the students can appreciate the real problems of the course. All experts bring to their analyses of social issues biases induced by their expert background. I am tempted to say that expertise and impartiality are often mutually exclusive. This is particularly true for physical scientists teaching arms-race courses. They bring with them into the course a background that includes all of their training that has made them successful physical scientists. This background includes all of the norms of science described by Merton.[4]

The consequence of these norms of science that is most relevant here is the very heavy emphasis on quantification. It can be very useful in sharpening one's thinking about the arms race. Consider the example of the neutron bomb.[7] If it were to be used in Europe as a tactical nuclear weapon, how many civilians might be killed? A true physicist would probably feel comfortable in estimating the area of lethality for each neutron-bomb warhead using a tank formation diagram, multiplying this area per bomb by an estimated number of required warheads, and finally multiplying that total area by the population density to obtain an estimate of the number of fatalities (e.g. 4 km^2 per warhead x 1000 warheads x 200 people per km^2 = 800,000 fatalities). The physicist would feel better for having a number to think about, even if that number might be irrelevant. This quantification approach is not necessarily bad. Once such a number exists the physicist might well realize all the weak assumptions involved in deriving it. However, the layman, who admires the physicist for this quantification talent, is likely to accept this number as far more than just a starting point for a discussion, he may accept it as having real significance. Students must learn to question experts' numbers.

Unfortunately, this ability to deal with numbers turns in the hand of experts often into an absolute demand for numbers even when they do not exist. The bias of technical experts is to have very little sensitivity for all non-numerical aspects of a problem, or, perhaps, to reject them altogether. This demand for quantification has been called by Yankelovich the "McNamara Fallacy" in honor of the former Secretary of Defense who tried to rationalize that Department's operations:

The first step is to measure whatever can be easily measured. This is okay as far as it goes. The second step is to disregard that which can't be measured or to give it an arbitrary quantitative value. This is artificial and misleading. The third step is to presume that what can't be measured easily isn't very important. This is blindness. The fourth step is to say what can't be easily measured really doesn't exist. This is suicide.[8]

We may now reexamine the problems likely to be encountered in an arms-race course taught by a physicist with a bias toward quantification. It is not so much a problem of distinguishing between quantifiable aspects and those that are qualitative. Both the physicist-professor and the students do know the differences between the technical aspects of issues and those that are political. The problem is the tendency to overemphasize the more quantifiable, and thereby make it seem more important than it is; this effect may creep in when a physicist teaches an arms-race course. (The alternative error is to lean so far over backward that the glibly qualitative aspects come to be overemphasized.) In the case of the neutron bomb I find myself much more inclined to compare the neutron bomb to "ordinary" fission bombs than to compare it with conventional warfare. I understand the effects of nuclear weapons

166

much better than I understand the tactics of conventional warfare, and my bias is toward analysis of the big bomb because its effects are (seemingly) much more quantifiable. Conventional war is so much less quantifiable that it seems unpredictable. If it is not quantifiable, my instinctive preference as a physicist is to ignore it altogether. I want to argue here for a careful balance between the quantitative and the qualitative.

Unfortunately this bias toward the quantitative has been accepted by the general public and political leaders as well. Students in arms-race courses also like certainties; they want to know the correct answer, even if they don't know how the answer is derived. Qualitative problems generally have no correct answers. The students, therefore, will abet the professor when he unconsciously exercises his quantitative biases. People on both sides of the lecture podium will recognize that there are qualitative aspects to the issue; but they will exorcise them through the McNamara Fallacy. My plea is for different behaviour. I ask that we not ignore issues we cannot definitely answer. We must discuss them as questions and explore their implications, in a manner as free of political bias as is possible.

Let me be clear. I do not oppose quantitative analysis of the nuclear arms race. I very strongly wish more good ones existed. However, calculations should be done to exhibit their limitations, not to provide answers. Students should be taught how to estimate arms-race numbers in order to know when they do not determine policy. We, as physicists, must acknowledge that we are professionally biased toward numbers, and that we may believe that they fix policies. Teaching quantification is necessary to reveal our biases, and to help students evaluate how meaningful such calculations are. Only by knowing when to ignore the numbers can the numbers be truly useful. Thus, as physicist-teachers it is often better to have revealed our professional biases than to have given correct answers. Truth is not always the same as correctness.

VI. REFERENCES

Ref. 1   A course on "War, Peace, Science, and Technology in the Atomic Age" has been described in the Am. J. Phys. 40, 1324-7 (1972), other courses are described in the Newsletter Countdown of the War Education Project of the Federation of American Scientists, obtainable from the FAS at 307 Massachusetts Ave., NE, Washington, DC 20002. See also D. W. Hafemeister's articles about the technologies of the nuclear arms race in the American Journal of Physics: "Science and Society Test for Physicists: the Arms Race,' 41, 1191-6 (1973); "Science and Society Test V: Nuclear Proliferation," 48, 112-120 (1980); and "Science and Society Test VIII: The Arms Race Revisited," Am. J. Phys. 51, 215-225 (1983).

Ref. 2   D. Schroeer and J. Dowling, "PNAR-1: Physics and the Nuclear Arms Race," Am. J. Phys. 50, 786-795 (1982); a resource bibliography.

Ref. 3   D. Schroeer, "Teaching about the Arms Race," Phys. Today, 36 (3), 52-59 (March 1983).

Ref. 4   See for example "The Normative Structure of Science", Chapter 13, pp. 267-78, in R. K. Merton, THE SOCIOLOGY OF SCIENCE, Chicago: University of Chicago Press, 1973.

Ref. 5   S. Glasstone and Philip J. Dolan, Eds., THE EFFECTS OF NUCLEAR WEAPONS, Washington, DC: U.S. Government Printing Office, 1977.

Ref. 6   B. A. Bolt, NUCLEAR EXPLOSIONS AND EARTH QUAKES: THE PARTED VEIL, San Francisco: W. H. Freeman and Co., 1976.

Ref. 7   F. M. Kaplan, "Enchanced-Radiation Weapons," Sci. Am. <u>238</u> (5), 44-51 (May, 1978).

Ref. 8   D. Yankelovich, as quoted in A. Smith, SUPERMONEY, New York: Popular Library, 1972.

CHAPTER 8
FILMS ON THE ARMS RACE

John Dowling, Physics Department
Mansfield State College, Mansfield, PA 16933

Physicists are seldom comfortable with physics films in the classroom, let alone films on the arms race and nuclear war. Part of this discomfort stems from the ego of physicists, most mistakenly think they can better any film with regard to most any topic. But partly, it stems from the lack of good films. While there are very few good physics films, there are now many excellent productions on the arms race and nuclear war.

Films convey the historical perspectives, the biographical stories, the insights of the participants, and the horror of nuclear war - far better than can any physicist. While films are not very efficient for covering details, derivation, or numbers, they can not be beaten in showing what really happens in a nuclear explosion, in getting across general concepts, in illustrating the parameters of a problem, and the problem itself. Most importantly, films and TV can reach the people who must be informed about these issues if we are to resolve the problems.

What I will point out is how films can contribute to an understanding of the issues of the arms race and nuclear war. Here is a crash course.

History   Building of the Bomb opens in the golden days of the 1930's in Gottingen. It covers the rise of Hitler and the subsequent emigration of Jewish physicists, the British and United States starts on the bomb due to a fear of the Hitler bomb, the Manhattan Project, and the dropping of the bombs on Hiroshima/Nagasaki (and the gruesome results). It concludes with Oppenheimer, Teller, Agnew, Tuck, etc., telling whether they would have done it all again - some 20 years after the fact. While Building of the Bomb does deal with some political aspects, Decision to Drop the Bomb gives them a much more detailed treatment.

The overall history of the arms race is very capably done via animation in Fable Safe. Fable Safe puts all the technological developments from the A-bomb to MIRV's in perspective. More details and all the historical and technological benchmarks are given in Nuclear Countdown.

Hiroshima/Nagasaki   No film covers the suffering and

0094-243X/83/1040168-16 $3.00 Copyright 1983 American Institute of Physics

the human tragedy of nuclear war as does Hiroshima/Nagasaki August 1945. It is must seeing for everyone who wants to comprehend what nuclear weapons do to people. Three other excellent films are Hiroshima: A Document of the Atomic Bombing which provides an indepth treatment of what happened at Hiroshima/Nagasaki, Hiroshima: The People's Legacy which examines the terrible legacy of the Bomb, and Bomb, which deals with both the decision to drop the Bomb and the results on Hiroshima/Nagasaki.

Effects of Nuclear War   There are several important films which discuss what would happen in a full scale nuclear war. One of the most explicit is The War Game - which has now achieved the status of being the "classic" film in this area. More details are graphically depicted in Nuclear War: A Guide to Armageddon, an excellent BBC production and Nuclear War Graphics, a 35 mm sound and slide show which also contains a short AEC 16 mm film showing how houses stand up to 5 psi overpressures. The medical aspects of nuclear war are well covered in The Last Epidemic and Race to Oblivion, both of which are from the Physicians for Social Responsibility.

Civil Defense   While films like The Last Epidemic and Nuclear War: A Guide to Armageddon paint a rather dim view of civil defense, there is one film that paints a rather rosy picture, Protection in the Nuclear Age. This film gives the official U.S. Government line on civil defense. If you are giving both sides of the civil defense story, show "Protection" first, and then show No Place to Hide, Under the Mushroom Cloud, or Seattle After World War III. If you don't show "Protection" first, you'll never get through it if the audience sees one of the other films first - they'll be rolling in the aisles laughing.

Weapons Technology   While very little "real" physics is discussed in these films, there is much applied physics shown which can be used as a springboard to develop concepts. The Bull's Eye War examines precision-guided munitions and is filled with examples of state-of-the-art electronics and optics. The Deep Cold War treats antisubmarine warfare and the enormous challenges physicists face there. The Nuclear Battlefield discusses, somewhat tongue-in-cheek, how nuclear war would be fought - we see all the marvelous technological gear controlled from an office where a soldier in full radioactive protective gear tries to cope with answering a telephone and using a typewriter. Proliferation problems, a plutoniun economy, and illicit bombs are discussed in The Plutonium Connection. Finally, Buck Rogers comes to life with lasers and charged-beam weapons blazing in The Real War in Space.

Political Issues   The MX Debate is a superb example of
an issue film. It examines whether the MX is needed; the
environmental, social, and economic questions; the technological and political issues; and finally how the public
views and reacts to such a weapon. Trillion Dollars for
Defense (now up by a factor of 1.6) looks at what all the
Pentagon money will buy over five years. How Much Is
Enough? examines what weapons numbers and postures should
suffice in the face off with the Russians. No First
Strike looks into all the ramifications of pledging not to
use nuclear weapons first, and what this policy entails in
the postures associated with conventional weapons. Call
to Arms and The War Machine look at what weapons are being
developed. The Rise of the Red Navy discusses the large
increase in the Russians' sea arm. Lastly, STOP Versus
START examines Reagan's Strategic Arms Reduction Talks.

The Public and the Arms Race   How the public perceives
the arms race and nuclear war has been well documented in
some excellent films. The two best are War Without Winners II and Thinking Twice About Nuclear War. Both examine what the general public really feels about all the
fuss. Films previously discussed, such as Trillion Dollars for Defense, The MX Debate, and How Much Is Enough?
also focus on the issues.

The Physicist's Perspective   Physicists have always had
a leading role in the debate, primarily because they had a
leading role in building the weapons and pushing the technology. Their unique insights and perspectives are given
adequate airing in A is for Atom, B is for Bomb (Teller
tells why he pushed nukes and muses on his role in the
Oppenheimer hearings), Building of the Bomb (a goldmine
on how physicists developed the bomb and what they thought
about it 20 years later), and The Day After Trinity (a
soul-searching look at Oppenheimer, the mystical guru of
nuclear weapons).

Conclusions   So what comes across in all these films?
I think that the public realizes that nuclear war at the
very best would be very, very bad; that scientists are on
all sides of every issue, that scientists share a great
part of the responsibility for the mess we are in, and
that weapons postures are most probably out of control.
Five years ago, there were few films available on these
issues, other than historical treatments of the events
leading to Hiroshima/Nagasaki. In November of 1982, over
20 films came into my office on these issues. PBS has
run many of these films and CBS did a five hour series of
productions on these issues on prime-time television.
This is encouraging in itself. People are becoming more
informed and more sophisticated about the issues. Physicists can contribute to public education by participating

in publicizing the films, and discussing them wherever and whenever they can. Films and TV are the only way to reach the full public.

Following is a list of some of the more important films on the arms race and nuclear war. A more complete list of about 150 films is maintained and updated weekly. The list is available for $2.00 from John Dowling, Physics Department, Mansfield State College, Mansfield, PA 16933.

FILM BIBLIOGRAPHY

A is for Atom, B is for Bomb
King Features Entertainment, 235 E. 45th St., New York, NY 10017. 3/4 in. videocassette, 60 min., 1980. $325 purchase, $75 rental.

Traces Teller's career from his youth in Hungary, student days in Germany, to Los Alamos, to Livermore, to the present. Features many interviews with Teller and covers the Oppenheimer controversy and hearings, questions of priority with Ulam, and his overwhelming drive to develop the hydrogen bomb. Includes interviews with contemporaries who knew and worked with Teller. Overall 7.

Bomb
Bonaventure Prod., Suite W219, 22555 Sheppard Ave. E., Willowdale, Ontario, Canada M2J 4Y1. 16 mm or 3/4 in. videocassette, 55 min., 1975. $785 (16 mm) or $645 (video) purchase. $24 rental from U. of Illinois Film Center, 1325 S. Oak St., Champaign, IL 61820.

Examines the background of the decision to drop the bomb on Hiroshima and Nagasaki, arguing that its use was primarily for diplomatic and political rather than military reasons. Outlines the complicated diplomatic maneuverings taking place in 1945 between the Japanese, the Soviets, and the Americans, studies Truman's options at the Potsdam conference in July, and features interviews with Averell Harriman, Alger Hiss, the commanders of the Hiroshima bombing mission and survivors of the attack. Truman brings news of the atomic bomb to Potsdam. In August 1945 it is dropped on Hiroshima. After a second bomb on Nagasaki, Japan surrenders. "A weapon is a weapon, and it really doesn't make much difference how you kill a man.

If you have to kill him, well, that's the evil to start
with and how you do it becomes pretty secondary." --Gen-
eral Curtis Lemay. Overall 8.

Building of the Bomb
Out of print. 16 mm, B/W, 72 min., 1969.

This important documentary is about the atomic bomb. It
starts in Germany in the early 30s with scientists Oppenheimer, Teller, Segré, Fermi and Heisenberg giving their
own interpretations of events that led to Hiroshima. The
film is absorbing because of its insistence throughout on
questions of human and scientific morality. Two particular points emerge from the film: one, Niels Bohr foresaw
a future nuclear arms race and unsuccessfully tried to
persuade Churchill and Roosevelt of its dangers; and two,
the project born out of fear that Hitler would develop
and use such a deady weapon unless the U.S. had one also,
took on a life of its own. In spite of evidence that the
Germans had not succeeded, the bomb became an offensive
weapon instead of a deterrent. Includes film interviews
with many of the scientists twenty years after Hiroshima
who discuss what they would have done from this later perspective. Overall 10.

Bull's Eye War, The
Films Inc., 1144 Wilmette Ave., Wilmette, IL 60091. 16
mm or 3/4 in. videocassette, 52 min., 1980. $700 (16 mm)
or $350 (video) purchase, $75 rental.

"Smart" bombs have changed the face of conventional warfare in the last few years. These precision guided
weapons make pinpoint accuracy possible for first-round
hits. Guidance systems work by wire, radio and radar.
They have given rise to electronic warfare for which much
of the growing defense budget is allocated. This program
examines the technology and intelligence applications and
analyzes implications for the future. Overall 6.

Call to Arms
CBS News, 524 W. 57th St., New York, NY 10019. 3/4 in.
videocassette, 52 min., 1981. $390 purchase, $60 rental.

This film assesses the strength of U.S. conventional
forces. It examines the recruit situation: how today's

compares to previous (favorably in general); women recruits; and the problems with getting, training, and
keeping recruits. The program covers some of the pecu-
liar demands of technological warfare: need for permanent and well-trained personnel, adequate parts and supplies, maintaining reasonable living standards - particularly on crowded Navy carriers, overworking the people
who do stay, etc. Overall 7.

Day After Trinity, The
Pyramid Films, P.O. Box 1048, Santa Monica, CA 90406.
16 mm or 3/4 in. videocassette, 88 min., 1981. $950 (16
mm) or $450 (video) purchase, $125 rental.

This film documents the life of J. Robert Oppenheimer,
the director of the atomic bomb project. Because of his
efforts to limit the amount of bomb technology, he was
dismissed from his position as director in 1954. Oppenheimer died in 1967 at the age of 63, a rather mystic
figure enveloped in the atomic cloud he helped to create.
Overall 6.

Decision to Drop the Bomb
Films Inc., 1144 Wilmette Ave., Wilmette, IL 60091. 16
mm or 3/4 in. videocassette, B/W, 82/35 min., 1965. $900
/$690 (16 mm) or $540/$415 (video) purchase, $125/$65
rental.

This film examines two issues - how was the decision made
and why was it made? It reviews the period from Roosevelt's
death to the bombing of Hiroshima. It is concerned with
the decision-making process in 1945 and an analysis of its
justification twenty years later. There are interviews
with many of the actual participants in that decision
which deal with the political and scientific problems of
the time. The moral questions involved are brought out,
on one hand by those who urged dropping the bomb without
warning on a city, to have maximum effect and thereby end
the war and save lives; and on the other by those who felt
that any military advantage would be outweighed by the
wave of horror and revulsion at such wholesale destruction of civilian life. Looking to an international agreement on the prevention of nuclear warfare, they urged a
demonstration of the bomb before UN observers on a barren
island. A very good film to start discussion of questions

of foreign policy decision-making, war and its alternatives. Overall 9.

Deep Cold War, The
Films Inc., 1144 Wilmette Ave., Wilmette, IL 60091. 3/4 in. videocassette, 52 min., 1980. $350 purchase.

Surface treaties and efforts at detente have not slowed the secret contest of wits and technology being waged beneath the waters of the world between the NATO allies and the Soviet Union. This program documents the anti-submarine warfare race, considered a crucial factor to the outcome of a major military conflict. Includes exciting scenes of air, surface and submarine tracking of the more than 300 Soviet diesel and nuclear-powered fleet. Overall 6.

Fable Safe
Museum of Modern Art, Film Program, 11 W. 53rd St., New York, NY 10019. 16 mm, 9 min., 1971. $150 purchase, $20 rental.

Animated satire pointing up the folly of the spiraling arms race among nations. It presents a comment on the arms race and national security using the drawings of cartoonist Robert Osborn and the music of Tom Glazer. This is an excellent short history of the arms race. Overall 9.

Ground Zero
CBS News, 524 W. 57th St., New York, NY 10019. 3/4 in. videocassette, 52 min. $390 purchase, $60 rental.

This film takes a close look at the terrifying prospects of nuclear war and all the preparations that must be made to wage it. It gives a simulation of what would happen when one 15 megaton bomb is dropped on SAC Headquarters near Omaha - complete with gory details and estimates of casualties.

Hiroshima: A Document of the Atomic Bombing
Michigan Media, 416 Fourth St., Univ. of Michigan, Ann
Arbor, MI 48104. 16 mm, 28 min., 1970. $12.50 rental
only.

Uses some of the same stock footage as in "Hiroshima/
Nagasaki August 1945." It adds a considerable amount of
contemporary footage and gives a sensitive historical
treatment of the holocaust's victims with some focusing
on the survivors of the blast. These is some concentration on the survivor's constant need for medical attention, their deep psychological wounds, and their inability to fit into everyday life. The contrast between current memorial services with the original black and white
film is startling, but well handled. Shows the "black
rain," teacups melted at 7000 degrees, human shadows
etched on concrete, and tells the story of the paper
cranes. There is significantly more medical footage than
in H/N. Document shows that life, tragically altered,
still goes on for the people of Hiroshima. Overall 10.

Hiroshima/Nagasaki August 1945
Museum of Modern Art, Film Program, 11 W. 53rd St., New
York, NY 10019. 16 mm, B/W, 17 min., 1970. $300 purchase.

Uses Japanese film that had been withheld from the public
for 20 years to show the results of the atomic bombings
of Hiroshima and Nagasaki. An informative and factual
narrative accompanies the view of absolute destruction
and the appalling agony that the Japanese experienced.
Very gruesome film but an extremely important one. It is
excellent to start discussion of questions about the in-
humanity of war, nuclear weapons, decision making, etc.
Overall 10.

Hiroshima: The People's Legacy
Electronics Arts Intermix, Inc., 84 Fifth Ave., New York,
NY 10011. 3/4 in. videocassette, 45 min., 1975. $200
purchase, $50 rental.

A newspaper in Japan asked survivors of the bombings of
Hiroshima/Nagasaki to send in their paintings and drawings
of what they saw or felt about the tragedy. This film
examines what was sent in and probes into the lives of
several of the people. It is extremely moving as a film
- and also quite heartbreaking. In Japanese with English
voiceover. Overall 7.

Hitler's Secret Weapons
King Features Entertainment, 235 E. 45th St., New York,
NY 10017. 3/4 in. videocassette, 59 min., 1978. $325
purchase, $75 rental.

Provides a history of German rocketry which led to the
V1 and V2 weapons of Hitler's war machine. Covers Von
Braun's work and developments and the struggle to support
rocketry in the competition to build the German war machine. Shows how the rocket weapons grew from a stephold
type support to a final desperate type as the ultimate
weapon. Ends with Von Braun and nearly all the Penne-
munde experts coming to the U.S. to further our rocket
research. The final impression is that of the old de-
rogatory poem concerning Von Braun's rocket work "Vee do
not care vere de rockets go down, vee only care vere de
rockets go up". Overall 6.

How Much Is Enough?
Documerica Films, P.O. Box 315, Franklin Lakes, NJ 07417.
16 mm or 3/4 in. videocassette, 60 min., $795 (16 mm) or
$595 (video) purchase, $85 rental.

Looks at the U.S. nuclear arms build up over the past two
decades. Interviews highly placed policymakers from the
Kennedy through the Reagan administrations. The thesis
of this production is that the American nuclear arsenal
grew not out of pure military necessity, but rather as a
part of a haphazard and capricious political process.
Includes first class animation scenes. Overall 7.

Last Epidemic, The
Resource Center for Nonviolence, P.O. Box 2324, Santa
Cruz, CA 95063. 16 mm or 3/4 in. videocassette, 36 min.,
1980. $350 (16 mm) or $125 (video) purchase.

This film is a record of a conference the Physicians for
Social Responsibility sponsored in San Francisco in November, 1980. The program consists of several prominent
physicists and arms controllers speaking on the more general consequences of nuclear war. They include Kosta
Tsipis of MIT, Bernard Feld of the Bulletin, Marvin
Goldberger of Cal Tech, Gene La Rocque of The Center for
Defense Information, and Herbert Scoville of The Arms
Control Association. They briefly discuss the destruction of the ozone layer, probable ecological catastrophes, civil defense impracticalities, destructive capability of a single submarine, the MX, the insanity of
nuclear war, etc. Physician Jack Geiger gives a detailed
explanation of the medical consequences involved. Overall
8.

MX Debate, The
WNET/13, 356 W. 58th St., New York, NY 10019. 3/4 in.
videocassette, 120 min., 1980. $600 purchase, $100
rental.

The MX Debate is a Bill Moyers' Journal program aired
nationwide on PBS. Overall it is an excellent program
which gives background on the facts and vital issues associated with the MX. It provides a forum for discussion
and debate on the issues by the many parties concerned
about this important weapons system. Moyers raises three
main issues: "Do we need the MX?", "How should the MX be
deployed and based?", and what was "...the issue of its
impact on Utah and Nevada." For each of these issues
there was a short filmed episode, narrated by Moyers,
which provided background on the issue. There were three
panels of three experts each, one for each topic. Moyers
directed questions on each topic to the appropriate panel.
The panel responded to Moyers' questions and to each
other's reponses. The audience had an opportunity to ask
questions after the second and after the third topic. A
special panel of four local citizens was set up and allowed to ask one question each. In summary, the program
provided a national forum for an MX debate. As Moyers
stated: "Obviously, we have not touched all the questions. We haven't even explored the big ones in the
depth that they deserve." - but it was a very good start.
Overall 7.

No First Strike
Union of Concerned Scientists, 1384 Massachusetts Ave.,
Cambridge, MA 02138. 16 mm or 3/4 in. videocassette,
27 min., 1982. $300 purchase, $25 rental.

This film was produced by UCS to present the reasons for
the U.S. to take a "no first use" position. It discusses
the history of the arms race, shows the results of nuclear
weapons at H/N, and shows many current weapons systems.
It features interviews with Robert MacNamara and several
military officers who have had extensive NATO experience.
Overall 7.

No Place to Hide
Direct Cinema Ltd., P.O. Box 69589, Los Angeles, CA 90069.
16 mm, 29 min., 1982. $495 purchase, $45 rental.

This is a documentary about growing up in the shadow of
the bomb. It chronicles the era of civil defense procedures, fallout shelters, and duck-and-cover exercises to

save yourself in an atomic bomb attack. It is composed
of vintage films, newsreels, cartoons, and TV programs
which tried to sell the American public on the idea that
nuclear war is survivable. Overall 7.

Nuclear Battlefield, The
CBS News, 524 W. 57th St., New York, NY 10019. 3/4 in.
videocassette, 52 min., 1981. $390 purchase, $60 rental.

This film examines the efforts of the superpowers to develop tactics for fighting nuclear war and what would
happen to Europe if it became a nuclear battlefield. It
provides an historical perspective on how and why nuclear
weapons are used to counter the threat of the Soviet-
Warsaw pact forces. Overall 7.

Nuclear Countdown
Journal Films, Inc., 930 Pitner Ave., Evanston, IL 60202.
16 mm, 28 min., 1978. $385 purchase, $40 rental.

Nuclear Countdown is a documentary designed to raise
public awareness to the necessity of nuclear disarmament.
It is excellent in sketching the history of the nuclear
race and gives an overview of the dangers of nuclear proliferation. Film clips show the important principals and
events, all the important milestones are mentioned and
there is some discussion of the successes, failures and
inadequacies of the treaties and agreements to date.
Overall 9.

Nuclear War: A Guide to Armageddon
Films Inc., 1144 Wilmette Ave., Wilmette, IL 60091. 3/4
in. videocassette, 30 min., 1982. For purchase or rental
inquire.

The first part of this superb production by BBC shows
what would happen to London if a one megaton bomb were
burst over St. Paul's Cathedral. It goes through all the
blast, heat, and radiation effects at different locations
for the blast. Its treatment of fallout patterns is particularly effective -- especially the area where a couple
stayed for 2 weeks inside a dark, damp, cramped hole.
The visual effects are extremely well done. Overall 7.

Nuclear War Graphics
Nuclear War Graphics Project, 100 Nevada St., Northfield,
MN 55057. 35 mm slides, audiocassettes, 16 mm film, 1981.

$45 purchase.

This package contains three separate items: a set of 130 35 mm slides, a set of 60 35 mm slides and 21 min. audio-cassette tape, and a 16 mm black and white film of the destruction of several houses in the 1953-55 nuclear weapons tests in Nevada. There is something in this package for everyone. The 130-slide set can be used as a library of slides to illustrate any talk on the arms race. For those who want a prepared talk, the sound and slide show is very good. It can stand alone or be used as a starter for discussion. Each comes with some descriptive material about how to make use of each item. All three sets are sold at cost. Together they comprise the best and by far the least expensive material around which illustrates the threat of nuclear war. Overall 8.

Plutonium Connection
Time-Life Films, 100 Eisenhower Dr., Paramus, NJ 07652. 16 mm or 3/4 in. videocassette, 59 min., 1975. $650 (16 mm) or $200 (video) purchase.

Deals with possibilities of diverting nuclear weapons-grade material from the nuclear fuel cycle to use in building unauthorized bombs. Traces an undergraduate student in a bomb making project. Interviews bomb expert Ted Taylor and elicits his opinions on safeguards for a plutonium economy. Overall 7.

Protection in the Nuclear Age
NAV Center, GSA, Washington, DC 20409. 16 mm, 24 min., 1980. $139 purchase.

This film is a prime example of overly optimistic estimates of survivability in a nuclear war. The picture portrayed of nuclear war is catastrophically misleading to the American public. The narrator explains that a "nuclear attack is possible but we can survive. Nuclear war does not mean the end of the world or civilization." The public should have the right to assess the consequences of a nuclear war. However, this film does not help much. Overall 3.

Race to Oblivion
Physicians for Social Responsibility, P.O. Box 35385, Los Angeles, CA 90035. 16 mm or 3/4 in. videocassette, 49 min., 1982. $295 (16 mm) or $75 (video) purchase.

While this film is quite similar to another PSR inspired

film "The Last Epidemic" (a different PSR conference with some old and some new faces talking about the medical effects of nuclear war and the arms race in general) there are some important plusses. The film centers around Burt Lancaster talking with Shigeko Sasamori, an A-bomb survivor of Hiroshima. Shigeko tells what she suffered then and now. Shigeko's story is interrupted periodically with scenes from the conference and H/N, and various clips of weapons tests, military hardware, and mushroom clouds. "Race to Oblivion" is a nice update to "The Last Epidemic" and the beginning and ending sound track of children singing "It's Up to You and Me" puts a touching cap on the whole thing. Overall 6.

Real War in Space, The
Films Inc., 1144 Wilmette Ave., Wilmette, IL 60091. 3/4 in. videocassette, 52 min., 1980. $200 purchase.

We monitor the Russian's satellites and they presumably monitor all of ours. Scientists have discovered that the Russians have been testing hunter-killer satellites, capable of seeking out and destroying other satellites in orbit. General George Keegan, former chief of U.S. Air Force intelligence, believes that the Russians are also developing the capacity to destroy inter-continental ballistic missiles by using high-energy lasers. While some disagree with Keegan's controversial assessment of Soviet intentions, others fear that both nations are on the verge of a costly new arms race for the domination of space. Overall 6.

Red Army
King Features Entertainment, 235 E. 45th St., New York, NY 10017. 3/4 in. videocassette, 58 min., 1981. $250 purchase, $60 rental.

This film examines the alleged Soviet superiority in four specific areas: tanks, planes, ships, and manpower. Past overblown perceptions of Soviet strength led us into the bomber, missile, and civil defense "gaps." This is a fine production which helps place the Soviet threat in perspective. It is a substantive contribution to an informed debate on whether we need to increase defense spending. Overall 7.

Rise of the Red Navy, The
Films Inc., 1144 Wilmette Ave., Wilmette, IL 60091. 16 mm or 3/4 in. videocassette, 57 min., 1980. $700 (16 mm) or $350 (video) purchase, $75 rental.

Since the Bolshevik Revolution, the Russian Navy has become very formidable. This film documents its growth,
focusing on warships and submarines, fishing fleets, an
oceanographic research fleet larger than that of all
other nations together, a merchant fleet second only to
that of Japan, and a unique flotilla of more than fifty
intelligence and surveillance vessels. Overall 6.

Seattle After WW III
KCTS-TV Programming, 4045 Brooklyn Ave. NE, Seattle, WA
91805. 3/4 in. videocassette, 30 min., 1982. $100 pur-
chase.

This film examines what nuclear war would mean to Seattle.
It treats evacuation, shelters, crisis relocation, the
physical and medical consequences of nuclear war on
Seattle, and what life would be like after nuclear war.
It covers a lot of ground and does it fairly well. Of
particular importance is how the film gets "ordinary"
people to talk about how they would handle such threats
and how they would deal, or attempt to deal, with the catastrophe of nuclear war. Overall 6.

STOP Versus START
Nuclear War Graphics Project, 100 Nevada St., Northfield,
MN 55057. 35 mm sound and slide show, 1982. $15 pur-
chase.

This slide show uses creative graphics to describe the
components and destructive capabilities of the U.S. and
Soviet strategic arsenals, and compares and contrasts the
nuclear weapons freeze proposal with the Reagan administration's proposal at the strategic arms reduction talks
(START). It concludes that START is an attempt to counter the freeze movement and begin a build up of new strategic weapons systems. Overall 6.

Thinking Twice About Nuclear War
Public Interest Video Network, 1736 Columbia Rd., NW,
Washington, DC 20009. 3/4 in. videocassette, 60 min.,
1982. $250 purchase, $40 rental.

This film is a look at how American families have adjusted to the realities of living in a world with nuclear
weapons and their response to disturbing new information
about the risk of nuclear war. It also presents several
different plans that individuals can take to reverse the
arms race. Overall 7.

Trillion Dollars for Defense
WNET Media Service, 356 W. 58th St., New York, NY 10019.
3/4 in. videocassette, 60 min., 1980. $385 purchase, $60
rental.

The program includes an annual arms fair where manufac-
turers like Singer, GE, Ford, RCA etc., exhibit their
latest high technology weapons systems, various promo-
tional clips of some dramatic weapons systems, interviews
with some of the exhibitors as well as then under Secretary of Defense Bill Perry. The film concludes with a
three-way discussion between Moyers and Senators Sam Nunn
and Gary Hart. Moyers questions both senators on how and
why and should the U.S. spend so much on the military.
Moyers is worried that the emphasis on such expensive and
complex weapons detracts from a leaner, simpler, and
tougher armed forces. The program appears to reach the
conclusion that there is no consensus on what the military should be doing or what U.S. foreign policy should
be. The trillion dollars for defense "could represent
the triumph of technology and the failure of wisdom,"
unless programs like these can stimulate the American
people into doing something about it. Overall 8.

Under the Mushroom Cloud: Civil Defense
Nuclear War Graphics Project, 100 Nevada St., Northfield,
MN 55057. 35 mm sound and slide show, 1982. $15 pur-
chase.

Reviews civil defense of the 1950's and 60's; discusses
why nuclear war civil defense is back again; explains the
crisis relocation concept; critically examines its fundamental assumptions and their invalidity; vividly describes
what nuclear war would really be like; and contrasts the
options of preparation and prevention of war. Overall 6.

War Game, The
Films Inc., 1144 Wilmette Ave., Wilmette, IL 60091. 16
mm or 3/4 in. videocassette, b/w, 49 min., 1968. $680
(16 mm) or $410 (video) purchase, $150 rental.

The War Game postulates a day when diplomacy fails, nations are backed into corners and bluffs are called. The
grim effects of a nuclear attack on Britain are shown in
horrifying detail, based on information supplied by experts in nuclear defense, economics and medicine. Use of
cinema verite heightens the impact and lends the authen-
ticity of a newsreel. Overall 10.

War Machine, The
CBS News, 524 W. 57th St., New York, NY 10019. 3/4 in.
videocassette, 52 min., 1981. $390 purchase, $60 rental.

This film asks whether our new weapons will work when the
time comes that we really need them. To find answers CBS
follows the F-18 fighter plane for nine months. The film
contains a simulated combat test which pitted sophisticated F-14s and F-15s against relatively crude F-5s. The
political game is played and the outcome is that the
Pentagon, Congress, and the defense contractors form "not
only a closed world...incestuous to a degree," but a
world where "...buyer and seller merge, perspectives narrow, alternatives disappear, and debate stops." Overall
7.

War Without Winners II
Films Inc., 1144 Wilmette Ave., Wilmette, IL 60091. 16
mm or 3/4 in. videocassette, 28 min., 1983. $400 (16 mm)
or $200 (video) purchase, $50 rental.

This is the sequel to War Without Winners. The film is
about people, auto workers, bomb assemblers and retired
government officials. The film is people expressing their
fears, thoughts and hopes about the future in an age of
nuclear weapons that can incinerate civilization in minutes. It is a fresh and delightful appeal to reason.
There is much new material in this version and it represents a very good update of an excellent film. Overall
7.

INTERACTIVE COMPUTER GRAPHICS: THE ARMS RACE

D. W. HAFEMEISTER
PHYSICS DEPARTMENT
CALIFORNIA POLYTECHNIC UNIVERSITY
SAN LUIS OBISPO, CA 93407

ABSTRACT

By using interactive computer graphics (ICG) it is possible to discuss the numerical aspects of some arms race issues with more specificity and in a visual way. The number of variables involved in these issues can be quite large; computers operated in the interactive, graphical mode, can allow exploration of the variables, leading to a greater understanding of the issues. This paper will examine some examples of interactive computer graphics: (1) the relationship between silo hardening and the accuracy, yield, and reliability of ICBMs; (2) target vulnerability (Minuteman, Dense Pack); (3) counterforce vs. countervalue weapons; (4) civil defense; (5) gravitational bias error; (6) MIRV; (7) national vulnerability to a preemptive first strike; (8) radioactive fallout; (9) digital image processing with charge-coupled devices.

I. INTRODUCTION

I.A. Methodologies: Words, Equations, Pictures, and ICG. In spite of the fact that so much has already been written about the arms race, some general confusion about the strategic importance of various missile systems and strategies continues. This paper explores the possibility of using the visual medium through interactive computer graphics (ICG) to learn about the arms race. ICG has three distinct advantages: (1) interactive, visual education; (2) debating policy; and (3) "networking" information.

We often have been told that "a picture is worth a thousand words;" the interactive pictures of ICG may be an improved medium for increasing our understanding of the arms race. The computer can relate both to our logical, mathematical left side of the brain as well as to the more intuitive right side of the brain as has been indicated by Sperry[1] and others. Gradel and McGill[2] have pointed out that "The graphical presentation of the results of complex computer model calculations is frequently as important as the computation, since it is generally through such presentations that the modeler and the modeler's audience derive the maximum amount of information." Since the deluge of words in our society has weakened the strength of verbal communication,

0094-243X/83/1040184-21 $3.00 Copyright 1983 American Institute of Physics

numbers have often taken on a greater political power than they deserve, both because they appear to be more reliable, and because they are not understood. ICG can correct this tendency by allowing both sides to explore and understand the variation in the data sets that are used to prove the "bottom line." Thus, ICG can be used both for educating and for debating national security policy. Of course, there is a limit to what can be quantified in these debates since the uncertainties of the arms race are larger than the certainties; nevertheless, we should try to be as accessible and accurate as possible when we do quantify. Lastly, ICG diskettes can be copied and mailed easily to those who wish to study and teach numerical aspects of the arms race; this ability to "network" information adds a new dimension to printed words and equations.

In this paper we will consider some simplified models of "war games" that can be used in interactive computer graphics. Most of these results can be obtained as easily with mathematical equations alone; but our purpose is to enhance the transfer of knowledge to those who are uncomfortable with the use of equations. In our ICG program "First Strike" (Sec. VI) we will use ninety adjustable parameters; only with ICG can one keep track of this plethora of parameters which seem to become further removed from reality the more we "talk" about them. Ultimately one can not use equations and parameters alone to describe the "action-reaction" escalation of the arms race and the degree of stability from mutually assured destruction (MAD), but these mathematical models do set some limits on what actually could be done, and they do give some meaningful insights into the interactions that effect the outcomes of these difficult issues.

I.B. Graphics. Computer graphics involves a number of tricks of the trade: the ability to "page-flip" between the two pages of high-resolution graphics; the use of toggle switches to input data to the computer; the ability to rapidly scale and rotate shape tables for animation; the use of "Easy Draw" to prepare large figures such as maps; the scaling and transformation of images; the use of light pens, joy sticks, graphics tablets, and graphics printers; and other processes. An excellent text on these graphics techniques has been been written by R. Myers[3]; those who would like to learn more about these topics can purchase a diskette with about 70 graphics programs that accompany the text. It is interesting to point out that undergraduate students are often more creative and faster with developing sophisticated graphics than many faculty; perhaps, this is because learning to use such terminology as peek and poke are very much like learning a new foreign language. At any rate, perserverance can overcome most graphics problems. Some of these graphics tricks are highlighted in the manual on ICG which is available to accompany our diskette (Sec. XI).

To demonstrate the power of interactive computer graphics, we will examine the following issues in the sections listed below: (II-IV) Target Vulnerability: The tradeoff between accuracy, yield, hardness, reliability, and gravitational bias error; Minuteman vulnerability; first and second strike weapons; Dense Pack MX; and civil defense. (V) Multiple Independently Targetable Reentry Vehicles (MIRV). (VI) National vulnerability to a preemptive first strike; is it possible? (VII) Distribution of radioactive fallout after a nuclear attack. (VIII) Digital image processing to enhance verification; the use of charge-coupled device cameras that can be used directly with computers.

II. TARGET VULNERABILITY

We shall begin by considering in detail an example that can be discussed with equations or ICG; the example is the case of an attack on the U.S. Minuteman force by Soviet SS-18 missiles. Sec. II will consider some equations, Sec. III will consider ICG, and Sec. IV will broaden the discussion to include the effects of a possible gravitational bias error.

II.A. Parameters. If the accuracy of a missile is increased, it follows that the yield necessary to carry out a mission against a hardened military target can be correspondingly reduced. As accuracy increased by a factor of twenty from about 5 miles in 1954 to 1/4 mile in 1970, the U.S. decreased the yield of its warheads by a factor of about 100 from 9 megatons (Mt) for the Titan ICBM to 50 kilotons (kt) for Polaris/Poseidon and 170 kt for Minuteman. Increased accuracy was the necessary precursor to the deployment of smaller warheads used with Multiple Independently Targetable Reentry Vehicles (MIRV) for counterforce purposes. The new technologies available to the cruise missile have further increased accuracy to less than 10 m. The tradeoff between accuracy and yield (for hardened targets) can be qualitatively understood by considering the empirical relationship[4] for blast over-pressure derived from nuclear testing (surface blasts):

$$p = (14.7)(Y/r^3) + (12.8)(Y/r^3)^{1/2} \qquad (1)$$

where p is the overpressure in psi, Y is the yield in Mt, and r is the distance in nautical miles (1 nm = 1852 m). For the case of a "silo-busting" attack on Minuteman where high pressures are needed, one need consider only the first term in Eq. 1. Since accuracy improved by a factor of 20 from 1954 to 1970, it follows that the yield could have been reduced by a factor of $(20)^3 = 8000$ in order to carry out the same military mission. Since the yield was reduced by a factor of only 100, the additional effective yield of Minuteman and Polaris/Poseidon can be used to overcome hardened missile sites and to increase the probability of a successful mission. The miniaturization of nuclear weapons has also enhanced the relative ability to destroy surface area (as well as point targets) since the total destructive area is increased (per Mt) with a larger number of smaller weapons.

II.B   Minuteman Vulnerability. In order to give some feeling for the numbers involved in Minuteman vulnerability, let us calculate[5] the single shot kill probability (SSKP = P_k) for a missile attacking a hardened silo:

$$P_k = 1 - e^{-(Y^{2/3}/B\ CEP^2\ H^{2/3})} \qquad (2)$$

where $B = 0.22$ when Y is in Mt, H is the silo hardness in psi, and CEP (circular error probable) is the accuracy in nm. We can determine the SSKP of destroying a Minuteman silo assuming the following parameters[5-8]: (1) Minuteman silos are hardened to about $H = 2000$ psi; (2) The Russian SS-18 warheads typically have a yield $Y = 0.75$ Mt and a CEP = 280 m (0.15 nm, at some point in the future); and (3) The reliability of an SS-18 is, perhaps, $R = 0.8$. Using these parameters, the SSKP for the SS-18 on a Minuteman silo is $P_k = 1 - e^{-1.05} = 1 - 0.35 = 0.65$. The SSKP should be multiplied by the reliability of the SS-18 to obtain the success rate for each SS-18 warhead; we obtain 52% for $R = 0.8$, and 59% for $R = 0.9$. For the case of aiming two SS-18 reentry vehicles from different launchers at a given silo, the kill probability is $P_{k2} = 1 - (0.48)^2 = 77\%$ for $R = 0.8$ and 83% for $R = 0.9$. Because one incoming warhead can destroy another incoming warhead (the fratricide effect), the two incoming warheads must arrive less than about 10 seconds from each other. For this reason we do not have to consider the case of three or more incoming warheads; nevertheless, for the case of 3 independent SS-18's, the success rate would be $P_{k3} = 1 - (0.48)^3 = 89\%$ for $R = 0.8$. Since there are 1000 Minuteman missiles, these results imply that, perhaps, 170 to 230 would survive two SS-18's, and 100 would survive 3 SS-18's. The latter case would consume the entire SS-18 force since there are about 308 SS-18 launchers and each could be MIRVed about 10 times. From this analysis we can conclude that "Minuteman vulnerability" means that the U.S. would have between 100 to 250 Minuteman launchers (200 to 600 warheads) remaining after a Russian first-strike attack. These calculations consider neither the possibility of systematic bias errors (Sec. IV) nor the other armaments that are available to deter such an attack (Sec. VI).

II.C Dense Pack. It has been proposed to base the MX missile in a very closely packed matrix (545 m apart) so that incoming missiles would destroy each other (fratricide). Let us determine the minimum value of the hardness (H in psi) of the MX silos that would prevent incoming warheads of yield Y = 0.75, 1, 5, and 25 Mt from destroying more than one MX silo. Since the nearest neighbor spacing is 545 m, an incoming warhead that landed halfway between two silos would be 273 m from each silo. Using the formula for overpressure from surface blasts (Eq. 1), we obtain the following values: H > 3500 psi for 0.75 Mt warheads; H > 5000 psi (1 Mt); and H > 22,000 psi (5 Mt), and H > 110,000 psi (25 Mt). In addition, one must consider the size of the craters from these warheads; if the radius of the crater[49] is greater than 275 m, both silos could be destroyed. By using the Rand Corporation "Bomb Damage Effect Computer" (1964), we have obtained the radii of the craters in rock: $r = 142$ m (0.75 Mt), $r = 158$ m (1 Mt), $r = 279$ m (5 Mt), and $r = 485$ m (25 Mt). These crater radii (in rock) can be approximately described by $r \cong 160\ Y^{0.3}$ where r is in meters and Y is in Mt. By building silos in more resilient rock media it is possible to reduce the effect of these craters somewhat, but it is clear that a very large warhead would create a large enough crater to destroy two MX silos. (There is some recent evidence that these crater radii for very large weapons must be reduced by about a factor of two.)

III. TARGET VULNERABILITY WITH ICG

III.A ICG. The complexity of the equations of Sec. II is usually enough to dissuade most people from moving from a discussion of the trees (equations) to a broader discussion of the forest (stability in the arms race as affected by numbers). To obtain a view of the forest we will use ICG to obtain a "physical feel" for the parameters and equations dealing with target vulnerability. We will briefly describe the ICG program "Bombs" which has five adjustable parameters: yield (Y), accuracy (CEP), hardness (H), number of warheads aimed at a silo (L), and gravitational bias error (Sec. IV). The program "Bombs" assumes 100% reliability for the incoming warheads; it could be modified with a random number generator to account for less than 100% reliability. Bombs does the following: It calculates the kill radius (r_k) of a warhead as a function of the yield of the warhead and the hardness of the silo; it scatters circles with a radius of r_k about the aim point with an accuracy of CEP using Gaussian statistics; it simulates gravitational bias error by shifting the aim point with a joy stick; the key "Q" can be used to call up the menu to scale the graphics and to vary Y, CEP, and H.

The program calculates the kill radius (in nm) of a surface blast as a function[5] of the yield of the weapon and the hardness (H = p in Eq. 1) of the target;

$$r_k = \{Y/(0.068H - 0.23H^{1/2} + 0.19)\}^{1/3} \ . \qquad (3)$$

For example, the kill radius of an SS-18 warhead (0.75 Mt) will vary depending on the hardness of the intended target: For a Minuteman silo (2000 psi), r_k = 335 m (0.18 nm); for superhardening Minuteman to H = 5000 psi for MX deployment, r_k = 245 m (0.13 nm); for a horizontal MX (600 psi), r_k = 510 m (0.28 nm); for cities with H = 5 psi, r_k = 6.7 km (3.6 nm); and for civil defense with H = 30 psi, r_k = 1.7 km (0.92 nm). Similarly, the kill radius will vary depending on the yield of the warhead that is aimed at a Minuteman silo (H = 2000 psi): for Y = 0.75 Mt, r_k = 335 m (0.18 nm), for Y = 0.35 Mt (MX), r_k = 260 m (0.14 nm); and for Y = 0.1 Mt (Trident), r_k = 170 m (0.093 nm). In Figure 1, the computer graphics has drawn the circles associated with these kill radii.

The accuracy of the incoming missiles is incorporated into the program in the following way: The program assumes that the individual missiles are randomly spread about the aim point with a normal Gaussian distribution;

$$P(r) = (1/2\pi\sigma^2)e^{-r^2/2\sigma^2} \ , \qquad (4)$$

with σ = CEP/1.17. This implies no systematic bias error; the distribution is centered about the aim point at r = 0. The program does its calculation on a rewritten form of Eq. 4 to obtain the random radius for this Gaussian distribution,

$$r(random) = -(2^{1/2}\sigma) \ln(Rnd) \qquad (5)$$

where Rnd is a random number between 0 and 1 generated by the computer. In addition, the program assumes random angles for the missiles with respect to the aim point. With these assumptions, "Bombs" can graph the case of SS-18 (Y = 0.75 Mt) aimed at Minuteman silos (H = 2000 psi) with an accuracy of CEP = 0.15 nm = 280 m (in 1985?). For the case of 100 independent SS-18 warheads falling on Minuteman, "Bombs" calculates (see Fig. 2) that 30 missiles survive and that 70 missiles are destroyed. This is statistically consistent with Eq. 2 which indicates that 35% of the Minuteman force would survive an attack of one SS-18 of 100% reliability on one silo. In 1981 and 1983 the U.S. government has proposed placing the MX missile in superhardened Minuteman silos

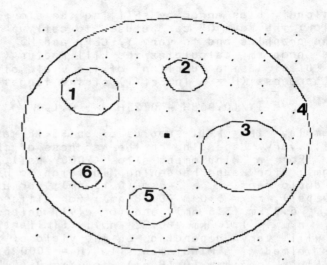

Fig. 1. KILL RADII: The kill radius of a missile is a function of the yield (Y) of the missile and the hardness (H) of the silo (Eq. 3). The dots in the figure are 0.2 nautical miles (1 nm = 1852 m) apart, and the matrix of dots is 1.6 nm (1.85 miles) on a side. For the case of the SS-18 missile (Y = 0.75 Mton) the ICG program "Bombs" has drawn the following circles: (1) H = 2000 psi for the Minuteman silo, (2) H = 5000 psi for the superhardened Minuteman silo, (3) H = 600 psi for the horizontal MX missile, and (4) H = 30 psi for the urban area with civil defense. For normal urban areas of H = 5 psi, the kill radius is offscale since it is 3.6 nm (6.7 km = 4.2 miles). For the case of the Minuteman silos with H = 2000 psi, circles have been drawn for (1) Y = 0.75 Mton (SS-18), (5) Y = 0.35 Mton for the MX missile, and (6) Y = 0.1 Mton for the Trident submarine. The circles appear elliptical in shape because of the Apple graphics.

Fig. 2. HARDENED MINUTEMAN SILOS: The pattern of missiles is distributed about the aim point according to Gaussian statistics (Eq. 4). In the upper portion of this figure we have considered the case of the SS-18 missile (Y = 0.75 Mton) with a circular error probable accuracy of CEP = 0.15 nm (perhaps, in 1985?) on the Minuteman silos (H = 2000 psi). In this example, 5 of the 7 silos were destroyed under the assumption of 100% reliability for the SS-18. The same parameters are used in the lower portion of the figure (Y = 0.75 Mton, CEP = 0.15 nm), but the Minuteman silos have been hardened to H = 5000 psi; 4 of the 7 missiles were successful (with 100% reliability). If the reliability of the SS-18 had been 85%, only 6 of the 7 missiles would have landed.

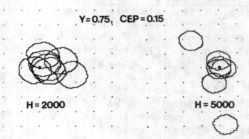

Y=0.75, CEP=0.15

H=2000   H=5000

Fig. 2   HARDENED MINUTEMAN SILOS:
(Caption on previous page)

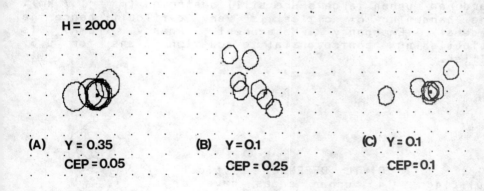

H = 2000

(A) Y = 0.35   (B) Y = 0.1   (C) Y = 0.1
     CEP = 0.05   CEP = 0.25   CEP = 0.1

Fig. 3.   U.S. COUNTER-FORCE WEAPONS: (A) The MX missile
(Y = 0.35 Mton, CEP = 0.05 nm) is aimed at silos with H =
2000 psi; the smaller, but very accurate MX could be used
as a first-strike weapon.   (B) The Trident I missile (Y =
0.1 Mton) may have modest accuracy (middle of figure, Y =
0.1 Mton, CEP = 0.25 nm), or (C) the Trident II may have
a very good accuracy (bottom of figure; CEP = 0.1 nm).
The former case of 0.25 nm could imply a second-strike
role against urban targets (or a first strike against an
airfield), while the later case of 0.1 nm could imply a
first-strike role against Soviet silos.

192

by increasing H from 2000 psi to 5000 psi. The ICG display in Fig. 2 shows that the consequence of this improvement are not very great. These results are consistent with the calculation of the SSKP which decreases from 65% for 2000 psi to 44% for 5000 psi for the case of R = 1.0.

III.B <u>Counter - Force Weapons</u>. Figure 3 shows quite graphically that the MX missile (Y = 0.35 Mt, CEP = 100 m = 0.05 nm, and let us assume H = 2000 psi for Soviet silos) is a counterforce weapon that could destroy a hardened target in a first-strike attack. In spite of the fact that the MX has a smaller yield than the SS-18 (0.35 Mt/0.75 Mt $\cong$ 50%), its greater accuracy (0.05 nm/0.15 nm = 1/3) allows the MX to have a considerably larger SSKP (99%) than the SS-18 (65%). Figure 3 shows that the Trident I submarine (Y = 0.1 Mt) would not be used as a first-strike weapon if its CEP = 0.25 nm because of its minimal kill ability, but that (depending on the hardness of the target) the Trident II could be a first-strike weapon if its CEP = 0.1 nm or less.

III.C <u>Civil Defense</u>. Lastly, one can examine the ability of warheads to destroy cities (H = 5 psi). In Figure 4 the ICG plot of a "random" attack by SS-18 warheads indicates that one warhead of 3/4 megaton can easily destroy (r_k = 6.7 km) the city in the figure (3 km by 3 km). If a civil defense policy is established to harden shelters with H = 30 psi, ICG shows that the hardened urban targets are still destroyed (r_k = 1.7 km). The hardening of cities is very difficult; hence the Federal Emergency Management Agency (FEMA) is establishing controversial evacuation plans for U.S. cities.

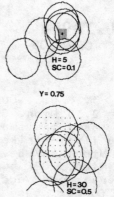

Fig. 4. CIVIL DEFENSE: Typical buildings in urban areas have a hardness of H = 5 psi; the SS-18 (Y = 0.75 Mton) can easily destroy large areas (top of figure; r_k = 6.7 km; circles seperated for visualization and a scale factor of 1:10). Using hardened, civil defense shelters (H = 30 psi) doesn't effectively mitigate the situation (bottom of figure; r_k = 1.7 km; circles seperated with CEP = 1 nm and a scale factor of 1:2).

IV. MISSILE ACCURACY AND GRAVITATIONAL BIAS ERROR

The issue of "bias error" in ICBM targeting has been raised recently[10] ; it has been suggested that systematic errors from gravitational uncertainties may reduce the accuracies of US/USSR missiles so that they may not attain their stated accuracies for trajectories over the poles. (This inaccuracy is probably less than the uncertainty in the listed aim point.) This reduction in accuracy could result because the ICBM's would experience a different gravitational field for ballistic trajectories over the poles (which are not tested) as compared to the trajectory to Kwajalein Atoll (which is often tested); the uncertainties in correcting for the different "g" force for the polar trajectory (as well as other gravitational shifts) might shift the "aim point" away from the intended target in an undeterminable way. The Executive Branch has indicated that the bias errors are included in the listed accuracies of the ICBM's.

However, it is important to consider the issue of gravitational bias since it questions whether either side could ever have confidence in its ability to successfully carry out a preemptive first strike. We will not be able to determine the uncertainties caused by these bias error corrections; however, we will simulate bias error with ICG in order to graphically describe the problems.

In order to avoid the mathematical complications[11] from an ICBM's Keplerian elliptical trajectory, let us consider[7] the trajectory of an ICBM that follows a parabolic path above a large flat earth. Since the earth's polar radius of 6357 km is 21 km (0.3%) smaller than its equatorial radius of 6378 km, we should expect different "g" values for the polar trajectory as compared to the trajectory to the U.S. missile target at Kwajalein Atoll. In the normal three dimensional problem one would consider the earth to be made up of a variety of "multipolar" shaped mass objects (hamburger shaped, oblate spheroidal, etc.), and then apply Newton's second law to a complicated force function which described this nonspherical, multipolar earth. We shall allow our pedagogical earth to be flat in order to estimate the uncertainty in the range

$$x = v^2 \sin(2\theta)/g \qquad (6)$$

that is caused by the uncertainty Δg in the different gravitational field of an untested trajectory. The uncertainty in the range will be

$$\Delta X/X = 2(\Delta v/v) + 2(\Delta\theta/\tan(2\theta)) - \Delta g/g. \qquad (7)$$

The difference in g between trajectories over the pole and along the equator can be approximately determined from the ratio of the mass in the bulge (ΔM) of the earth to the mass of the earth (M). For trajectories which stay reasonably close to the earth (within about 1000 miles), we can crudely approximate $\Delta g/g \cong \Delta M/M$. Since $\Delta R/R \cong 0.3\% \cong 10^{-3}$, we can expect[12] approximately that $\Delta g/g \cong \Delta M/M \cong 10^{-3}$. From this we see that the potential uncertainty in the ICBM flight path from an uncorrected gravitational bias error (10^{-3}) is considerably larger than the uncertainty in the range due to uncertainties in velocity and angles which are about $10^{-5} = (0.1 \quad km/10,000 \quad km)$. Thus, it is clear that gravitational corrections (due to a nonspherical Earth, local gravity anomalies, and the Earth's rotation) must be made for polar trajectories; the only question is the degree of accuracy with which one must be able to perform these corrections in order to reduce the uncertainty in $\Delta g/g$ to better than 10^{-5}. Since neither the US, nor the USSR will be allowed to test the quality of their calculations with ICBM launches over the pole to the territory of the other side, a government would have to believe that their aerospace experts are capable of measurements and calculations with better than 1% accuracy in the correction for the g forces in order to have any kind of confidence that a strike on land-based missile silos would have any chance of success. ICG can be used to illustrate these effects; in Figure 5 we have simulated a substantial gravitational bias error by shifting the pattern of circles from the aim point with the "joy stick" of the computer. In this figure we have set the magnitude of the bias error equal to 0.3 nm in both the x and y directions. Since the magnitude of the total bias error in this example is 0.45 nm (or three times the CEP of 0.15 nm used in this ICG example) all of the missiles (in this figure) missed the target. The actual magnitude of the inaccuracy in the bias error correction is less than this figure, but ICG descriptively indicates how small the bias error has to be in order to neglect its consideration.

△G/G BIAS: X=Y=0.3
CEP = 0.15

Y = 0.75
H = 2000

Fig. 5 GRAVITATIONAL BIAS ERROR: Corrections for different trajectories ($\Delta g/g \cong 10^{-3}$) must be carried out to about 1% if the bias error is not to degrade the CEP of missiles (CEP/Range $\cong 10^{-5}$). In Fig. 5, the aim point of the SS-18 (Y = 0.75 Mt, CEP = 0.15 nm, H = 2000 psi) was moved from the silo with a very large bias error of 0.4 nm with the "joy stick" of the computer.

V. MIRV/ASW

In this section we will briefly describe two ICG programs that were developed by Kent Norville, a student in my class on the arms race; these programs on MIRV and ASW (anti-submarine warfare) are intended to be provocative "teaching tools" rather than sophisticated "policy making tools."

The MIRV technology is capable of spreading its individual warheads over a distance of about 1000 km; the actual calculation of the elliptical, ballistic trajectories of the reentry vehicles can be approximated as a shift in the range and tracking position from the original ballistic trajectory. In order to simplify the mathematics, the program uses the "flat earth" approximation (Fig. 6) for the ballistic projectiles; the program allows the user to vary the number of reentry vehicles, and their individual velocities (magnitude and angle) with respect to the bus. This approach allows only a variation in the range direction, and not the tracking direction; nevertheless, this approach has assisted those students who are unfamiliar with MIRV to understand it more completely.

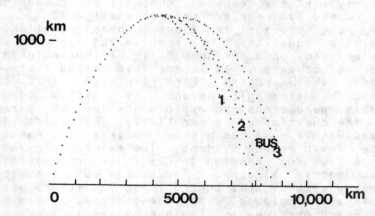

Fig. 6. MIRV: The equations for the "flat earth" (Sec. IV) are used to obtain the trajectories for the individual reentry vehicles and the missile bus. An initial velocity of 10^4 m/s is used for the ICBM. The number of reentry vehicles and their velocities and angles (in range only) can be varied in the program. Momentum is conserved in the system as the reentry vehicles are released. The range, height and spread in range for the re-entry vehicles from the "flat earth" calculations are very similar to those for the actual "round earth."

The ICG program "ASW" applies an ASW technology that determines the position of a submarine by comparing the time delays of signals reflected from a submarine to several hydrophones; more recent technologies[13] use sonobuoys that transmit and receive their own sonar signals, infrared observations from satellites, Fourier transforming sonar signals, lasers, and other technologies. The "ASW" program stimulates consideration of these more modern technologies as well as making clear the uniqueness of the ocean-going leg of the triad.

VI. FIRST STRIKE

The question of vulnerability to a first-strike is continually being debated. The issue of "vulnerability" usually means "will we have enough nuclear weapons to assure destruction on those who would attack us." The debates on this issue often use words like "superiority, parity and sufficiency." In spite of the fact that the uncertainties in first-strike situations are likely to be greater than the certainties, both sides in this debate do use numerical data to buttress their arguments. (For example, it would be difficult to quantify the uncertainty in the command, control, and communications systems caused by the electromagnetic pulse.)
If one chooses to answer these questions with numbers and equations, one soon learns that the size of the data base, and the complexity of the manipulations, requires a computer; it is for this reason that we have developed the program "First-Strike." This program analyzes the strategic balance by letting us vary 90 adjustable parameters to explore how the many parameters affect the question of vulnerability. Each side (America = A and Russia = R) has a 6 by 6 matrix of parameters to describe their weapon systems. The matrix formulation is merely a way of keeping track of the parameters; we do not use matrix algebra. Three data sets (the present, the future without arms control, and the future with arms control) are included with the program. Table I shows the A and R matrices as they approximately appear at the present time (1980-85):

Table I. The A and R matrices for the present time (1980-85). The rows are the 6 different missile classes (land, sea, air, intermediate range ballistic missile, tactical weapons, and cruise missile). The columns are the average values of the parameters which describe each system (yield, average number of reentry vehicles per launcher, number of launchers, circular error probable accuracy (nautical miles), reliability (0-1), silo hardness (psi)).

	AMERICAN A MATRIX				(1985??)		RUSSIAN R MATRIX					
	1=Y	2=M	3=N	4=CEP	5=R	6=H	1=Y	2=M	3=N	4=CEP	5=R	6=H
1=LAND	.335	2.1	1052	.12	.8	2000	.75	3.2	1398	.15	.7	2000
2=SEA	.1	8	656	.25	.8	5	.5	2	950	.5	.7	5
3=AIR	1	4	346	.1	.8	3	1	4	156	.1	.8	3
4=IRBM	.1	1	1500	.1	.8	5	.1	1	4500	.1	.8	5
5=TAC.	.02	1	20000	.1	.8	5	.02	1	10000	.1	.8	5
6=CM	.2	1	1000	.01	.8	5	0	0	0	.01	0	0

This data set can be updated easily to account for changes in the parameters. For example, we might want to take into account that the 52 Titan ICBM's are being decommissioned; by typing 1,3 (row,column), and 1000, we will have changed the number of American land-based missiles from 1052 to 1000. If we are content with the data set, we would type 0,0 to accept it. One can easily take issue with the averaging process that we have used; for example, we have properly taken into account the MIRVing of Minuteman (II, and III) by using M = 2.1, but we have further assumed that they will have an average yield comparable to the Mark 12-A warheads of 335 kilotons. The main point of "First-Strike" is to create a method of changing the parameters as one discusses (and updates) the data base; after the data base is acceptable to the debators, the numerical operations can be carried out. In addition, there are other parameters which deal with the number of warheads sent to attack a silo (L), gravitational bias error (BE = CEP(with)/CEP(without)), submarine duty factor, effectivness of ASW, and various other parameters.

One can examine the one-on-one match-ups between various systems by calculating the single shot kill probability (P in Section II) with reliability (T_k = R x P_k), and then the kill probability with L warheads (from seperate missiles),

$$K_p^L = 1 - (1 - T_k)^L \qquad (8)$$

After we have examined these initial one-to-one results, "First Strike" matches up the land-based missiles against each other. In the example discussed below we have used 2 warheads (L = 2) against each silo. In the next version of this program, we intend to add further flexabilities[6] to allow any system to attack any other system, to allow for the attack of cities, and to allow for a "launch on warning" situation and other operational factors.

We have allowed for broad uncertainties in "First Strike" by allowing for fractional (0-1) parameters; for our "base case" we have used the following:

FSUB = submarine duty factor = 0.6 for A; 0.2 for R.
FASW = effectiveness of ASW = 0.5 for A and R.
FIRBM = FTACTICAL = FCRUISE = FBOMBER
      = fractions destroyed = 0.5 and 0.5 for A and R.

It would not be difficult to show that all the parameters (other than FSUB) should be smaller than 0.5; thus, the destructiveness of the results from "First Strike" are actually an upper bound estimate. The number of surviving launchers are merely the products of the appropriate fractional probabilities and the number of launchers, except for the case of an attack on the land-based missiles where we have used the probability formulation from Eq. 2 and 8. The resultant numbers of remaining warheads after a first strike were obtained from the base case of the "present" data set and they are displayed in Fig. 7. We see that approximately 50% (an upper bound) of the first triad (land/sea/air) and the second triad (IRBM/tactical/cruise missile) would be destroyed[14] if either side should be tempted to carry out such a scenario. Since the number of warheads surviving would be rather large, there is still a large second-strike force to retaliate. Since the fractional probabilities are not certain; debators can change these parameters on the next run of the program. "First Strike" also keeps track of the total yield, lethality (K), and the numbers of launchers before and after.

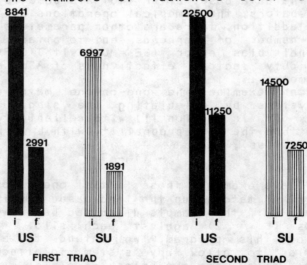

Fig. 7. FIRST STRIKE: The number of US/USSR warheads are compared before and after a "first strike." The present (1980-85) data base and the parameters described in the text were used to obtain these results.

VII. FALLOUT.

If a nuclear attack ever occurred, there would be radioactive plumes extending from the target points. The ICG program "Fallout" calls up a map of the United States (Fig. 8) and allows the user to locate the targets and the number of weapons used on each. Parallelograms are used to simulate the more complex shapes[15] of the plumes; the area of the parallelograms agree with the more accurately calculated affected areas by about 25%. The final shapes for the plumes are highly dependent on the wind velocities; our ICG program allows for two different wind velocities, 20 mph and 60 mph where the later is approximately 50% of the stratospheric velocity. Two plumes are located at each target; the inner plume represents a minimum dose of 1350 REM outside of buildings and 450 REM inside the building; the outer plume represents 450 REM outside and 150 REM inside. These dose levels would vary considerably within the plumes and within the buildings. A radiation dose of about 450 REM would kill about 50% of the population. The present version of this program does not determine the number of fatalities; a future version will include population density and a biological coupling factor in order to do this even though there would be great uncertainties in the absolute numbers of fatalities.

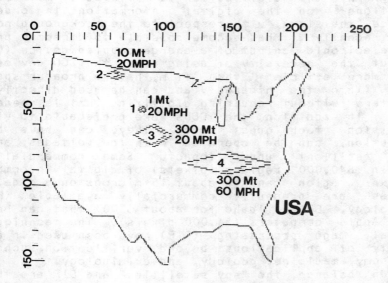

Fig. 8. FALLOUT: (Figure caption on next page.)

Fig. 8. FALLOUT: Four different attacks are considered with "Fallout;" (1) one megaton weapon (50% fission), (2) 10 one megaton weapons, (3) 300 one megaton weapons with a wind velocity of 20 mph, and (4) 300 one megaton weapons with a wind velocity of 60 mph (about 50% of stratospheric velocity). The inner plume represents an area in which a minimum dose of 1350 REM is received for those who are outside of buildings, and a minimum dose of 450 REM inside of buildings; the outer plume represents an area with a dose of 450 REM outside and 150 REM inside. The bases attacked are (1) Ellsworth (x=82,Y=50), (2) Malmstrom (55,28), (3) Warren (78,67), and (4) Whiteman (115,85).

VIII.   DIGITAL IMAGE PROCESSING.

In order for the Senate to ratify an arms control agreement, it must have confidence in its "national technical means" to verify that the various conditions of the treaty are being upheld by the other party. Ultimately, the Senate must decide how likely it would be that the other party could carry out a significant infraction (in terms of national security) before we could discover the infraction. One of main techniques for verification is photography from spy satellites. The information in the photographic images can be enhanced by digitizing the intensity at various regions of a photograph; the digitization is followed by mathematical operations[16] on the digital information in order to enhance the signal with respect to the background noise. More recently it has become possible to directly process the electronic output of a charge-coupled device (CCD)[17] without the necessity of using film; the CCD devices are much more efficient than film, have a broader spectral range (into the infrared), and can be used directly with computers without requiring digitizing that is needed for film. In addition the CCDs are preferable to vidicon television techniques since they can have better resolution, can be operated with low voltages, and are more resilient under impact. Some commercial CCDs contain 640,000 regions (pixels) of digital information; a pixel region can be about 15 microns on a side. CCD cameras are presently commercially available (Micron Technology, Boise, Idaho for about $300) that can be used with Apple computers (54,000 pixels). The techniques of digital image processing (DIP) have been used in a wide variety of applications beyond verification such as in astronomy, medicine, geology, and criminology.

On balance, the "spy satellites" and DIP are thought to be stabilizing for the arms race because they (1) enhance verification giving greater confidence to the SALT/START process, and (2) because more information from satellites can reduce the tendency towards "worst possible case analysis." Some Senators objected to SALT II because of the loss of verification facilities in Iran. President Johnson made the comment in 1967 when he indicated that the $35-40 billion spent on space was worth it because "I know how many missiles the enemy has."

This section of our paper will do the following: (1) Briefly discuss some mathematical applications of DIP, and (2) demonstrate some initial results of CCD with Apple computers. In Fig. 9 we have created a one dimensional silo (20 pixels wide) that is partially blurred by noise developed from a random number generator. When the signal-to-noise ratio (S/N) is very high, one does not need to process the data to see the silo, but when the S/N ratio is less than one it becomes necessary to process the image in order to be able to "see" the silo. In Fig. 9, we have subtracted the image from a version of the same image that has been shifted by 1 to 20 pixels. This technique allows us to examine the data for an auto-correlation within the data; if we expect the silo to be 20 pixels wide, we would look for a figure that looks like a square wave with a wavelength of 40 pixels. In order to work with lower S/N ratios, one could then Fourier analyze the difference image and look for Fourier components consistent with the size of the silo.

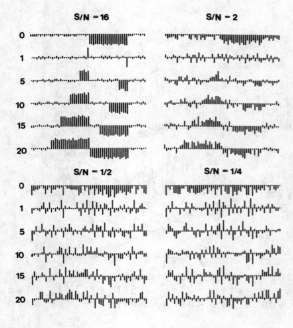

Fig. 9. DIGITAL IMAGE PROCESSING: A random number generator is used to partially mask a missile silo with S/N ratios between 16 and 1/4. The silo (20 pixels wide) can be discovered with auto-correlation techniques by subtracting the data from the original data after shifting it by 1 to 20 pixels. If the resulting data was Fourier analyzed, one would obtain a strong component with a 40 pixel wavelength.

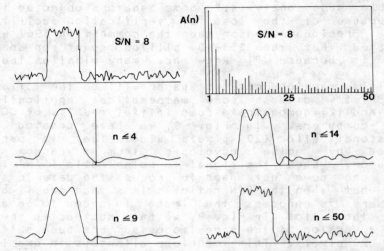

Fig. 10. FOURIER TRANSFORM: The spatial data can be Fourier analyzed to obtain its fequency spectrum; by reversing the process the spatial representation can be reestablished. The silo appears more clearly when only the 4 lowest frequency components are used and the 46 higher components containing the "noise" are rejected.

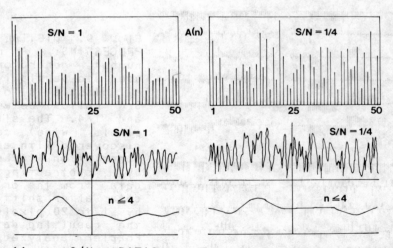

Fig. 11. S/N RATIO. The approach of Fig. 10 successfully identifies a silo with a S/N of 1/4. Two dimensional transforms would use the geometry of the silo to improve the results.

Another approach of DIP is to Fourier transform the digital information of the image, remove the high frequency components that are associated with the noise, and then Fourier transform the remaining low frequency (long wavelength) signal back to the spatial image. By analyzing the data we convert it from a spatial representation to a freqency spectrum; the second Fourier transform then synthesizes (reconstructs) the spatial image from the truncated frequency spectrum. In Fig. 10, we have obtained a 101 pixel array in one dimension from the equation $y = (NN)(Rnd(1)) + S$ where NN is the peak noise and S is the intensity of the signal. The image in Fig. 10 has a favorable signal-to-noise ratio of $S/N = S/(NN/2) = 8$; when this image is Fourier analyzed, one obtains large components in the low frequency region which is associated with the silo signal, and small components at the high frequency region which is associated with random noise. When the image is reconstructed from these components, we observe that only the 4 lowest components are necessary to easily detect the silo, and that a greater number of components tends to blurr the reconstructed image. In Fig. 11 we have increased the noise level to give $S/N = 1/1$ and $1/4$; the reconstructed images with the lowest four components clearly show the location of the silo. These examples must next be extended to the regime of two dimensions; the convolution theorem must be applied to remove distortion by the viewing system. Other transforms can be used in order to apply mathematics that is more "tailor-made" to the geometry of the object. Finally, we have obtained some grey level presentations of pictorial images by using a charge-coupled device, an opticRAM with 128 × 256 pixels. The grey levels in the final picture are obtained by comparing images of the same object that were obtained with different exposure times. Since the cost can be as low as 3 cents/pixel, it is clear that the CCD technology will have a tremendous impact on verification and digitial image processsing.

IX. THE SCIENCE AND SOCIETY PUBLIC POLICY DISKETTE

A diskette and manual of 15 programs is available from us (553 Serrano, San Luis Obispo, CA 93401) on a nonprofit, noncopyright basis for $10. Please send us your results for possible inclusion on future diskettes so that we can create an imformation network. I would like to thank James Hauser, Fred Jaquin, Alan Lyon, Kent Norville, Dietrich Schroeer, and Walt Wilson for assistance on this project.

REFERENCES

1. R. Sperry, Science <u>217</u>, 1223 (1982).

2. T. Graedel and R. McGill, Science <u>215</u>, 1191 (1982).

3. R. Myers, <u>Microcomputer Graphics</u>, Addison-Wesley, Reading, MA, 1982.

4. M. Brode, Ann. Rev. Nuc. Sci. <u>18</u>, 153 (1968).

5. K. Tsipis, Science <u>187</u>, 393 (1975); <u>206</u>, 510 (1979).

6. J. Steinbruner and T. Garwin, Int. Sec. <u>1</u>, 138 (summer, 1976).

7. D. Hafemeister, Am. J. Phys. <u>51</u>, 215 (1983).

8. P. Nitze, <u>Military Implicatons on the SALT II Treaty</u>, U.S. Senate Armed Services Comm., Washington, D.C., 1979, p. 891.

9. S. Glasstone and P. Dolan, <u>The Effects of Nuclear Weapons</u>, Dept. of Defense and Dept. of Energy, Washington, D.C., 1977.

10. E. Marshall, Science <u>213</u>, 1230 (1981). J. E. Anderson, Bull. At. Sci. 37, 6 (Nov., 1981).

11. D. Hoag, p. 19 in <u>Impact of New Technologies on the Arms Race</u>, Ed. by B. Feld, T. Greenwood, G. Rathgens, and S. Weinberg, MIT Press, Cambridge, MA, 1971.

12. W. Kaula, <u>Theory of Satellite Geodesy</u>, Blaisdell, Waltham, MA, 1966. M. Caputo, <u>The Gravity Field of the Earth</u>, Academic, NY, 1967.

13. K. Tsipis, A. Cahn, and B. Feld, <u>The Future of the Sea-Based Deterrent</u>, MIT Press, Cambridge, MA, 1973.

14. R. Forsberg, Sci. Am. <u>247</u>, 52 (Nov., 1982).

15. S. Drell and F. von Hippel, Sci. Am. <u>235</u>, 27 (Nov., 1976). K. Lewis, Sci. Am. <u>241</u>, 35 (July, 1979). S. Fetter and K. Tsipis, Sci. Am. <u>244</u>, 41 (April, 1981). <u>The Effects of Nuclear War</u>, Office of Technology Assessment, Washington, D.C., 1979.

16. B. Hunt, Chapter 6 of this book.

17. J. Kristian and M. Blouke, Sci. Am. <u>247</u>, 67 (Oct., 1982).

LEWIS F. RICHARDSON AND THE ANALYSIS OF WAR

Paul P. Craig
Department of Applied Science
University of California
Davis, CA 95616

and
Mark D. Levine
Lawrence Berkeley Laboratory
Berkeley, CA 94720

ABSTRACT

We show that since about 1970 the strategic nuclear weaponry of the US and the USSR have been following the classic exponential growth pattern first discussed by Lewis Fry Richardson, the founder of mathematical analysis of war. We also present data suggesting a correlation between war deaths and two-generation, half century long waves (Kondratieff cycles) in human systems.

INTRODUCTION

Lewis Fry Richardson (1881-1953), a British physicist and meteorologist (D.Sc., Physics, University of London, 1929; Fellow, Royal Society, 1929) was one of the founders of mathematical analysis of war. Richardson's personal experiences as a Quaker and a pacifist gave him strong views, which he took pains to clarify. "...the first duty of a social scientist is to declare his prejudices. So I had better mention those of mine which are relevant to the arms race. I have a prejudice that the moral evil in war outweighs the moral good, although the latter is conspicuous. This prejudice is derived from the Quakers, who brought me up. I am not ashamed of it; indeed it has been one of the two principle motives for writing this book. The other principle one is that the scientific method is more trustworthy than rhetoric" (Richardson 1960a).

Richardson's efforts resulted in a few publications and two posthumously published books. Arms and Insecurity (Richardson, 1960a) focuses on the development of a mathematical theory of war. In it he introduces and tests a pair of coupled differential equations which are today known as "Richardson's equations", which we will discuss. Statistics of Deadly Quarrels (1960b) is a compilation of data on some 315 wars occurring between about 1810 and 1952.

0094-243X/83/1040205-11 $3.00 Copyright 1983 American Institute of Physics

Our purposes in this paper are 1) to provide a brief introduction to Richardson theory and apply it to the nuclear arms race and 2) to show how those times when the probability of war is relatively small coincide with mimima in the half century Kondratieff waves of economics, demeography and culture.

RICHARDSON EQUATIONS APPLIED TO THE NUCLEAR ARMS RACE

Richardson used a pair of coupled first order differential equations to discuss arms races. He tested his equations in Europe in the period preceeding World War I. (The equations are traceable to Frederick William Lanchester, 1914). Let the expenditures on armaments of two nations be denoted by x and y. The Richardson equations express the rate of change of armament expenditures:

$$dx/dt = ky - ax + g$$

$$dy/dt = cx - by + h$$

The terms are to be understood as follows: k and c are "defense coefficients" representing the response in one country to military expenditures in the other; a and b are "fatigue coefficients" representing the (exponential) decay in arms expenditures if there are no threats; g and h are "grievance" or "unsatisfied ambition" coefficients which lead to an increase in arms expenditure (when positive) or a decrease (when negative) due to "deeply rooted prejudices, standing grievances, old unsatisfied ambitions, wicked and persistant dreams of world conquest, or, on the contrary, a permanent feeling of contentment" (Richardson, 1960a, p.16). Richardson justifies his use of first order equations, and his decision to treat all the coefficients as constant using Occam's razor: "Mathematical physics has progressed by trying out first the simplest formulae that described the broad features of the early experiments and by leaving complicated formulae to wait for more complicated experiments" (1960b, p.14).

Richardson was well aware of stability theory, and provided an extensive technical analysis of the stability characteristics of these equations. What is of interest to us, though, is his test of steady state growth. The first major arms race of this century occurred from 1909-1914. France was allied with Russia, and Germany with Austro-Hungary. Britain, Italy and the United States were not allied. Since the two alliances had similar technologies available and were roughly equal in size, Richardson assumed symmetry and set k=c and a=b (without justification). He added the two equations to obtain:

d(x+y)/dt = (linear function of (x+y))

Figure 1a, from Richardson (1960a), examines this relation using finite differences - the year to year variation in arms expenditures in the arms race which preceded World War I (he uses U and V, which differ from x and y by additive constants). While the expenditures on neither side grow exponentially, the total expenditures do - as demonstated by the linearity of the graph. The agreement is spectacular. "Since I first drew this diagram...I have been incredulous about the marvelously good fit" (1960a, p33). An early reviewer of this data wrote:

> The agreement with the theory is remarkably close, and if the question were one of physics, it would be regarded as settled. But in physics we believe that laws hold, in politics we may not unreasonably doubt it, and many more confirmations would be needed to produce conviction. (H. G. Forder, 1940 quoted in Richardson 1960a).

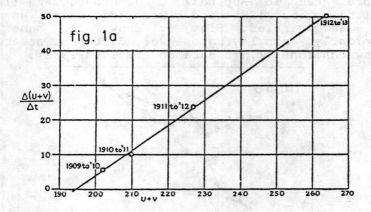

Fig. 1a. Lewis Fry Richardson's demonstration that arms expenditures prior to World War I followed an exponential growth law. The factors U and V are (to within an additive constant) the expenditures by the two competing powers.

AN APPLICATION OF RICHARDSON EQUATIONS

Does the current arms race fit the Richardons model? Constant dollar expenditures by both NATO and by the WTO (Warsaw Treaty Pact) have grown only modestly in the past two decades, and certainly do not fit the theory (see data in NATO (1981), Jones (1982) and SIPRI (1981)). However the data problem is severe, and any conclusions must necessarily be viewed with much suspicion because of the complexity of developing meaningful cost comparisons between Eastern and Western bloc nations, a matter that has been extensively discussed (Cusack and Ward (1981), Cockle, (1978)).

Dollar figures are not even obtainable for the nuclear component of the arms race. Even for the US it is hard to disentangle nuclear costs (about 15% of the total) from total military costs; for the USSR this disaggregation is impossible.

We therefore decided to focus on technical instead of fiscal factors. What emerges is the conclusion that since about 1970 the US and the USSR have been engaged in a classic Richardson arms race. Current Administration requests for nuclear armaments are consistent with the continuation of this pattern.

Figure 1b presents the (base 10) logarithm of the number of warheads of the US (squares) and the USSR (circles). The DoD figures and DoD extrapolations are those presented in Appendix C of Ground Zero (1982). Extrapolations to 1985 and 1990 are shown as open squares and circles. Fig. 1c shows the same data parametrically, displaying the very different strategies of the two nations. (This data covers 1960-1982).

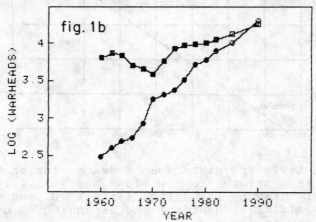

Fig. 1b. Logarithm of the number of warheads of the US (squares) and the USSR (circles). Open points are Department of Defense forecasts.

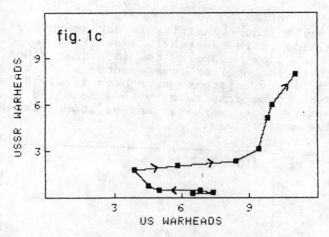

Fig. 1c. Parametric plot of US versus USSR warheads.

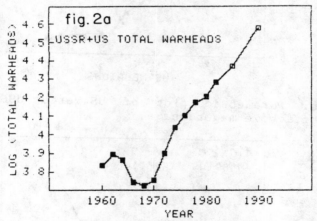

Figure 2a is a Richardson logarithmic plot of the sum of US and USSR warheads. It shows a striking exponential growth pattern after 1970, with a doubling time of about eight years.

Figure 2b is a parametric plot of US versus USSR effective megatonnage from 1960-1982. (Equivalent megatonnage weights low yield devices according to the square root of the yield and warheads above one MT as the cube root of the yield). Figure 2c is the same data in a Richardson plot. The forecasts call for the escallation of the last decade to continue, doubling each eleven years. Figure 3a is the historical data from fig. 2c presented in the format Richardson used: the derivative of the sum of the effective megatonnage versus the sum of the effective megatonnage (1970 to

210

1980) While the linearity is not so good as that which he obtained, it is nevertheless striking and disturbing.

These data support the idea that since about 1970 the strategic nuclear components of US and the Soviet Union military systems have been involved in a classic Richardson type arms race, with a doubling time of about eight years for warheads and of about eleven years for effective megatonnage.

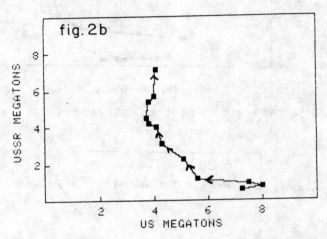

Fig. 2b. Parametric plot of US versus USSR strategic weapon effective megatonnage.

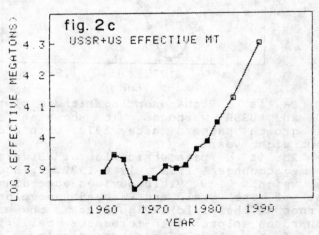

Fig. 2c. Semi-logarithmic plot of US + USSR effective megatonnage. The doubling time is about eleven years.

Exponential solutions to the Richardson equations occur only when rather special conditions on the coefficients are satisfied. One set of suitable conditions occurs when there is symmetry between the competing powers. The early stages of the nuclear arms race were dominated by the US. Around 1970 considerable symmetry developed. It may be that the military-industrial bureauracies in the two nations have much in common. One type of similarity that would encourage a self-perpetuating arms race is a tendency of military analysts to assume high reliability for enemy systems, while using "conservative" (i.e. low) reliability assumptions for their own systems.

The critical challenge of our era is to find a way to find an escape route from this exponentially escallating arms race.

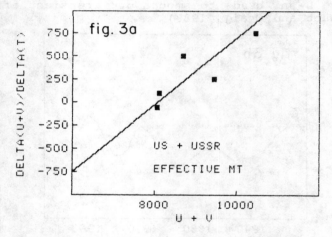

Fig 3a. Richardson plot (similar to that of Fig. 1a) of the rate of change of US + USSR effective megatonnage versus the US + USSR effective megatonnage (1970 - 1982). This is a particularly sensitive way to show the exponential character of the escallation.

STATISTICS OF DEADLY QUARRELS

Lewis Richardson had an experimentalist orientation. His data base on wars, Statistics of Deadly Quarrels (1960b) is one of the most important extant sources of detailed information. Richardson characterized some 315 wars over the period 1810 until just prior to his death in 1953. His data have been reviewed and summarized by Wilkinson (1980). This data can be used to examine questions having to do with the timing of wars relative to other events - as evidenced for example in the Long Cycles of Kondratieff.

The Kondratieff Long Cycle lasts about a half century, or two human generations. (For a brief review see Dickson 1983). The cycle has been observed in many aspects of human affairs. A number of causes have been adduced for the Kondratieff cycle. A two generation cycle in birth rates can arise from cycles of optimism and pessimism in job prospects, leading to low birth rates when times are bad (e.g. the 1930's) which results in low cohort size, a spirit of optimism, and high birth rates in the next generation. Cycles in innovation and the natural lifetimes of economic goods have also been found to lead to half century periodicity. One of the best pieces of evidence for Kondratieff cycles is the wholesale price index, relative to gold. This cycle is shown in figure 3b, where a two decade arithmetic average has been used to smooth out transient effects and noise (Watt and Craig, 1983).

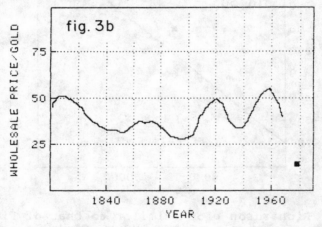

Fig. 3b. The two generation long wave (Kondratieff wave) in US wholesale prices relative to gold. Nadirs are times of economic stagnation.

Richardson's data on wars include an estimate for each war of the number of people killed, and of the war duration. The estimates of war dead are necessarily rough, and so Richardson used as an indicator the base-10 logarithm of the number of people killed. Figure 3c is calculated from this data, with a one decade arithmetic average. The cyclic nature of war deaths is apparent. The solid and open boxes at the bottom of the curve show, respectivly, the minima and maxima in the Kondratieff wholesale price index. (Minima: war deaths: 1840, 1887, 1926; wholesale/gold: 1845, 1895, 1936.

Maxima: war deaths: 1828, 1861, 1915, 1945;
wholesale/gold: 1813, 1869, 1919, 1959). At times of
Kondratieff minima (which occur when economic times are
bad) war deaths are low. Maxima in war deaths occur a
few years prior to the Kondratieff maxima.

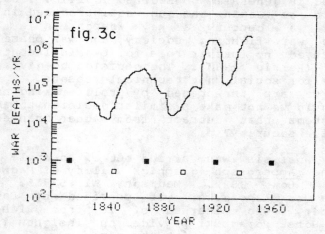

Fig. 3c. Logarithm of war deaths versus time.
Richardson's data set includes about 315 wars. There is
an apparent correlation between economic activity and
war.

The work of Lewis Richardson serves to raises
profound questions regarding the nature of arms races,
and the relation between wars and economics. To what
extent are the correlations indicative of causal
connections between economics and war? If there are
causal connections, what are the primary drivers? Are
these drivers still relevant in an age of nuclear
weaponry? And, most important, what policy measures can
be used to decrease the likelihood of war, especially
during the next Kondratieff upswing which may occur in
the 1990's?

These considerations may be relevant to the
competition between the US and the Soviet Union. The
next few decades are likely to see nuclear weapons in
the hands of many nations. As the French historian
Fernand Braudel observed in his review of the
preindustrial world (1975): "Those who succeeded usually
did so ruthlessly at the expense of others". The
pressures to pull nations out of recessions may well
create an environment in which nuclear conflict becomes
more likely. Should this be the case, the security of
our nation may be connected with our ability to minimize
large amplitude global economic swings. These

considerations can provide linkages between the
"East-West" conflict between the US and the USSR, and
the so-called "North-South" conflict between developed
and developing nations.

The idea that a weapons system, once invented, can
be abandoned is generally considered idyllic. Yet in the
island world of Japan the gun was introduced in 1543,
became for a century a key weapon in a highly
militaristic war fighting society, and was phased out.
The gun played no further role for two full centuries,
until 1879 (Perrin 1980). The decision to give up the
gun was made for social, not technical reasons.

Today we are threatened by proliferating nuclear
weapons. Could we not make a similar decision to abandon
weapons systems that detract from rather than enhance
our collective security?

The analysis was carried out on an Apple II,
using the Ampergraph graphics utility of MadWest
Software, Box 9822, Madison, WI 53715. The
reference to Lanchester was provided by David
Hafemeister, who learned of it from Richard
Garwin. Horst Porembski provided the insights into
the Japanese experience with the gun.

REFERENCES

Braudel, Fernand, "Capitalism and Material Life: 1400-1800" Harper and Row, N.Y., 1973.

Cusack, Thomas R. and Ward, Michael Don, "Military Spending in the United States, Soviet Union, and the Peoples Republic of China". J. Conflict Resolution Vol. 25 No.3, 429-469 (1981).

Dickson, David. "Technology and Cycles of Boom and Bust". News and Comment section of Science Magazine V. 219, 933-936, February 25, 1983.

Ground Zero, "Nuclear War, What's in it for You". Pocket Books, N.Y. 1982.

Jones, David R. ed. Soviet Armed Forces Review Annual, Gulf Breeze, Academic International Press, 1982.

Lanchester, Frederick William, "Aircraft in Warfare: The Dawn of the Fourth Arm". Engineering, October 14, 1914, pp 422-423.

NATO Facts and Figures, NATO Information Service, Brussels, 1981.

Perrin, Noel, "Giving up the Gun: Japan's Reversion to the Sword, 1543-1879". Shambhala Press, Boulder, CO., 1980.

Richardson, Lewis F. "Arms and Insecurity", Boxwood Press, Pitsburgh, PA. 1960a.

Richardson, Lewis F. "Statistics of Deadly Quarrels", Boxwood Press, Pittsburgh, PA. 1960b.

Richardson, Stephen A., "Lewis Fry Richardson (1881-1953): A Personal Biography". Conflict Resolution Vol 1, 300-304 (1960).

SIPRI. Stockholm International Peace Research Institute. "World Armaments and Disarmaments Yearbook", Taylor and Francis, Ltd. London, 1981.

Wilkinson, David, "Deadly Quarrels: Lewis F. Richardson and the Statistical Study of War". University of California Press, 1980.

Watt, K.E.F. and Craig, Paul P. "The Ecology of Long Waves: Resources, Technology, Markets, Demography, Politics and Culture". submitted to Science, 1983.

CHAPTER 11

COMPREHENSIVE TEST BAN NEGOTIATIONS

G. Allen Greb
University of California, San Diego
La Jolla, Ca. 92093

Warren Heckrotte
Lawrence Livermore National Laboratory
Livermore, Ca. 94550

ABSTRACT

Although it has been a stated policy goal of American and Soviet leaders since 1958 (with the exception of Ronald Reagan), the world today is still without a Comprehensive Test Ban Treaty. Throughout their history, test ban negotiations have been plagued by a number of persistent problems. Chief among these is East-West differences on the verification question, with the United States concerned about the problem of possible Soviet cheating and the USSR concerned about the protection of its national sovereignty. In addition, internal bureaucratic politics have played a major role in preventing the successful conclusion of an agreement.

Despite these problems, the superpowers have concluded several significant partial measures: a brief (1958-1961) total moratorium on nuclear weapons tests; the Limited Test Ban Treaty of 1963, banning tests in the air, water and outer space; the Threshold Test Ban Treaty of 1974 (150 KT limit on underground explosions); and the Peaceful Nuclear Explosions Treaty of 1976 (150 KT limit on individal PNEs). Today, the main U.S. objections to a CTBT center is the nuclear weapons laboratories, the Department of Energy, and the Pentagon, who all stress the issues of stockpile reliability and verification. Those who remain committed to a CTBT emphasize the potential political leverage it offers in checking both horizontal and vertical proliferation.

0094-243X/83/1040216-14 $3,00 Copyright 1983 American Institute of Physics

I. INTRODUCTION

The idea of a ban on nuclear weapons tests arose almost thirty years ago, a product of the beginning of the age of thermonuclear weapons. It is an idea that remains with us, regarded by some as unfinished business in the attempt to inhibit the spread of nuclear weapons and by others as a potential threat to national security. In this article, we trace the history of the negotiations to achieve a complete ban on the testing of nuclear weapons. It is a long, many faceted, and complicated story; in the brief span of these pages, we can at best present a selective and condensed summary.

The first test of a thermonuclear or hydrogen weapon was the Bravo test carried out by the United States in the Pacific in 1954. The yield was about fifteen megatons, almost twice that anticipated. Radioactivity from the fallout fell on a Japanese fishing boat that had unwittingly crossed the path of Bravo's radioactive cloud. By the time the vessel arrived in Japan two weeks later, the entire crew had fallen ill; one sailor later died. Suddenly there was worldwide consternation over fallout and over the existence of weapons of such enormous destructive power. Calls for the cessation of nuclear weapons tests came from all corners of the globe.

The United States examined the idea of a comprehensive test ban (CTB) and rejected it -- government officials, especially those in the Atomic Energy Commission (AEC) and the Pentagon, insisted that the continued development of nuclear weapons was needed to counter the Soviet Union. By contrast, the Soviet government adopted the test ban as an item in its extensive arms control agenda in mid-1955. The Soviets, however, offered this proposal as part of a vague package of general arms control and disarmament measures which had no chance of being accepted by the United States.

II. THE TEST MORATORIUM

From this initial call for a test ban until the beginning of serious negotiations in 1958, a complex web of problems -- involving domestic and alliance politics, nuclear strategy, national security questions, and East-West rivalry -- stood as a formidable barrier to negotiations. There can be little doubt, however, that the widespread public concern over radioactive fallout produced by atmospheric testing continued to push and fuel the test ban cause.

During this period both superpowers suggested that nuclear tests be stopped for some finite duration, but each proposal had conditions regarding verification and other matters which were unacceptable to the other side. Then in the spring of 1958, the Soviets announced a unilateral test suspension if the United States would agree to follow suit. The Kremlin made this proposal following an extensive test series of its own and with the knowledge that Washington had a similar series planned. Whether disingenuous or not, this measure clearly put the Soviets ahead in the international propaganda battle.

At the same time, the Eisenhower administration began to gradually shift its internal views on a CTB. Scientists brought into the government and the decision-making process in the aftermath of Sputnik -- including such illuminaries as James Killian of MIT, George Kistiakowsky of Harvard, and Hans Bethe of Cornell -- began to counter the strong anti-test ban views of officials and scientists in the AEC and the military. These scientists stressed the value of a test ban not in preventing fallout but as an initial control on the Soviet American arms race. Bethe, for example, argued that "if we once get one controlled disarmament agreement [a CTB], I believe that others may follow and that the principle will thereby be established." This was, he maintained, the "overriding argument."[1] By the spring of 1958, moreover, Secretary of State John Foster Dulles had reversed his earlier opposition to a ban in response to the mounting international pressures against nuclear testing. The President himself remained somewhat ambivalent on the test ban issue but made clear his commitment to somehow diminish the threat posed by nuclear weapons.

Although Eisenhower continued to move cautiously, a breakthrough came when Secretary Dulles drafted and the President agreed to send two letters to the new Soviet Premier Nikita Khrushchev calling for technical studies of the question. The letters proposed that a group of scientists from East and West meet to develop a CTB control system, a necessary precondition in Eisenhower's eyes for implementing any political decision on a test ban. The Soviets agreed.

The resulting Conference of Experts met in Geneva in the summer of 1958. By far the most challenging problem faced by this prestigious group was how to detect and identify underground nuclear explosions under a possible CTB regime. After examining this and other questions for nearly seven weeks, the experts concluded that it was technically feasible to set up an effective control system and described the elements of such a system. Significantly, they included in this system provisions for on-site inspection of events that could not be identified as earthquakes. The seismic analysis underlying the experts' final recommendations rested on a single datum: the low yield underground Rainier test conducted by the United States in 1957. It was, as it turned out, a slender base for the proposed edifice.

With the successful conclusion of the Conference of Experts, Eisenhower announced that the United States was prepared to enter formal treaty negotiations and proposed October 31, 1958, as the opening date. He also stated that the United States would stop further testing, unless the Soviets tested, for a period of one year. This moratorium, he continued, would be extended on a year-by-year basis if an agreed control system could be installed and if progress could be made on major arms control agreements. The Soviets agreed to begin talks but objected to the linkage of a CTB with other arms control measures. The negotiations began on schedule in Geneva with Britain, the only other nuclear power of the day, as a third negotiating party. A day or two later, the Soviets carried out several more nuclear tests but then stopped. Curiously, the Soviet leadership made no formal statement at this time that it would observe the moratorium.

Moscow initially took the stance that a ban should be established by treaty exclusive of any control provisions. U.S. and British negotiators, however, insisted that discussion of verification procedures take place first and the Soviets acquiesced. But it soon became clear that the two sides' conception of a control organization was far apart. The West (the United States and Britain) envisaged an international organization, internationally staffed, with decisions made by majority vote. Most important, on-site inspections (OSIs) could be made whenever a questionable event occurred. From the Soviet perspective, such an organization could be manipulated by the West for intelligence purposes. The Soviets thus sought a verification system under which they would in effect retain control of all operations on their own territory. For the West this simply spelled veto and self-inspection. It was evident that negotiations would be much more difficult than at first anticipated, and some believed resolution of this impasse would prove impossible.

Meanwhile within the Eisenhower administration, opposition to a test ban remained strong. AEC and military critics, led by AEC Chairman Lewis Strauss and supported by such influential scientists as Edward Teller and E. O. Lawrence of the Livermore Laboratory , argued that tests must proceed to develop new strategic and tactical weapons (the latter the "clean" or neutron bomb). A test cessation, Strauss warned, would be a "very fateful step" for the United States.[2] Strauss, Teller, and other nuclear spokesmen in addition argued that verification was not adequate to assure against Soviet cheating. This view was reinforced when assessment of the data from underground experiments conducted before the moratorium (the U.S. Hardtack tests of October 1958) led scientists to conclude that the experts' control system was substantially less capable than had been estimated from the Rainier event. At the Geneva Conference, the Soviets dismissed this analysis (suggesting even that it had been fabricated) and reasserted that the technical problems had been settled by the agreed experts' report. To offset the negative impact of the new technical findings, the United States announced it would no longer require progress toward nuclear disarmament as a condition for a test ban. But in light of the Hardtack results and the development of the new theory of "decoupling" (the muffling of the seismic signal by firing an explosion in a very large underground cavity), Eisenhower decided to turn away from a total ban to either a ban on atmospheric tests alone or a threshold ban.

Despite these problems, the Geneva negotiations proceeded during 1959 and 1960; each side made concessions, some substantial, and the form of a treaty gradually began to take shape. Yet there was little real optimism because difficult verification issues remained

unresolved. In November 1959, Washington, still beseiged by internal
test ban opposition, proclaimed it would extend the moratorium only
until the end of the year. The following month, Eisenhower announced
that "the voluntary moratorium on testing will expire on December
31," but that "we shall not resume nuclear weapons tests without
announcing our intention in advance."[3] Early in January 1960, Khrushchev declared that Eisenhower's statement obviously meant that the
United States could resume nuclear tests at any moment. But he also
pledged that the Soviet Union would not resume tests unless the
"Western powers" did so.[4]

With the official basis for the moratorium thus gone but with it
preserved in fact, the U.S. delegation presented a threshold treaty
-- banning underground tests above a specified seismic magnitude
(estimated at twenty kilotons) and atmospheric tests -- at the Geneva
Conference in the spring of 1960. The Soviets accepted with one pro-
viso -- a voluntary moratorium on tests below the threshold. The
United States in turn accepted this proposal if the moratorium were
only for one or two years; the Soviets countered with four or five
years. Both sides also agreed in principle to an annual quota of
OSIs, a significant compromise, although neither party advanced a
specific quota number.

With optimism again high, both sides hoped to settle the duration and quota questions and set the stage for the resolution of all
remaining issues at the Big Four summit conference to be held in
Paris in May. The U-2 incident, however, undermined the summit and
ended any expectation of achieving a negotiated ban on the part of
the Eisenhower administration.

Even without the U-2 incident, it is at best problematical that
agreement could have been achieved on the substantive test ban
issues. Given the difference between the two sides on the numerical
value of the OSI quota as it later emerged -- three on the Soviet
side; twenty on the U.S. side -- one may doubt the possibility of a
successful resolution of the question. In fact, this OSI "numbers
game," as one participant in the debate calls it,[5] came to symbolize
the difference between the two sides over arms control verification,
underscoring the basic East-West conceptual divergence about the
rights and functions of a control organization.

The Kennedy administration came into office determined to try to
solve the test ban impasse. It did not have the internal divisions
on this subject that had existed within the Eisenhower bureaucracy.
But after several months of meetings with the Soviets, negotiations
were clearly stalemated. Moscow in effect reintroduced the idea of a
veto and despite U.S. demonstrations of flexibility there was no
indication of a Soviet desire to compromise. In time Khrushchev --
under pressure from constituencies within the Soviet bureaucracy, in
particular the powerful military elite -- began to downplay the
importance of a test ban altogether. On several occasions, he suggested that nuclear weapons tests by the French, a "Western power"
and NATO member, might compel the Soviet Union to resume testing. On

August 30, 1961, the Soviet government announced it would resume nuclear tests and began on the very next day. It was the start of an extensive test series, which included the largest nuclear explosion ever carried out, equivalent to fifty-seven megatons. The negotiations in effect were over.

This first set of CTB negotiations was the SALT of its day; it was seminal both for the future of the test ban in particular and arms control in general. The debate focused almost exclusively on verification, and this issue has been a dominant element in arms control discussions to this day. The Soviet unwillingness to accept the verification procedures the United States sought strengthened the belief that the Soviets were unwilling to accept reasonable control constraints of any kind. Their abrupt resumption of testing in 1961 reinforced this view. The Soviet tests, whose extent and magnitude necessitated long and careful preparation, caught the United States by surprise and emphasized the difficulty of knowing what the other side might or might not be planning to do.

III. THE LIMITED TEST BAN TREATY

On March 4, 1962, the Soviet-American test ban negotiaions entered a new forum, the Eighteen Nation Disarmament Conference (ENDC). For the first time nonnuclear, nonaligned states actively engaged in the arms control and disarmament process with the superpowers, bringing new concerns and complications into the test ban debate. The ENDC (today the Committee on Disarmament or CD) in effect became the focus of international pressure for a test ban. By the spring of 1963, the United States had simplified and reduced its verification requirements for a CTB, but the impasse over OSIs remained.

Personal diplomacy again broke the logjam. A letter from Kennedy to Khrushchev proposed that very high level representatives from both sides meet in Moscow to try once more to achieve a treaty. Khrushchev agreed. A potent underlying factor in this renewal of contacts was the Cuban missile crisis of October 1962. Taking both sides to the brink of a nuclear confrontation that neither wanted, the resolution of the crisis represented a major turning point in superpower relations. As Kennedy's letter to Khrushchev put it, "perhaps now, as we step back from danger, we can make some real progress in this vital field (nuclear arms control). I think we should give priority to questions relating to the proliferation of nuclear weapons... and to the great effort for a nuclear test ban."[6] Early in 1963, Soviet Foreign Minister Andrei Gromyko confirmed the desire for conciliation and flexibility that prevailed in both Moscow and Washington after the missile crisis. "History has so developed," Gromyko told the Supreme Soviet in January, "that without mutual understanding between the USSR and the USA not a single serious international conflict can be settled; agreement cannot be reached on a single important international problem."[7]

At Moscow, the U.S. intention was to seek a CTB. But when the U.S., British, and Soviet delegations met in July, the Soviets expressed interest only in a partial ban, leaving untouched the question of underground tests. With the major problem of verification thus set aside, the parties reached agreement in ten days. The Limited Test Ban Treaty (LTBT) of 1963 brought to a close the second chapter in attempts to ban all nuclear weapons tests. By prohibiting tests in all environments except underground, the LTBT met the concern that had fueled the initial public outcry -- fallout from atmospheric explosions. With this pressure removed and with the sense of accomplishment of a treaty, efforts toward a complete ban subsided.

IV. NONPROLIFERATION AND THE CTB

The principal arms control measure that followed the LTBT was the Nonproliferation Treaty (NPT). Recognizing that it was in their mutual interest to check the spread of nuclear weapons, the superpowers eschewed their traditional confrontational stance and sought agreement within the ENDC on this issue during the mid and late 1960s. Although a CTB remained high on the agenda of most members, the United States and Soviet Union accorded highest priority to the proliferation question and pushed through the NPT in 1968. Most if not all of the nonnuclear states perceived the treaty as discriminatory and sought to balance in part their renunciation of nuclear weapons with a superpower commitment to negotiate a CTB. The two superpowers refused such an explicit commitment, but accepted as a compromise Article VI of the NPT which calls upon the "parties of the treaty to pursue negotiations in good faith on effective measures relating to cessation of the nuclear arms race."[8] The nonnuclear nations took this to mean a CTB. They could also point to the NPT preamble which "recalled" the preambular passage in the 1963 LTBT expressing the determination of the signatories to continue negotiations for the "discontinuance of all test explosions of nuclear weapons for all time."* Since then the CTB has been inextricably linked with the NPT. The NPT Review Conferences of 1975 and 1980, for example, heavily criticized the nuclear powers for their feet-dragging in this matter. The "non-proliferation regime," Sigvard Eklund of the IAEA concluded at the 1980 meeting, "can only survive on the tripod of the Non-Proliferation Treaty, effective international safeguards, and a Comprehensive Nuclear Test Ban treaty. The vital third leg is still missing as it was 5 years ago."[10]

V. THE THRESHOLD TEST BAN TREATY

After 1968, the main focus in East-West arms control discussions shifted from a test ban to limits on offensive and defensive strategic systems. But following the successful negotiation and

_____

*Preambular language comes before the "have agreed" phrase of the treaty and thus does not represent a legal commitment.[9]

ratification of SALT I in 1972, these talks, too, ran into major roadblocks. In 1974, President Nixon and Soviet Secretary General Brezhnev faced the prospect of a summit meeting with no new strategic arms agreements ready for signing. High level exchanges led to the propoal to negotiate a threshold test ban treaty (TTBT), which would not require working out intrusive verification measures. Without this burden, and with pressure to meet the summit deadline, negotiators finished their work in five weeks, concluding a treaty prohibiting underground nuclear tests of more than 150 KT yield in time for signing at the July summit. The TTBT provided for verification of the yield limit by national technical means (NTMs), but also provided for the exchange of data to assist in the determination of yield, the first agreed step in bilateral arms control going beyond NTMs.

VI. THE PEACEFUL NUCLEAR EXPLOSIONS TREATY

The TTBT postponed consideration of the troublesome issue of nuclear explosions for peaceful purposes (PNEs). The United States originally had sought a complete ban on PNEs. The Soviets, however, who had an extensive PNE program, objected. Accordingly, the two sides agreed to take the subject up in a separate negotiation. The PNE negotiation began in October 1974 in Moscow and lasted until April 1976. The resulting treaty restricts the yield of individual explosions to less than 150 KT, but permits salvo or group explosions in which the aggregate yield is less than 1.5 MT under circumstances in which the individual yields can be determined to be no greater than 150 KT. The treaty and protocol have a technical content that far exceeds any other arms control agreement. They provide for extensive data exchange of all PNE events and on-site inspection for group explosions. Under specific and well-defined conditions, designated inspectors are permitted to bring equipment into the Soviet Union to determine the individual yield of explosions.

Many CTB proponents were immediately critical of the TTBT and PNET. They claimed that a threshold ban would divert attention from a CTB, and that the threshold was so high that it did not represent a real restraint on weapons development. The PNET, they argued, endorsed PNEs and thus could prove an obstacle to a CTB and provide an excuse for potential proliferators to label weapons development as peaceful. Finally, they criticized the OSI provisions because the circumstances were unlike those of OSIs under a CTB where the location of the inspection is fixed by the event. A provision for OSIs in this case could be used to strengthen the demand for OSIs in the context of a CTB which could prove to be a stumbling block as in the past.

Neither the TTBT nor the PNET have been ratified by the U.S. Senate. Washington and Moscow agreed to observe the yield restrictions, but the verification provisions have not gone into effect. These provisions nonetheless did break important ground. Many of the OSI procedures that negotiators painstakingly developed could carry over to a CTB and possibly could have parallel effect in other arms control treaties requiring on-site inspection. They represented, moreover, a significant evolution in Soviet thinking about cooperative and intrusive forms of verification.

VII. RENEWED NEGOTIATIONS

Promptly after taking office, the Carter administration directed an interagency study of the test ban and other arms control issues. In March 1977, at the ill-fated Moscow meeting in which the Soviets sharply rejected the administration's proposal to negotiate reductions in strategic arms, the United States also proposed to open CTB discussions. The Kremlin accepted. This was not surprising since the Soviets had at the time of the TTBT talks indicated a renewed interest in the CTB. In 1975, they presented a draft CTB treaty to the UN General Assembly. It provided for PNEs and verification by NTMs. In 1976, they amended the text to include the possibility of voluntary challenge inspections. Although worded in very general terms, the draft treaty offered additional evidence of a new Soviet approach to the verification question.

Full tripartite negotiations began in Geneva in the fall of 1977. As anticipated, PNEs and verification became the major points of controversy. With regard to the former, the United States and Britain wanted a ban on all nuclear explosions; the Soviets wanted to preserve the right to carry out PNEs. As one Defense Department official explained, the Western position rested on the premise that "PNE technology is indistinguishable from that required for weapons tests and that a state undertaking PNE's [sic.] inevitably derives weapons-related information."[11] On November 2, 1977, Brezhnev announced in a public speech that the Soviet Union would accept a moratorium on PNEs for the duration of a CTB. This outcome early in the negotiations on a question that many perceived would be a major obstacle seemed to indicate a genuine commitment on the Soviet side to achieve a CTB.

Although the United States had learned a great deal about the seismology of underground nuclear test detection since the signing of the LTBT in 1963 and had substantially increased its capabilities in this field, a detection and identification threshold still existed. Verification and OSIs, therefore, remained a central issue for U.S. negotiators in 1977. As noted earlier, the Soviets had accepted the idea of voluntary or challenge inspection the previous year. Briefly, a voluntary OSI works as follows. If a seismic event occurs in Country A's territory and the event appears ambiguous to Country B (and thus possibly an explosion), Country B can request permission to inspect the region surrounding the event. Country A then can either accept or reject the request. If rejected, Country B could choose to withdraw from the treaty.

The Swedes had first introduced the voluntary OSI concept at the ENDC in the 1960s. The Soviet Union had rejected it then, and the United States had roundly criticized the idea. In 1977/78, however, with the change in Soviet attitude, the U.S. government reassessed its position and concluded that challenge OSIs could have as much deterrent value (to cheating) as mandatory OSIs* if the rejection of a request could be established as a serious matter (in effect little different from thwarting a mandatory inspection) and if effective procedures could be worked out for the conduct of an OSI. This was a significant political step and, whether intended or not, a response to the Soviet concession on PNEs.

A separate verification issue taken up at the tripartite Geneva talks was that of setting up permanent national seismic stations within the boundaries of each negotiating party which would contribute data to the other parties for the purpose of verification. This had constituted an important element of the first CTB negotiations. In the U.S. view, these national stations would be automatic and tamper-detecting therefore not requiring foreign personnel in constant attendance (staffing of control posts had been a constant problem in the first CTB negotiations).** In their 1980 tripartite report to the Geneva Committee on Disarmament, it is evident that the three parties had made significant progress on implementing OSI procedures. But with regard to national seismic stations, the report simply notes that the parties "are negotiating" the issue.[12]

Even while detailed negotiations on the verification and other provisions were under way, the Washington bureaucracy that had gone along with the decision to open CTB discussions began to question the wisdom of concluding a treaty. The fundamental objection, based on assessments by the nuclear weapons laboratories and given bureaucratic clout by the Department of Energy (DOE) and Joint Chiefs of Staff (JCS), related to stockpile reliability. Under a permanent CTB, the laboratories maintained, confidence in the reliability of the nuclear weapons stockpile would inexorably erode without nuclear tests. This issue had simmered over the years*** and with the possibility of a CTB on the doorstep, it became central. Laboratory, DOE, and JCS representatives made an exposition of their concerns at congressional hearings on the CTB in 1978. In summary, they argued that testing would be required "in the long run" to protect against

---

*Mandatory does not mean unconditional. A mandatory OSI would be circumscribed by various criteria which would establish if and when an inspection could be carried out. Provisions for voluntary OSIs need not require precise criteria.

**The United States has developed high quality, automatic seismic stations and has installed five such stations in the United States and Canada; these are now operating. The data from each is sent via satellite to a control station. A group of British and Soviet scientists visited the U.S. development program in July 1979.

***Carl Walske, special assistant to the secretary of defense for atomic energy, seems to have been the first to introduce the issue officially in congressional testimony in 1971. Walske called stockpile reliability considerations at that time "sort of new."[13]

"stockpile aging" and to facilitate modifications of warheads for new military requirements.[14] Under a CTB regime, the technical expertise necessary to carry out these tasks would be lost both through attrition of skilled laboratory designers and engineers and the deterioration of skills on the part of those who stayed.

Three members of the nuclear weapons technical community -- Norris Bradbury and J. Carson Mark, formerly of Los Alamos, and Richard Garwin, a prominent general consultant on weapons technology -- supplied a dissenting opinion. In a letter written directly to President Carter at the time of the hearings, they argued that the issue of reliability could be met by foregoing changes in existing stockpiles and by remanufacturing or replacing those weapons in which defects appeared. Through these and other measures, the DOE and its contractors and laboratories could "provide continuing assurance for as long as may be desired of the operability of the nuclear weapons stockpile."[15]

CTB critics also challenged in 1978 the contention of some advocates that whatever stockpile degradation occurred would affect both sides equally. Because of U.S./Soviet "asymmetries," one could not be assured that the reliability of the Soviet weapons stockpile would decrease as the U.S. stockpile did. Moreover, according to DOE spokesman Donald Kerr (now director of Los Alamos), there would be "no guarantee that Soviet tests at the level we find necessary for maintaining long-term confidence in our nuclear weapons stockpile can be detected and identified with confidence."[16] Verification thus continued as a basic issue despite the technical progress that had been made over the years.

Although these internal doubts about a CTB continued to mount, the Carter administration remained committed to the negotiations primarily because of the political gains they offered in checking proliferation. Rear Admiral Thomas Davies of the Arms Control and Disarmament Agency (ACDA) and Leslie H. Gelb of the State Department outlined the government's position at the 1978 hearings. They identified three ways they believed a CTB would contribute to U.S. nonproliferation efforts: (1) it would directly bolster the NPT; (2) it would "inhibit testing by threshold states;" and (3) it would give the United States greater bargaining leverage in general "nuclear matters." "Of course, the CTB is not a panacea," Gelb concluded. "I cannot quantify . . . exactly how much a CTB would help our proliferation efforts. But I am confident that it will be of substantial benefit."[17] Opponents, however, claim that a CTB would have little effect on proliferation, arguing that a potential proliferator is motivated by its perceived political-military needs and not by the presence or absence of testing by the major powers.

The internal CTB opposition had its effect. In the spring of 1978, the administration changed its position from an indefinite duration treaty to a finite treaty of five or three years, with an option to resume tests after the accord lapsed. The DOE favored an explicit commitment to this option. The House subcommittee which conducted the 1978 hearings further recommended that any treaty provide for testing up to a 10 KT limit.

VIII. CTB AND SALT

While controversy thus stirred around a CTB in Washington's inner circles, Carter's principal arms control item, SALT II, also ran into unexpected troubles and delays. The administration did not need another contentious subject interjected into the already difficult strategic arms control debate. A CTB, officials decided, would have to wait until SALT II was safely out of the way. The Soviet view of the CTB/SALT II relationship is more difficult to discern. According to Carter, however, he and Gromyko met in September 1978 and "agreed to conclude the SALT II Treaty first, followed by a comprehensive test-ban agreement."[18] Moreover, there is no indication that the Soviets took any initiative to move matters along in Geneva. The CTB became hostage to SALT. SALT II problems and then the invasion of Afghanistan put to rest any immediate expectation of ratifying SALT and with it any expectation of successfully concluding the CTB negotiations.

IX. CONCLUSION

The negotiations recessed in November 1980, several weeks after the U.S. presidential election. They have not resumed. In July 1982, President Reagan tabled consideration of a CTB indefinitely. "It is fair to say," an administration spokesman reported, "that we won't go on with it until we get verification terms we can live with." According to this official, the first order of business was "to seek Soviet agreement to negotiate improved verification procedures which would make the Threshold Test Ban Treaty and the PNE Treaty adequate for submission for ratification."[19] Even earlier, in October 1981, ACDA Director Eugene Rostow had explained the administration's overall view on a nuclear test ban in negative terms to the First Committee of the UN General Assembly:

> Holding nuclear tests has been an issue before this Committee, the Committee on Disarmament and its predecesors for many years. High hopes have been attached to the proposal [for a test ban], and no one can question the goal it is designed eventually to achieve. Of course, the United States Government supports that long-term goal. But a test ban cannot of itself end the threat posed by nuclear weapons. We shall co-operate fully in appropriate procedures to examine the problems which the proposal presents. However, international conditions have not been propitious and are not now propitious to immediate action on this worthy project.

> As we consider the question of a nuclear test ban, we should keep in mind that, in order for such a ban ultimately to be effective, it must be verifiable and it must be concluded

under conditions which ensure that it will enhance, rather than diminish, international security and stability.[20]

The long history of the test ban debate and negotiations indicates that it is an issue which is not likely to fade away. But whether a time and formula can be found to bring a test ban about remains a matter for the future.

FOOTNOTES

1. Senate Committee on Foreign Relations, Hearings on Disarmament and Foreign Policy, 86th Cong., 1st sess. (1959), p. 178.

2. "Notes by Chairman Lewis L. Strauss, U.S. Atomic Energy Commission, before Subcommittee on Disarmament, U.S. Senate" (April 17, 1958), p. 7, E. O. Lawrence Papers, Bancroft Library, University of California, Berkeley.

3. U. S. Department of State, Documents on Disarmament, 1945-1959 (2 vols.; Washington, D. C., 1960), II, pp. 1590-91.

4. U. S. Department of State, Documents on Disarmament, 1960 (Washington, D. C., 1961), pp. 4-6.

5. Robert Neild and J.P. Ruina, "A Comprehensive Ban on Nuclear Testing," Science, 175 (January 14, 1972), p. 141.

6. Glenn T. Seaborg, Kennedy, Khrushchev, and the Test Ban (Berkeley, Ca.: University of California Press, 1981), p. 176.

7. Current Digest of the Soviet Press, 15 (January 20, 1963), p. 11.

8. ACDA, Arms Control and Disarmament Agreements: Texts and Histories of Negotiations (Washington, D. C., 1982), p. 93.

9. Ibid., p. 41.

10. ACDA, Office of Public Affairs, Special Report: 1980 Review Conference of the Treaty on the Non-Proliferation of Nuclear Weapons (November 1980), p. 10.

11. House Intelligence and Military Application of Nuclear Energy Subcommittee of the Committee on Armed Services, Effects of a Comprehensive Test Ban Treaty on United States National Security Interests, 95th Cong., 2d sess. (1978), p. 57.

12. "Tripartite Report to the Committee on Disarmament," CD/130 (July 30, 1980), p. 5. Also NPT Review Conference Document, NPT/CONF. II/13 (August 13, 1980).

13. Senate Subcommittee on Arms Control, International Law, and Organization of the Committee on Foreign Relations, Prospects for a Comprehensive Nuclear Test Ban Treaty, 92d Cong., 1st sess. (1971), p. 101.

14. House Committee on Armed Services, Effects of a Comprehensive Test Ban Treaty on U.S. National Security Interests, pp. 3-27, 175-76.

15. Norris E. Bradbury, Richard L. Garwin, and J. Carson Mark to President Jimmy Carter, August 15, 1978; released through Senator Edward Kennedy's office, August 17, 1978. Reproduced in ibid, pp. 181-82.

16. House Committee on Armed Services, Effects of a Comprehensive Test Ban Treaty on U. S. National Security Interests, p. 8.

17. Ibid., pp. 107-08.

18. Jimmy Carter, Keeping Faith: Memoirs of a President (New York: Bantam, 1982), p. 231.

19. Washington Times, July 21, 1982; New York Times, July 20, 21, 1982; The White House, Office of the Press Secretary, "Background Briefing on Test Ban Treaty" (July 20, 1982), p. 3.

20. UN General Assembly, First Committee, "Verbatim Record of the 6th Meeting," A/C.1/36/PV.6 (October 21, 1981), p. 31.

CHAPTER 12

ON LIMITING TECHNOLOGY BY NEGOTIATED AGREEMENT

Albert Carnesale
John F. Kennedy School of Government
Harvard University
Cambridge, MA

The substance of my remarks tonight will be far narrower in scope than the prescribed title of my talk would indicate. This reflects two considerations: first, this topical meeting is focused on technologies associated with nuclear weapons systems; and, second, President Reagan recently (i.e., on March 23, 1983) called for "a program to counter the awesome Soviet missile threat with measures that are defensive." In light of these considerations, I will concentrate tonight on the case of anti-ballistic missile (ABM) systems as an example of a continuing effort to limit technology by negotiated agreement.

Why limit ABM systems? After all, such systems are defensive in nature, not offensive. Defensive systems are intended to protect people and the things of value to them. It is the offensive systems that cause death and destruction. Why don't we just go ahead and deploy the best available ABM system, and develop and test even better systems for deployment in the future?

The ABM Treaty

There are several reasons for not simply charging ahead with ABM systems, most prominent among them being the ABM Treaty ratified by the United States and the Soviet Union in 1972. Some of the basic provisions of the Treaty appear in the following chart:

SOME PROVISIONS OF THE ABM TREATY

1. DEFINITION OF ABM SYSTEM

    ". . . an ABM system is a system to counter strategic ballistic missiles or their elements in flight trajectory, currently consisting of . . . ABM interceptor missiles, . . . ABM launchers, . . . and ABM radars . . ." (Article II)

2. NO NATIONWIDE DEFENSE

    "Each Party undertakes not to deploy ABM systems for a defense of the territory of its country and not to provide a base for such a defense . . ." (Article I)

0094-243X/83/1040230-09 $3.00 Copyright 1983 American Institute of Physics

3. WHAT'S NOT PERMITTED IS PROHIBITED

"Each Party undertakes not to deploy ABM systems or their components except that . . ." [one site with 100 ABM interceptor missiles, 100 ABM launchers, and specified ABM radars] (Article III and 1974 Protocol)

4. ONLY FIXED LAND-BASED ABM IS PERMITTED

"Each Party undertakes not to develop, test, or deploy ABM systems or components which are sea-based, air-based, space-based, or mobile land-based." (Article V)

5. NO DEPLOYMENT OF "EXOTIC" ABM SYSTEMS

"In order to insure fulfillment of the obligation not to deploy ABM systems and their components except as provided in Article III of the Treaty, the Parties agree that in the event ABM systems based on other physical principles and including components capable of substituting for ABM interceptor missiles, ABM launchers, or ABM radars are created in the future, specific limitations on such systems and their components would be subject to discussion in accordance with Article XIII [i.e., in the Standing Consultative Commission] and agreement in accordance with Article XIV [i.e., by amendment] of the Treaty." (Agreed Statement D)

6. NO RELABELING

" . . . each Party undertakes . . . not to give missiles, launchers, or radars, other than ABM interceptor missiles, ABM launchers, or ABM radars, capabilities to counter strategic ballistic missiles or their elements in flight trajectory . . . " (Article VI)

7. UNLIMITED DURATION

"This Treaty shall be of unlimited duration." (Article XV)

8. RIGHT OF WITHDRAWAL

"Each Party shall . . . have the right to withdraw from this Treaty. . . . It shall give notice of its decision to the other Party six months prior to withdrawal from the Treaty." (Article XV)

Embodied as the basic principle of the ABM Treaty is the undertaking by the Soviet Union and by the United States "not to deploy
ABM systems for the defense of the territory of its country and not
to provide a base for such a defense." An ABM system is defined as
"a system to counter strategic ballistic missiles or their elements
in flight trajectory, currently consisting of . . . ABM interceptor
missiles, . . . ABM launchers, . . . and . . . ABM radars."

Under the Treaty (as modified by a 1974 protocol), each side is
permitted one geographically, quantitatively, and qualitatively
constrained ABM deployment—either at its national capital or at an
ICBM deployment area. Technological innovation is inhibited by
provisions banning the development, testing, and deployment of "ABM
systems or components which are sea-based, air-based, space-based,
or mobile land-based" and of "ABM interceptor missiles for the
delivery by each interceptor missile of more than one independently
guided warhead." Moreover, the agreement provides that if exotic
new "ABM systems based on other physical principles and including
components capable of substituting for ABM interceptor missiles, ABM
launchers, or ABM radars are created in the future," their deployment would be prohibited unless specific limitation were agreed upon
through consultation and amendment of the Treaty.

The ABM Treaty is of unlimited duration. The two sides are
formally to review the agreement at five-year intervals, but the
Treaty remains in force unless one or both of the parties takes
positive action to terminate it. Either party can propose amendments at any time, and it can withdraw upon six-months notice "if it
decides that extraordinary events related to the subject matter of
this Treaty have jeopardized its supreme interests." The formal
review is in no sense a "window" for modification of or withdrawal
from the ABM Treaty.

BMD Dilemmas

The world is not as it was in 1972. America has recovered from
its "Vietnam syndrome" and detente is seen in retrospect as a dis-
appointment. The Soviet military threat has increased markedly, and
their large force of ICBMs equipped with multiple independently-
targetable reentry vehicles (MIRVs) now threatens or soon will
threaten the survivability of America's land-based retaliatory
forces. But BMD technology also has moved forward over the past
decade. The nuclear world has changed.

What should be the role of BMD in America's national security?
Would modification or termination of the ABM Treaty be in the interest of the United States? My primary objective here is to identify
and clarify the basic issues which must be addressed if reasoned
answers to these questions are to be provided. Each of these issues
constitutes a dilemma. Any resolution incurs both costs and benefits. There are no cost-free alternatives. We find such dilemmas
in the following areas: the nature of the BMD choice, technology,
economics, deterrence, nuclear warfighting, relations with allies,
and arms control.

The Nature of the BMD Choice. The ABM Treaty, with its severe constraints on BMD, represents a joint Soviet-American selection of a world in which neither side has a meaningful defense against the nuclear arsenal of the other. To modify or to terminate the Treaty is to modify or to terminate it for both sides, not just for the United States.

The benefits to the United States derived from the ABM Treaty are the constraints upon Soviet BMD. These constraints help to ensure the effectiveness of America's retaliatory forces by essentially precluding Soviet deployment of defensive barriers to the penetration of our ICBMs and SLBMs. By undermining the capability of United States retaliatory forces, Soviet BMD could pose a real threat to American security. Deployment by the Soviets of a capability to counter the nuclear warheads on an American missile nullifies the deterrent value of that missile. It is essential to keep in mind that modification or termination of the ABM Treaty would free the Soviets (and not just the Americans) from some or all of the constraints on BMD. A world in which the Soviet Union is constrained by the present Treaty while the United States is permitted significant BMD deployment is attractive to Americans, but in all likelihood is unattainable. The real choice is between a world in which both nations can have significant BMD deployments or a world in which neither nation can have them.

Technology. The lead times characteristic of strategic weapons systems are quite long. The decision to deploy BMD probably would precede full operational capability by about a decade. In a sense, then, one faces the following, somewhat loaded, question: would today's American BMD technology be able to deal with the Soviet threat ten years from now? Standard projections of Soviet offensive capabilities are inadequate to answer this question. It must be assumed that the threat is a reactive one; i.e., that the Soviets would develop tactics and deploy systems designed to counter the American defenses. One must also ask the mirror-image question: would America's offensive forces ten years from now be able to penetrate Soviet BMD (if the ABM Treaty were to be modified or terminated to permit one)?

Care must be taken not to rely upon either best-case or worst-case analyses. By making assumptions which are sufficiently defense-conservative, it can invariably be demonstrated that the BMD task for us is a hopeless one. On the other hand, unrealistically offense-conservative assumptions attribute to even the most Rube Goldbergian of conceivable Soviet defenses a near-perfect level of performance. Neither model should provide the basis for decisions by policymakers.

Economics. BMD, like all modern strategic weapons systems, is expensive. But that observation by itself argues neither for nor against BMD deployment. The relevant economic questions are these: how would the cost of the BMD system compare with the cost of offensive measures (e.g., additional reentry vehicles, maneuvering reentry vehicles, decoys, electronic jamming devices) required to

offset the BMD system? In other words, would the cost exchange
ratio favor the defense or the offense? How would the cost of the
BMD system compare with the costs of other means (e.g., improvements
and/or additions to our offensive forces) for providing the same
degree of enhancement of our nation's security? How would the costs
of the offensive measures required to offset a Soviet BMD system
compare with the benefits of a United States BMD system? Where
would the money for a new BMD system come from?

   Questions of this type call for analyses of trade-offs among
systems associated with different military services. Such questions
are more easily asked than answered, or even addressed. To whom can
we turn for objective and credible analyses of whether it is better
to spend money for new ICBMs for the Air Force, or new SLBMs for the
Navy, or a new BMD system for the Army, or a real rapid deployment
force for the Marines? For such comprehensive cost-benefit analyses
we must rely primarily upon the Office of the Secretary of Defense
and agencies of the United States Congress (e.g., the Office of
Technology Assessment and the General Accounting Office), recognizing that political factors will restrain freedom of both thought and
action.

   Deterrence. To be useful for deterrence, our strategic offensive forces should be able to survive an attack and subsequently
penetrate to designated targets in the U.S.S.R. Deployment of a BMD
system to defend our ICBMs would enhance their survivability, but
their ability to penetrate to their targets would be degraded by the
presence of a Soviet BMD system. Moreover, a Soviet BMD system also
would degrade the penetrability of our SLBMs and, indirectly, our
bombers and cruise missiles. (The penetrability of the air-
breathing delivery systems normally would be enhanced by precursor
attacks by ballistic missiles upon the adversary's air defenses.)

   The tension between survivability and penetrability is unavoidable. Its significance depends strongly on what it is that is
defended (or is perceived to be defended) on the other side. If a
signficant portion of the other side's territory is defended, deterrence could be jeopardized. If only strategic offensive forces are
defended, deterrence is unimpaired. Indeed, BMD deployments which
enhance the survivability of one's offensive forces without
impairing the ability of the other side's forces to deliver a
retaliatory blow would, in principle, enhance deterrence. Any
assessment of the effect of BMD on deterrence must include estimates
not only of the effects of BMD on ICBM survivability, but also of
the effect of the adversary's BMD on the penetrability of the
surviving offensive forces. Would BMD on both sides be advantageous
to the side that strikes first, or would it favor the retaliator?
The former situation would undermine deterrence; the latter would
strengthen it.

   Nuclear Warfighting. Thinking about fighting a nuclear war is
irresponsible to some, is anathema to others, and is an unpleas-
ant necessity to those responsible for the nation's security.
Maintenance of a nuclear warfighting capability can serve to deter

nuclear war by convincing potential adversaries that they could not achieve "victory" at any level of aggression. If deterrence should fail, the same capability could be employed actually to deny such "victory." (Victory in nuclear war is generally understood to have been achieved when the outcome is less unfavorable to your side than to the other side.)

As part of a warfighting capability, one seeks survivable and enduring forces which can be used to destroy elements of the adversary's strategic offensive forces, his other military capabilities (e.g., theater nuclear forces and general purpose forces), his political and military leadership and control, and his industrial and economic base. Ideally, the ability to attack the adversary's forces would be augmented by the ability to defend against those elements of his forces which survive. Defense of forces and of population both contribute to one's warfighting capability, while defense on the other side degrades one's warfighting capability. Thus, in warfighting, as in basic deterrence, the trade-off to be examined is between the benefits of United States BMD and the costs (to the United States) of Soviet BMD.

Relations with Allies. How would the deployment of a United States BMD system be viewed by America's allies? They might see reduced American vulnerability as enhancing extended deterrence by increasing our willingness to risk escalation of a war that might start in Europe, the Persian Gulf region, or elsewhere. On the other hand, they could view an American defense as a return to "Fortress America" and as reinforcing their concerns about the possibility of a nuclear war "limited" to their territory.

Our allies' views of expanded BMD deployments in the Soviet Union would be unequivocally negative. The missile forces of the United Kingdom and France now can penetrate with high confidence to virtually any target in the Soviet Union. If the Soviets were to deploy BMD at levels substantially higher than now permitted by the ABM Treaty, the deterrent value of the British and French independent strategic forces could be diminished markedly. And if it is a United States initiative which leads to modification or termination of the ABM Treaty, blame for the expansion of the Soviet BMD and corresponding diminution of the British and French deterrents would (rightfully) be attributed to us. (The utility of the People's Republic of China's nuclear deterrent vis-a-vis the Soviet Union also would be threatened by an expanded Soviet BMD deployment. Would that be in the U.S. interest?)

It is hard to see how modification or termination of the ABM Treaty to permit expanded BMD deployments in the U.S. and in the U.S.S.R. could have other than a negative effect on America's relations with its allies. How large or important the effect would be is not nearly so clear.

Arms Control. The ABM Treaty remains the principal enduring accomplishment of over a decade of Strategic Arms Limitation Talks (SALT). (The SALT I interim agreement on offensive arms expired in 1977, and the SALT II agreements have not been ratified by the

United States; however, both the U.S. and the U.S.S.R. continue to avoid irreversible actions inconsistent with these accords.) Termination of the Treaty probably would signal the end of the SALT process (or, as it is now known, the START process)--a process whose future is clouded in any event.

More difficult to assess are the likely effects on arms control of attempts, either successful or unsuccessful, to modify the Treaty. A frustrated attempt to amend the Treaty could be followed by a return to the conditions--legal and political--prevailing before the amendment attempt; or, at the other extreme, it could lead to withdrawal from the Treaty by the frustrated party. A successful amendment attempt might result in modifications to the Treaty far different than those originally proposed--a phenomenon not unheard of in international negotations. Modifications which permitted expanded, but still strictly limited and relatively unambiguous, deployments of BMD for defense of ICBM silos would be far less detrimental to the future of arms control than modifications which permitted extensive area defenses. And the nature of the survivability versus penetrability trade-off associated with the expanded BMD deployments on both sides would influence the prospects for control of strategic offensive weapons.

The Treaty restrictions on development and testing of BMD systems and components confront decisionmakers with another dilemma. It would be difficult for any president to decide to withdraw from the Treaty before he was confident that the U.S. had a workable BMD system ready for deployment. Do we have such a system now? Apparently not.

The only kind of BMD system that could be available for deployment within this decade has as its principle components radars, nuclear-tipped interceptor missiles, and elaborate data processing equipment, and is designed to destroy the incoming reentry vehicle (RV) in the terminal portion of its flight (i.e., at low altitude after the RV has reentered the atmosphere). In time, this so-called "terminal" or "endoatmospheric" defense could be augmented by another layer of defense. The "overlay" or "exoatmospheric" system is designed to intercept RVs much earlier in their trajectories (i.e., well before they reenter the atmosphere) and makes use of rocket-borne optical sensors and long-range missiles carrying multiple independently-guided non-nuclear devices for intercepting and destroying RVs in space.

But the endoatmospheric and the exoatmospheric systems are in the testing phase of development, with the endoatmospheric systems being considerably further along. It is my understanding that, for the next few years, there is no conflict between the technologically determined (rather than politically determined) testing programs and compliance with the ABM Treaty. After that, however, the Treaty provisions would begin to constrain the testing programs. For example, radars and interceptor missiles could not be built or tested in mobile or deceptive-basing modes. Nor could the exotic exoatmospheric system, especially the multiple independently-guided devices for killing RVs, be fully tested. Without the testing needed to acquire confidence in the systems, one would be hesitant to withdraw

from the Treaty; but, without withdrawing from (or modifying) the
Treaty, one cannot do the necessary testing. Unfortunately, I can
only describe this dilemma. I do not yet know how to resolve it.

In light of all this, it is my judgment that the U.S. should
not seek to modify or terminate the Treaty, especially since the
Treaty does not yet impose any important constraints on our BMD
programs. It appears that that situation will prevail for at least
a few years. We should use that time to conduct intensive research
and development on BMD and to think hard about the dilemmas I have
posed. This is essential if we are to understand better the
implications of alternative future actions.

"Star Wars"

Let us turn now to President Reagan's remarks of March 23--the
so-called "Star Wars" speech. It was in that speech that the
President called for "a program to counter the awesome Soviet
missile threat with measures that are defensive." He expressed hope
that we might reach a state in which "we could intercept and destroy
strategic ballistic missiles before they reached our soil or that of
our allies." The President went on to say:

> I know that this is a formidable technical task, one that
> may not be accomplished before the end of this century.
> . . . It will take years, probably decades, of effort on
> many fronts.

Then came the substance:

> Tonight, consistent with our obligations under the ABM
> Treaty and recognizing the need for close consultation
> with our allies, I am taking an important first step. I
> am directing a comprehensive and intensive effort to
> define a long-term research and development program to
> begin to achieve our ultimate goal of eliminating the
> threat posed by strategic nuclear missiles.

Wasn't there more substance than that? No. Didn't the President
call for something more dramatic than an "effort to define a long-
term research and development program"? No. Didn't he say anything
about space stations, or particle beams, or lasers? No. The rhetoric may have been bold, but the proposal was really quite modest.

Would the effort called for by the President be in conflict
with the ABM Treaty? No, because he called for no more than a
definition of an R&D program. Would implementation of the program
be in conflict with the Treaty? Probably yes. Research itself is
not constrained by the Treaty, but development is. Of particular
relevance is the undertaking (in Article V) "not to develop, test,
or deploy ABM systems or components which are sea-based, air-based,
space-based, or mobile land-based." The meaning of the restrictions
on development were explained (in testimony before the Senate Armed

Services Committee on June 28, 1972) by Ambassador Gerard Smith, chief U.S. negotiator of the ABM Treaty:

> The prohibitions on development contained in the ABM Treaty would start at that part of the development process where field testing is initiated on either a prototype or breadboard model. It was understood by both sides that the prohibition on "development" applies to activities involved after a component moves from the laboratory development and testing stage to the field testing stage, wherever performed.

These restrictions on development were intended primarily to enhance verification of the limitations on testing: if, through our national technical means of verification, including reconnaissance satellites, we "see" in the field an ABM component of a type anything other than fixed land-based, that is evidence of a violation of the Treaty, even though we may not have seen any test being conducted.

On Limiting Technology

The ABM Treaty demonstrates that (at least one kind of) technology can be limited by negotiated agreement. The Treaty works: neither party to it could deploy, or even prepare to deploy, an effective defense against ballistic missiles. The earlier portion of these remarks makes clear my view that, for the foreseeable future, BMD technology should be constrained. But not every technology that <u>could</u> be controlled <u>should</u> be controlled, nor vice versa. Implementation of some technologies can reduce the likelihood of war. For example, reconnaissance satellites reduce uncertainties about the capabilities of potential adversaries; permissive action links on nuclear weapons reduce the chances of accident or unauthorized use; and ballistic missile submarines reduce the incentives to launch pre-emptive strikes in times of crisis.

There is a tendency to blame the nuclear arms race on "runaway technology." This is a transparent attempt to shift responsibility for our problems from ourselves to the products of our ingenuity. But technology has no mind or character of its own. It can be used, by us, for good or for evil. The choice is ours to make, and the responsibility is ours to bear.

CONGRESS AND NATIONAL SECURITY

PETER SHARFMAN*
OFFICE OF TECHNOLOGY ASSESSMENT

The starting point for any serious discussion of Congress'
role in matters of national security is the recognition that
Congress does some kinds of things very effectively, but generally
fails when it tries to do other kinds of things. Consequently, a
citizen with a desire to shape national policy may find Congress
to be the focal point of national decision, or largely irrelevant,
depending almost entirely on the nature of the issue. As a
political scientist, I am tempted to relate this to the provisions
of the U.S. Constitution and to the differing structures of the
Executive and Legislative institutions; since I am addressing an
audience of physicists, I will confine my explanations of causes
to the observation that you cannot easily push on a string.

This paper advances and elaborates on two pairs of
propositions:

1. Congress can be very effective in influencing or even
determining what weapons will be procured, but is largely
ineffective in attempting to influence the military strategies
that underly actual deployments or use of weaponry.

2. Congress is generally ineffective in exerting pressure in
favor of any particular kind of arms control agreement -- or even
for any arms control agreement at all -- but reasonably effective
in making particular agreements which it dislikes impossible.

I shall illustrate my propositions regarding weapons and
strategy by discussing the case of the MX missile, which we at OTA
have studied in some depth.[1] The history of this weapon has
displayed continuing controversy over two issues -- issues that
have been much too important to ignore, and much too difficult to
resolve to anybody's complete satisfaction. The first issue has
been the question of survivable basing: how should this very
large and capable intercontinental ballistic missile (ICBM) be
stored and deployed so that the Soviets could not destroy our MXs

---

* Peter Sharfman is Program Manager for International Security
and Commerce at the Office of Technology Assessment, U.S.
Congress. The views expressed herein are his own, and do not
reflect the views of the Office of Technology Assessment or the
Technology Assessment Board.

[1] U.S. Congress, Office of Technology Assessment, MX Missile Basing,
Washington, D.C.: Government Printing Office, 1981.

0094-243X/83/1040239-07 $3.00 American Institute of Physics

in a first strike? There has been continuing strong Congressional
influence over this issue for at least seven years, and as I write
the issue has just been returned to Congress for what the Air
Force and the President hope will be a final resolution. The
second issue has been the question of how this unprecedentedly
accurate ICBM should be used in deterrence or in warfare -- in
other words, the question of why the U.S. needs the MX in the
first place. This is an issue on which Congress has been largely
silent, and efforts made by individual Members to influence this
matter have been essentially ineffective.

Consider the following passage from a magazine article:

> In debate so far, few congressmen have quarreled
> over evidence that supports the need for a Minuteman
> replacement. But they do criticize the Air Force's plan
> to place its early production versions of Missile X in
> present Minuteman silos and to defer a decision on
> whether to make the new missile air-mobile or land-
> mobile until later in the 1980's.
>
> "The Air Force is trying to tell us that our silos
> will be safe enough from Soviet attack for the next 10
> years, but it knows better," says one key congressional
> source. "It is taking that line because it wants to
> hurry the new missile into production sooner than it can
> hope to finish research" [on basing].[2]

This passage appeared in <u>Business Week</u> of October 25, <u>1976</u>. In
the spring of 1983, the Air Force is still trying, and Congress is
still resisting. And in the intervening 6-1/2 years, clearly the
Air Force has been doing its best to do Congress' bidding.
Equally clearly, at no time in this 6-1/2 years has Congress had a
serious debate over the question of whether the MX is needed in
the first place.[3]

Why, if Congress is so powerful on procurement matters, and
if Congress has consistently resisted silo-basing (that is, basing
in a way that would be vulnerable to an attack by the most
accurate Soviet ICBMs), is the Administration proposing yet again
to proceed with MX in Minuteman silos? The answer is that
Congress has been demanding something which is apparently

[2] <u>Business Week</u>, October 25, 1976, p. 63.

[3] In this connection, it may be of interest that the request for
OTA's MX study (note 1 above) explicitly <u>excluded</u> any
consideration of whether or why the MX itself was needed.

impossible -- a basing mode for the MX that is fully survivable
and which meets the Air Force's criteria of military effectiveness
as well. A basing mode which provides survivability, instant
readiness, land-basing, tolerable dollar costs, and availability
by 1986 (or anyway 1987) does not seem to exist. Congress can
hardly legislate away the laws of physics or the facts of
engineering. But it is a striking fact that Congress has accepted
at face value the Air Force criteria. Only a tiny minority of
Congress has challenged, or even seriously discussed, the military
strategy from which these criteria are derived.

Does this mean that the Members of Congress generally accept
and approve the evolution in U.S. strategic nuclear strategy that
has occurred over the last ten years? Do Members of Congress
believe that their constituents accept and approve the change?
Not at all. What it means is that the Members of Congress
understand very well that the best means at their disposal to
influence strategy is control of weapons procurement, while their
constituents are so horrified by the very idea of nuclear war that
they have little interest in issues of how one should be fought.

Imagine a Congressional debate over strategy. It might
reasonably revolve around two questions: (1) Is it true that the
best way to deter a Soviet attack in Western Europe (or elsewhere)
is to be visibly able, and possibly willing, to start a limited
strategic nuclear war in response? (2) Is it true that the best
way to deter the Soviets from starting a limited strategic nuclear
war is to be visibly able, and possibly willing, to fight one?
For some years the Executive Branch answer to both questions has
been "yes," although Harold Brown as Secretary of Defense added
the nuance that a threat of limited strategic nuclear war was more
fearsome (and hence more effective as a deterrent) because the
limited nuclear war would most likely escalate to a massive
nuclear war. And for an equal number of years many Members of
Congress have had their doubts. But the general feeling (which
this author would agree with) has been that it is quite unlikely
that a Congressional debate could resolve the issue, while an
unresolved debate (which would surely become a national issue)
would tend to undermine whatever effectiveness our deterrent and
our alliance system have. For the most part Congress has been
quick to refute anything that sounded like a claim that a limited
strategic nuclear war could be tolerable or acceptable or (by any
definition) successful, and has left it at that.

Instead, Congress has acted through the procurement
process. A vulnerable MX would be a valuable asset in a limited
strategic nuclear war, but a highly questionable one in the role
of retaliating against a massive attack. Congress has repeatedly
made it clear in the past that it considered a vulnerable MX to be
somewhere between a menace to U.S. security and a waste of
money. If Congress in the coming debate over the recommendations
of the Scowcroft Commission holds to its past position, it will

kill the MX program. In so doing, it will oblige the Executive to
shift U.S. strategy back in the direction of Mutual Assured
Destruction. If Congress accepts the proposal to base MX in
Minuteman silos, it will in effect be acquiescing in, and perhaps
somewhat reinforcing, the shift to a strategy of flexible
strategic nuclear response. Either way, Congress will have fought
the battle out on its own ground: as a procurement issue rather
than as a strategy issue.

Turning now to arms control, I restate my proposition that
Congress can be very effective in preventing agreements which it
dislikes, but has little or no ability to promote or encourage
agreements which it desires. The example of prevention which
comes most readily to mind is the question of a nuclear test ban
treaty more comprehensive than the one which was signed and
ratified in 1963. More than one President has been conscious of
the advantages of a full ban on nuclear testing -- though to be
sure we have also had Presidents since 1963 who have quite
genuinely believed that such a treaty would be disadvantageous for
the United States. But no President has felt that Congress would
ratify such a treaty in the teeth of opposition from the military
and from the weapons R&D establishment. Consequently, none has
been negotiated. On the other hand, a so-called threshhold test
ban treaty was negotiated -- and a skeptical Senate (or rather
several successive skeptical Senates) has not even brought its
ratification to a vote.

Another example is Congress' insistence on a quite stringent
standard of verifiability. In its most extreme version, this
viewpoint appears to believe that the Soviets would gain a
significant advantage from the mere fact of cheating and getting
away with it -- hence, the possibility of cheating is itself
sufficient motivation to cheat, and sufficient reason for the U.S.
to reject a possible agreement. An alternative view would be that
the Soviets would weigh the risks of detection against the
concrete advantages obtainable from undetected cheating, and would
cheat only if the military benefits promised to be quite
substantial. In this view, the importance of high confidence
verification varies with the military importance of the particular
provision of the agreement whose observance is being verified; in
other words, we need high confidence of detecting violations of
great military significance, and only low confidence of detecting
violations of low military significance. This latter view has
never prevailed in the United States government, primarily because
it was clear that a vocal minority in Congress would insist on the
need for high confidence verification of every detail of every
treaty provision, and the majority in Congress would be unwilling
to resist this pressure.

In contrast, consider the fate of Congressional efforts to
promote particular arms control objectives. The classic case is
the fate of the so-called Jackson Amendment, which was attached to

the Senate action to ratify the SALT I Agreements in 1972. This amendment was eventually supported by the Nixon Administration, and then adopted by an overwhelming margin: the resulting passage of the law states that "the Congress...requests the President to seek a future treaty that, inter alia, would not limit the United States to levels of intercontinental strategic forces inferior to the limits provided for the Soviet Union."[4] Now what this means is not altogether clear, but we know that Senator Jackson and many of his colleagues were and remain worried about the Soviet advantage over the United States in the size (and therefore, once technological parity is attained, the capability) of their ICBMs. Senator Jackson, like the rest of the United States, would be very pleased if the Soviets would dismantle and not replace their largest missiles, which in 1972 were SS-9s and have since been replaced by SS-18s. If that is not going to happen, it is not necessarily sensible for the United States to design, build, and deploy an ICBM of comparable size. However, the Jackson Amendment plainly sought future arms control agreements which would give the United States the right to build ICBMs as large as the SS-9 or SS-18, and to deploy as many of these monsters as the Soviets were allowed. This was a "request" of the Congress, but it was as noted above accepted by the Nixon Administration. However, the subsequent course of negotiations resulted in two agreements -- the abortive Vladivostok Accords and the signed but unratified SALT II Treaty -- which allowed the Soviets but not the United States to deploy so-called "heavy" ICBMs. Why? As far as one can determine, successive administrations have concluded that there is no practical possibility that the Soviets would agree to a treaty which obliged them to dismantle their most formidable weapons, and have also concluded that it would be unwise for the United States to build anything comparable. Therefore, the most that could be obtained would be purely symbolic: if the Soviets had the right under the treaty to deploy 308 heavy ICBMs, the U.S. would have a similar right even though we had no intention of exercising that right. It would seem that neither the Nixon, Ford, nor the Carter Administration was willing to bargain, to give up some substantive point, for this symbolic equality. In the face of Presidential judgments that Jackson's arms control objective was not in the national interest, the vote of the Senate in 1972 counted for little.

A similar fate no doubt awaits any so-called "freeze" resolution which may pass the Congress. Regardless of what the resolution may say, and regardless of how great the vote in favor, it is most unlikely that the Administration will negotiate any agreement that it does not believe is good for the United States. Indeed, the debates in the current Congress over a nuclear weapons freeze and over the staffing of the Arms Control

[4] Public Law 92-448, Section 3.

and Disarmament Agency are more nearly events in the 1984
Presidential campaign than they are a process of deciding critical
national policies. That is why the Administration takes them so
seriously, even though they have no direct effect on the
President's programs or policies. (Congress could, of course,
impose a unilateral freeze -- but that would be weapons
procurement policy as a means of affecting strategy; it would not
be arms control as we generally understand it.)

It is not easy for some Members of Congress to accept this,
and every year or two one or more Members seek to launch an arms
control proposal. Indeed, if one believes that arms control is at
once the most difficult and perhaps the most important item on the
national agenda for the coming years, it is frustrating to face
the respects in which arms control differs from virtually every
other major national issue. In most national issues, Congress and
the Administration hammer out a policy that both can accept,
though it is not usually anybody's first choice. In the area of
arms control, unless the Administration and the Congress see eye-
to-eye at the outset, nothing happens. Once a treaty is
negotiated, Congress can really only choose to ratify or not.
(This is because it is generally quite unlikely that the other
party or parties would accept reservations or amendments attached
by the U.S. Congress after negotiations were supposedly
concluded.) It is extremely difficult to find an opportunity for
bargaining and compromise between legislative and Executive
positions. (Between the delivery of this paper and the printer's
deadline, there was a serious effort at bargaining; the President
modified his START negotiating position in exchange for critical
Congressional votes on the MX. It remains to be seen whether this
bargain will have any real effect on the course of
negotiations.) There is a second difference. On most national
issues, Congress has every opportunity to initiate specific
provisions, though comprehensive programs are generally shaped by
the Administration. On matters of arms control, specific
provision must be negotiated with one or more foreign countries,
and the Congress frequently is not privy to detailed and up-to-
date information about these other countries' positions. There
really is no way to advance a bright idea regarding the terms of a
prospective arms control treaty except to get a key job in the
Executive Branch, or to persuade (not pressure, but genuinely
persuade) one of a handful of Executive Branch officials of the
merits of the idea. In the entire United States, there are today
probably fewer than a dozen individuals who are in a position to
affect directly the U.S. position in any arms control negotiation;
the rest of us, whether in or out of the Government, can be
effective only if we persuade one of that handful to change his
own preferences about what to do. This is surely one of the
reasons why arms control is so difficult, and why so many years of
effort have produced so few agreements of real significance.

Does this mean that citizens who are interested in national

security issues other than weapons procurement should ignore the
Congress? Not at all. On all national security issues, including
the selection of military strategies and the negotiation of arms
control agreements, Congress can make it easier or more difficult
for any Administration to do what it wants. Even though Congress
may not be able to influence military strategy or the evolution of
arms control negotiating positions directly, it can slow or speed
their implementation, and can encourage or discourage a President
who must conserve his political capital. Put differently,
Congress may not be able to change certain national security
policies, but Congressional support might make the difference to a
President who was pursuing an arms control agreement that was
attainable, but just barely.

Moreover, it is generally the Congress rather than the
Executive that sets the level of a debate. It is the Congress
that provides the most subtle and up-to-date assessment of public
opinion, and in this way sets and then alters the bounds of what
is politically possible -- that is, what the American people will
or won't accept. On both the issues of MX basing and strategic
arms control, it has been the Congress that has posed the key
issues for debate. Perhaps in the end a national debate that
actually focuses on the real issues and their real significance is
the achievement we should expect and demand from the Congress, and
a national debate that never takes place or that misses the point
is the Congressional failure we should fear.

CHAPTER 14

STRATEGIC ARMS LIMITATION

G. Allen Greb
University of California, San Diego
La Jolla, California 92093

Gerald W. Johnson
TRW Corporation
Redondo Beach, Ca. 90278

ABSTRACT

Following World War II, American scientists and politicians pro-posed in the Baruch plan a radical solution to the problem of nuclear weapons: to eliminate them forever under the auspices of an interna-tional nuclear development authority. The Soviets, who as yet did not possess the bomb, rejected this plan. Another approach suggested by Secretary of War Henry Stimson to negotiate directly with the Soviet Union was not accepted by the American leadership. These ini-tial arms limitation failures both reflected and exacerbated the hos-tile political relationship of the superpowers in the 1950s and 1960s.

Since 1969, the more modest focus of the Soviet-American arms control process has been on limiting the numbers and sizes of both defensive and offensive strategic systems. The format for this effort has been the Strategic Arms Limitations Talks (SALT) and more recently the Strategic Arms Reduction Talks (START). Both sides came to these negotiations convinced that nuclear arsenals had grown so large that some form of mutual restraint was needed.

Although the SALT/START process has been slow and ponderous, it has produced several concrete agreements and collateral benefits. The 1972 ABM Treaty restricts the deployment of ballistic missile defense systems, the 1972 Interim Agreement places a quantitative freeze on each side's land based and sea based strategic launchers, and the as yet unratified 1979 SALT II Treaty sets numerical limits on all offensive strategic systems and sublimits on MIRVed systems. Collateral benefits include improved verification procedures, working definitions and counting rules, and permanent bureaucratic apparatus which enhance stability and increase the chances for achieving addi-tional agreements.

0094-243X/1040246-16 $3.00 Copyright 1983 American Institute of Physics

I. INTRODUCTION

SALT I, the Vladivostok accord, SALT II, START -- -- more has been said and written about these superpower forays into the world of arms control than any other set of agreements or negotiations of the nuclear era. Included in this vast literature are several book-length accounts by participants, volume upon volume of Congressional testimony and exhaustive analyses by government experts and scholars. Yet while they are familiar with the acronyms, most Americans (and for very different reasons nearly all Soviet citizens) would be hard pressed to discuss the substance of these efforts to limit strategic nuclear arms or what they mean in the context of U.S. and international security.

In the following pages, we hope to help demystify this historic set of negotiations. How and why did the talks begin? How has the negotiating style and apparatus of each side evolved? What have been the major agreements concluded? What do they contain? What do they mean for each side? What issues do they leave unresolved? What lessons do they teach us for the present and the future of arms control?

II. BARUCH AND STIMSON PLANS

Although the Strategic Arms Limitation Talks (SALT) constitute the most important, most intense, and longest running attempt to curb the arms race, they do not stand alone. In fact, SALT would not have come into existence at all if its most famous antecedent, the so-called Baruch plan had been implemented. On June 14, 1946, at the very outset of the nuclear age, the United States introduced this plan to ban forever all atomic weapons and place nuclear technology under international control. The men behind the plan were a mixed bag of scientific and government experts: the atomic scientist Robert Oppenheimer, who originally conceived it; U.S. Atomic Energy Chairman David Lilienthal and Secretary of State Dean Acheson, who substantially modified it; and the financier Bernard Baruch, who presented it before the United Nations. Specifically, the proposal called for the creation of an international nuclear development authority with the power to control all atomic energy activities, including the establishment of inspection procedures not subject to UN Security Council veto. Once this agency was established, the United States promised to dismantle its nuclear bombs. The Soviet reply, presented by Deputy Foreign Minister Andrei Gromyko, came only five days later. Because America enjoyed a nuclear monopoly, the Soviets called for the destruction of existing atomic arsenals before any discussion of inspection and control procedures could take place. This fundamental difference blocked all progress toward any form of arms control or disarmament for a dozen years.

The roots of this Soviet-American failure went beyond disagreement on the specifics of the Baruch plan. Any agreement on atomic energy was highly unlikely at this juncture because of the growing tension between East and West. The authors of the plan, in particular Oppenheimer, were well aware of this hostile political climate. They believed however, that the uniqueness and magnitude of the problem demanded extraordinary measures. As they put it:

> The program we propose will undoubtedly arouse skepticism when it is first considered. It did among us, but thought and discussion have converted us.

> It may seem too idealistic. It seems time we endeavor to bring some of our expressed ideals into being.

> It may seem too radical, too advanced, too much beyond human experience. All these terms apply with peculiar fitness to the atomic bomb.

> In considering the plan ... one should ask oneself, "What are the alternatives"? We have, and we find no tolerable answer.[1]

These words proved all too prophetic. As many anticipated, the Soviet Union soon acquired the bomb, the Cold War deepened and intensified, and the nuclear arsenals of both sides grew in both numbers and sophistication. Only after two stormy decades were the superpowers again able to directly address the issue of arms limitation, if only partially, through the medium of SALT.

An early approach to negotiation proposed by Secretary of War Henry Stimson to President Harry Truman and the Cabinet one month after the nuclear attacks on Hiroshima and Nagasaki has recently come to light. Although Truman did not follow Stimson's suggestion, it did outline a bold and possibly attractive alternative to the Baruch plan. Stimson suggested that the United States in concert with the United Kingdom enter into direct discussions with the Soviet Union, warning that a multinational approach could create the impression of "ganging up" on the Soviets. He noted that the atomic bomb was a uniquely dangerous weapon that must be brought under control, and that traditional diplomatic procedures might not suffice to meet the challenge. In his view, when the Soviets acquired the bomb -- whether it be four or twenty years -- was less important than the superpowers having a peaceful, cooperative relationship when they did.

III. INITIAL NEGOTIATIONS

In November 1969, the United States and Soviet Union in essence adopted such an approach, formally initiating bilateral negotiations on strategic nuclear arms. This followed a number of abortive overtures and initiatives made during the 1950s and 1960s, one by Soviet Premier Nikolai Bulganin in 1955 and several by U.S. President Lyndon

B. Johnson -- in 1964 for a "verified freeze" on strategic weapons, in 1966 for a limit on antiballistic missile (ABM) deployment and again in 1967 to restrict both offensive and defensive systems. Finally, in mid-1968, each side publicly committed itself to the negotiating table. But as so often has happened in the history of SALT, an unrelated political action -- the Soviet invasion of Czechoslovakia -- rudely interrupted the course of events. Then in November, the Washington elections occasioned even more delay. After extensive review, the new President, Richard Nixon, and his special assistant, Henry Kissinger, approved the SALT idea and the United States and Soviet Union set November 17, 1969 as the official date to open talks.

The immediate question that comes to mind is: why? What put arms limitation on the political and diplomatic agendas of each side at this particular point in time? Analysts generally agree that one point stands above all others: stockpiles of nuclear weapons and the means to deliver them (the so-called strategic nuclear delivery vehicles, or SNDVs) had reached such levels by 1968/69 that Soviet and American leaders independently concluded that their national security interests would best be served by some form of mutual arms restraint. A nuclear-maturing Washington and Moscow cited both the risks of nuclear war and the overburdening costs of carrying on an unbridled strategic competition.

Of course, each side had its own unique variations and perspectives on this common theme. For the United States, arms limitation seemed more attractive because of a fundamental change in the global strategic environment.   For the first time, American nuclear dominance was clearly and irrevocably at an end, the product of a determined Soviet buildup following the Cuban missile crisis of 1962. Each side now possessed (or more accurately perceived each other to possess) a nuclear retaliatory force that could inflict unacceptable damage after a first strike. This tenuous balance -- what U.S. defense analysts call a situation of mutual deterrence -- seemingly offered the promise of new possibilities in the arms control field.

The individual primarily responsible for rethinking and retooling U.S. strategic goals in light of this new international reality was Johnson's defense secretary, Robert S. McNamara.   Backed by an army of "systems analysts," Secretary McNamara persuasively demonstrated to the President and (perhaps more important) to the Joint Chiefs of Staff the absurdity of engaging in a strategic arms competition that no one could win. He also successfully conveyed his concerns about ABMs. This new and developing technology, he argued, offered no sure guarantees of safety and had the potential to completely destabilize the deterrent system by acting as a stimulant to offensive countermeasures. ABM curbs, therefore, became the prime target of the U.S. negotiating position.

For the Soviet Union, catching up with the United States in nuclear weaponry also made negotiations attractive but for different reasons.   The Soviets had always been haunted by feelings of inferiority with their capitalist adversaries. The leaders who took over in the wake of the 1962 Cuban fiasco, Leonid Brezhnev and Alexei Kosygin, were, in the words of several prominent Kremlinologists, particularly "obsessed by the notion of inequality."[2] For such leaders, SALT offered above all the promise of confirming their country's status as a legitimate world and superpower. The Brezhnev/Kosygin regime also saw the opportunity to check the U.S. technological lead in new weapons systems, such as ABMs (even though the Soviets had been first to deploy these) and multiple independently-targeted reentry vehicles or MIRVs. U. S. advances in multiple warheads in particular threatened to undermine the hard-fought Soviet gains in SNDV strength.

Beyond these strategic-defense considerations, domestic constraints played a major role in both the U.S. and Soviet decisions to negotiate. In the United States, popular negative reaction to the Vietnam War and a growing public and Congressional interest in national security matters produced demands for the scaling down of defense budgets and programs. Some officials even worried that Congress would legislate unilateral restraints on strategic forces in the absence of mutual agreements. Similarly, discussion and negotiation become an important way for the Soviet supreme leadership to avoid costly global confrontations while consolidating power at home and within the Soviet sphere. Specifically, the deepening Sino-Soviet breach (opened in the late 1950s and fully confirmed after 1964), persistent nationalism in Eastern European bloc countries, and the allure of trade and credits with the West caused the new Kremlin regime to initiate what Brezhnev and Kosygin termed an "opening to the West."

Finally, the desire to build upon a series of secondary but important arms control agreements achieved during the previous decade -- the Antarctic Treaty, the Hot Line agreement, the Limited Test Ban Treaty, the Outer Space Treaty, and the Nonproliferation Treaty (NPT) -- helped spur the superpowers on. Article VI of the 1968 NPT, for example, specifically committed the two giants "to pursue negotiations in good faith on effective measures relating to cessation of the nuclear arms race."[3] In this sense, all arms control negotiations are interrelated, each successful agreement setting the stage for others and providing building blocks toward a more stable international community.

Thus, with both the strategic forces and psychology of the two leaderships "in phase," as the chief U.S. negotiator, Gerard Smith, put it, SALT began. Neither side, however, entered into these initial SALT negotiations with a great deal of optimism. "Although overall political relations had mellowed slightly in the 1960s," Ambassador Smith remembers, "American and Soviet SALT negotiators were under no illusions that they were involved in anything but an adversary relationship, representing two superpowers tentatively searching for a less risky and costly way to maintain the balance of nuclear terror."[4] Nevertheless, after two and a half years of hard bargaining involving both formal negotiations and important behind-the-scenes maneuvering between Henry Kissinger and Soviet Ambassador A. F. Dobrynin, Washington and Moscow forged two sets of accords, collectively known as SALT I.

IV. SALT I

The 1972 ABM Treaty (and its companion 1974 Protocol) severely restricts ballistic missile defense systems, allowing each side only one deployment site and placing qualitative limits on ABM technology. Described typically as the "most significant" or "most impressive" result of SALT, one expert writes that the ABM Treaty is "a classic case of both sides agreeing not to build something that neither side wanted but that each would have probably ended up building in the absence of an agreement."[5] Of unlimited duration, the pact still comes up for review every five years. Each side reaffirmed its support for the Treaty at the 1982 review, but ballistic missile defense is again being considered in light of the theoretical vulnerability of the U.S. land-based missile force.

The SALT I Interim Agreement on Offensive Arms essentially imposes a five-year quantitative freeze on the superpowers' intercontinental ballistic missles (ICBMs) and submarine launched ballistic missiles (SLBMs) -- the first such set of actual limits on strategic hardware ever achieved. Much more controversial and difficult to negotiate than the ABM Treaty, the Interim Agreement is best viewed as a series of compromises or "tradeoffs" between adversaries. On the one hand, the agreement left the Soviets with several hundred more launchers than the United States -- 2347 to 1710. Within this total or aggregate, moreover, Russia was permitted to retain about 300 "modern large" or "heavy" ICBMs while the United States had none. The U.S. research and development community much earlier had turned away from large missiles in favor of a smaller, more accurate, more efficient rocket (the Minuteman). But this discrepancy worried many Pentagon planners all the same.

On the other hand, the Soviets could point to a number of important compromises which in many ways compensated for the aggregate force level asymmetries. First, it was significant that the Soviets accepted any restrictions on their offensive missiles at all. The Kremlin had resisted including such forces in the negotiations for a full year and a half, reflecting internal pressure from a strong military establishment. Second and more important, the government agreed to defer consideration of what it saw as a number of key issues: U.S. technological superiority; U.S. forward-based systems (FBS) capable of delivering a nuclear attack on Soviet territory from outside the United States; and the U.S. approximately three-to-one advantage in long-range strategic bombers.

Despite not being fully satisfied by its terms, both sides recognized that the Interim Agreement was just that: a temporary measure with the potential to produce more meaningful limits on strategic forces in the future. In addition, both governments welcomed it and the ABM Treaty as highly significant political achievements. During the course of the negotiations, in fact, SALT acquired an institutional life of its own that has fundamentally altered the superpower political relationship. As one first-hand observer describes this SALT "process" or "experience":

> Soviet and American officials sat across from each other at long tables, sipped mineral water and discussed military matters that used to be the stuff that spies were paid and shot for.... The process was the product. There emerged SALT bureaucrats in Washington and SALT apparatchiks in Moscow. SALT became a career in the civil service. The process acquired an institutional mass ... that served as a kind of deep-water anchor in Soviet-American relations.[6]

The American public and Congress also greeted SALT I with a high degree of initial enthusiasm, reflected most vividly in the Senate ratification vote of 88-2. Then Congressman Gerald Ford perhaps best summarized the national mood. "What it all comes down to is this," Ford told a middle-America audience in June 1972. "We did not give anything away, and we slowed the Soviet momentum in the nuclear arms race."[7] Support was not unconditional, however. Congress overwhelmingly approved the SALT I accords in September but also added an amendment sponsored by Senator Henry Jackson requesting "the President to seek a future treaty that ... would not limit the United States to levels of intercontinental strategic forces inferior to the limits provided for the Soviet Union."[8]

V. STALEMATE: TECHNOLOGY AND THE SALT PROCESS

Foreshadowed by the Jackson amendment and the debate over Soviet "headroom," the problems rather than the promise of SALT were to dominate the next phase of U.S.-Soviet discussions. Between November 1972 and November 1974, the negotiations completely deadlocked over several familiar issues: what systems to include in a permanent, comprehensive agreement (for example, FBS and bombers); how to deal with qualitative as well as quantitative limits, especially with regard to MIRV; and how to keep outside political events -- "linkage" -- from influencing the negotiations.

By far the greatest impediment to progress in this period was the continued, seemingly relentless improvement of Soviet and U.S. strategic nuclear forces within the limits set by SALT I. Members of the military, industrial and scientific bureaucracies on both sides kept constant pressure on policymakers to modernize strategic systems during the 1970s. Working from worst-case analyses, they warned about the possibility of technological "breakout" and worried out loud about the uncertainties and unknowns of a possible SALT II agreement. Naturally, this created a very difficult and uncomfortable situation for SALT negotiators.

The Soviet military buildup was particularly dramatic, especially in the eyes of U.S. defense planners. Of gravest concern to these officials was a new generation of hardened Soviet ICBMs -- the SS-18, SS-19, SS-17, and SS-16 -- each with improved accuracy, increased throwweight (payload capacity), and MIRV capabilities. Moscow also introduced new SLBMs; a new intermediate-range ballistic missile (IRBM), the SS-20; a new bomber, the Tu-22M Backfire; and continued to improve its air defense and civil defense systems. According to the political analyst, Thomas Wolfe, during the 1970s the "Soviet R&D establisment was continuing to operate at its own cyclical rhythm, little affected by SALT."[9]

Reacting to these provocative moves, President Nixon's defense secretary, James R. Schlesinger, suggested in 1974 that the Soviets were creating a "major one-sided counterforce (hard-target) capability against the United States ICBM force" that was "impermissible from our point of view."[10] This became the major justification for the ongoing modernization of U.S. nuclear forces. In addition to proceeding with MIRV deployment (which helped maintain the substantial U.S. advantage in numbers of warheads), Washington expanded the existing Trident SLBM and B-1 intercontinental bomber programs and initiated numerous other R&D projects. These included the MX missile, the long range cruise missile, the increased yield MK-12A warhead, and MARV, a maneuverable reentry vehicle that was the technological successor of MIRV. Thus, Wolfe's cogent observation about the Soviet R&D establishment could as easily apply to developments within the U.S. R&D community.

Besides attempting to keep pace with these rapid technological advances, negotiators also found themselves adversely affected by the deterioration of international relations. Just as SALT I helped to bolster U.S.-Soviet political accommodation, SALT II fell victim to the first signs of political trouble between the two adversaries, a problem that would plague the SALT process throughout the 1970s. Specifically during the 1972-1974 period, Soviet backing (and possible instigation) of the October 1973 Arab attack on Israel and Soviet adventurism elsewhere in the Third World noticeably strained the negotiations in Geneva.

VI. BREAKTHROUGH AT VLADIVOSTOK

Still, both sides had a great deal invested in SALT and the preservation of detente. Hence the talks continued, surviving another blow in 1974 when the U.S. presidency changed hands in the wake of Watergate. Finally in November, the new president, Gerald Ford and Secretary General Brezhnev met at Vladivostok where they hammered out an "agreed framework" for SALT II. Henry Kissinger, again utilizing his talents at personal diplomacy and his new position as secretary of state, deserves the lion's share of credit for engineering this major SALT breakthrough.

Like SALT I, the 1974 Vladivostok executive accord contained a mixture of superpower compromises. Meeting a central U.S. concern, Brezhnev and the supreme leadership accepted the principle of equal overall ceilings on strategic launchers and agreed to drop the contentious FBS issue. However, the specific force level aggregate he and Ford set -- 2,400 launchers to include ICBMs, SLBMs, and for the first time, strategic bombers -- was higher by about 300 than the United States would have preferred. Within this aggregate, moreover, each side would be permitted to mix and modernize its forces in any way it saw fit (which left unresolved the issue of Soviet heavy missiles). Also bowing to U.S. pressure for "parity," the two leaders established a subceiling of 1320 MIRVed missile launchers for each side, but placed no specific limits on throwweight in which the Soviets had a large advantage.

Despite having a treaty "90 percent" completed by the end of 1975, according to Kissinger, and despite a January 1976 "near miss" by Kissinger to find that 10 percent solution, SALT II would not be signed for another three years.[11] There were a combination of reasons for this renewed stalemate in the SALT process. For the sake of agreement, the principals at Vladivostok once more left the hardest strategic questions for future discussion. They underestimated the dynamism of new technology. And lastly, they met a formidable roadblock in the form of the American political process.

The two particular new technologies which created a storm of controversy following Vladivostok were the Soviet Backfire bomber and the American long-range cruise missile. The Backfire is a modern supersonic airplane intended for use primarily as a medium bomber but which under certain circumstances could be employed for strategic missions. The modern cruise missile is a small, pilotless aircraft. It can be launched from the air, land, or sea, and its sophisticated guidance system permits it to fly at very low altitudes. The versatility of these "gray area" systems -- that is, those capable of strategic as well as tactical roles -- made them attractive to the military establishment of each side at the same time it posed almost insurmountable difficulties for negotiators. How or even whether to count them in a new SALT agreement became the all but impossible arms control task.

VII. CARTER AND SALT II

Nevertheless in 1977 a new administration came to power in Washington deeply committed to arms control in general and determined to resolve the outstanding SALT issues in particular. In March, President Jimmy Carter sent Secretary of State Cyrus Vance to Moscow with a comprehensive package of SALT proposals Carter felt confident would break the current impasse. In a typically bold but rash move, Carter and his national security advisor, Zbigniew Brezezinski, included in this package several provisions that went far beyond the Vladivostok equal-aggregates formula. As the administration's chief arms negotiator, Paul Warnke, later recalled, the Carter/Brezezinski team hoped "to shortcut the arms control negotiating process and move in one single giant step toward very significant reductions in numbers and toward a whole series of qualitative restraints."[12] Predictably, the proposal package shocked Soviet bureaucrats who had a natural distaste for surprises or changes of any sort. Brezhnev summarily rejected it.

Thus, what seemed at first as a giant leap forward for arms limitation turned out to be in fact a temporary step backward. But despite continuing Soviet uneasiness over Carter's idealism, his open style of diplomacy, and his unpredictability, the two sides narrowed their differences and adopted a compromise negotiating framework in September 1977. At the same time, they issued a set of unilateral statements agreeing to abide by the SALT I Interim Agreement, set to expire on October 3, while the SALT II talks continued. This persistence ultimately paid off. On June 18, 1979, Carter and Brezhnev met in Vienna and penned their names to the three-part SALT II accords.

What is SALT II? At first glance, the accords appear to be a complex mass of documents describing the minutiae of nuclear arsenals. They include the basic Treaty on the Limitation of Strategic Offensive Arms to run through 1985, a protocol of three years' duration, and a bewildering jumble of Joint and Agreed Statements, Common Understandings and Memoranda totalling more than 100 pages. Buried in this necessary legalistic and technical jargon, however, is a working, viable agreement.

At the heart of the Treaty are the aggregate numerical limits and sub-limits: on total launcher vehicles (2400 initially, 2250 within a year, which not insignificantly would require the Soviets to dismantle 250 existing systems); on total MIRV launchers, including missiles and cruise-missile-carrying bombers (1350); on MIRVed missiles alone (1200); on MIRVed ICBMs alone (820); and on "heavy" ICBMs (308). Other ceilings are established on numbers of warheads or reentry vehicles (RVs) per missile (10 for each MIRVed ICBM, 14 for each SLBM); and on the number of long-range cruise missiles per heavy bomber (an average of 28). The parties in addition established several important bans: on the flight-testing or deployment of new types of ICBMs, except for one new type of "light" missile for each side; on heavy mobile ICBMs and heavy SLBMs; on the construction of additional fixed ICBM launchers; and on rapid reload systems. Restrictions on the troublesome Backfire bomber are less clear, but the Soviets did add an official statement which clarifies the airplane's status as a medium bomber and limits its production rate to 30 per year. Finally, the agreement includes detailed definitions and counting rules for the weapons systems it covers as well as an agreed data base.

The spirit of compromise is again very evident in the SALT II Treaty. No less than the Joint Chiefs of Staff explicitly recognized this in 1979 and consequently supported SALT II as a "modest but useful step in a long-range process which must include the resolve ... to maintain strategic equivalence coupled with vigorous efforts to achieve further substantial reductions."[13] Many former national security officials and members of Congress did not see it this way, however. Led by Senator Jackson and Paul Nitze, former SALT I negotiator and co-founder of the Committee on the Present Danger, domestic critics mounted a massive campaign against SALT II during 1978/79 which intensified during the ratification process.

Nitze became the intellectual driving force behind the anti-SALT movement. He and his Committee on the Present Danger attacked the Treaty itself as "fatally flawed" and the government official most responsible for it, President Carter, as soft on national security issues. Nitze zeroed in on the persistent question of the Soviet heavy missile advantage, stressing the hypothetical threat it posed for U.S. ICBMs and creating ingenious scenarios in which the Soviets might use this advantage in a post-SALT II world for political blackmail. Nitze, Jackson, and other SALT detractors noted one other Fatal Flaw --- the noninclusion of the Backfire bomber as a strategic system --- and also raised questions about the ability to monitor Soviet compliance with the treaty.

This barrage of criticism notwithstanding, political analysts and Senate-watchers still gave SALT II at least a 50-50 chance of ratification when a rapid fire series of world events during the latter half of 1979 threw the agreement into limbo. In August, the United States discovered a Soviet troop brigade in Cuba; in November, Iranians seized the U.S. Embassy in Teheran; and in December, Russia invaded Afghanistan. The overall effect of these actions was to both divert attention from SALT and deepen an already strong public suspicion of Soviet intentions. In January 1980, President Carter saw no choice other than to ask the Senate to "delay consideration" of the Treaty.[14]

VIII. REAGAN AND START

Today, with SALT II still on the Senate shelf, we have yet another administration with yet another approach to the problem of arms limitation. Ronald Reagan came to office with a strong public mandate to "get tough" with the Russians. The President initially took this mandate to heart, adopting the policy line "arms now - talk later" in his dealings with the Soviets. For nearly a year and a half, high-level adminstration officials, including former Arms Control and Disarmament Agency (ACDA) Director Eugene V. Rostow, paid almost no attention to the issue of arms control in any form. But responding in part to a growing international and national antinuclear movement, Reagan's advisors have taken a renewed interest in arms control while at the same time emphasizing the need to bolster the defense establishment.

What does this mean for SALT? Nearly all members of the Reagan national security team, including the President himself, have been extremely critical of the SALT agreements negotiated by previous administrations. Rostow, for example, has stated that the only "successful" treaty in "modern history" was the Rush-Bagot Agreement of 1817. The benefits of SALT I and SALT II, he claims, "have turned to ashes in our mouths."[15] Thus when forced to formulate their own policies and goals, these advisors not surprisingly proposed going beyond the SALT regime to achieve "truly substantial" reductions in nuclear arsenals.[16] SALT became START --- Strategic Arms Reduction Talks --- and was carried on in conjunction with the new Intermediate Range Nuclear Forces (INF) talks to deal with European theater systems.

The START deep-cut strategy envisions downgrading the importance of equal launcher numbers, emphasizing instead restraints on the "destructive capacity" of strategic weapons -- throwweight, megatonnage, warheads, and missile accuracy. Whether this approach can actually lead to agreement or not is problematical. As a Carnegie Endowment study notes, the task is formidable: "Any scheme must deal with the wide differences in current forces and consider how to move from the current regimes toward a significantly different one without creating new instabilities."[17] Still, the Soviets have not rejected the Reagan proposals outright (as they did Carter's in 1977), and talks have again begun.

In the meantime, SALT II survives in the form of de facto observance by both the United States and the Soviet Union. At the same time, however, Reagan's defense officials continue to explore ways to deal with the ICBM vulnerability problem that threaten to unravel existing elements of SALT -- for example, using ABMs to protect the MX, requiring the abrogation or "technical adjustment" of the 1972 ABM Treaty; or "dense pack" basing for the MX.

IX. THE SALT LEGACY

What do we stand to lose if SALT I and SALT II are gradually eroded or even scrapped in this way? To be sure, SALT has left each side with huge nuclear arsenals. Moreover, certain negotiating tactics that can be traced back to the Nixon/Kissinger years have actually accelerated weapons procurement. The utilization of "bargaining chips" -- new weapons systems employed to gain negotiating leverage -- is perhaps the most disturbing of these practices. "The bargaining chip," one study notes, "cannot always be cashed in."[18] Or as another puts it, "weapons initially justified as bargaining chips soon become building blocks -- weapons systems which become permanent parts of the arsenal."[19] SALT gains, however, far outweigh SALT costs. Apart from the specific restrictions of individual agreements, which act as a control mechanism (however imperfect) on the East-West arms race, the SALT process as a whole has produced a number of general benefits that enhance U.S. security. Heading this list are advances in verification procedures and apparatus which both increase our knowledge about the other side and build mutual confidence. The 1972 agreements explicitly legitimize (in Article XII of the ABM Treaty and Article V of the Interim Agreement) and SALT II reaffirms (in Article XV) the unimpeded use of "national technical means of verification," a euphemism for satellite reconnaissance and other noninstrusive information gathering techniques (e.g. radar and electronic monitors). SALT II builds upon this basic achievement with several new verification measures: "counting rules" for hard-to-inspect-and-categorize systems, such as MIRVed missiles (adopting the standard "once tested {as a particular type}, always counted"); "type definitions" of systems, for example, of "heavy" missiles; and an "agreed data base" on ten categories of strategic forces to be updated at regular intervals.

Together these provisions represent a significant gain for U.S. intelligence and a small but marked change in Soviet attitudes about secrecy. After agreeing to the data base exchange, for example, chief Soviet negotiator Vladimir Semyenov reportedly said to Paul Warnke that 400 years of Russian history had just been "repealed." "But on reflection," he added, "maybe that's not a bad thing."[20] Not only have Soviet leaders agreed to tell more to the West, they have in the process opened up the flow of information within their own bureaucracy. Apparently, the military is no longer the sole custodian of strategic facts and doctrine. New constituencies, in particular foreign policy experts, have taken their place in the internal decision-making process through the negotiations themselves and such special SALT bodies as a jointly-run ad hoc group within the Ministry of Defense.

Another valuable product of the SALT process that deserves special mention is the Standing Consultative Commission (SCC). Set up originally by the ABM Treaty to monitor problems and complaints about SALT compliance, the SCC has quietly evolved into an important force for continuing dialogue and accommodation on nuclear issues. Since it began operations in 1973, the SCC has met regularly at least two times a year in Geneva in sessions completely separate from the formal negotiations. The U.S. delegation has been represented in the past by a civilian commissioner (Sidney Graybeal followed by Robert Buchheim, both of ACDA), a military deputy, and a staff of advisors from various government agencies. Today both top officials are military officers. The Kremlin also has a mix of civilian and military personnel on its delegation, led since the outset by Major General G. I. Ustinov of the General Staff and Viktor P. Karpov of the Ministry of Foreign Affairs (who now also heads the START negotiations). According to one SALT scholar, the SCC's most significant contribution to date has been the "successful ... defusing {of} public accusations of Soviet cheating by American critics of the SALT process."[21] This confidence-building role will expand under SALT II and START since the SCC will be responsible for maintaining the data base inventories, supervising any dismantling of strategic weapons that may be called for, and even considering new arms limitation strategies.

X. CONCLUSION

In sum, then, each side is now playing the strategic armaments game by a new set of rules. These rules make the game more predictable and therefore enhance stability. They should not be sacrificed or discarded in the interests of short-term military goals. SALT is not a panacea for all our national security problems, neither is it the main cause of U.S. strategic deficiencies. Rather, it provides a valuable framework in which the superpowers can work toward further progress in negotiated arms limitation.

REFERENCES

1. Chester I. Barnard, J. R. Oppenheimer, Charles A. Thomas, Harry A. Winne, David E. Lilienthal, A Report on the International Control of Atomic Energy, prepared for the Secretary of State's Committee on Atomic Energy (Washington, D.C.; March 16, 1946), p. 31.

2. Helmut Sonnenfeldt and William G. Hyland, Soviet Perspectives on Security, Adelphi Paper No. 150 (Spring 1979), p. 14.

3. ACDA, Arms Control and Disarmament Agreements: Texts and Histories of Negotiations (Washington, D.C.; August 1980), p. 92.

4. Gerard Smith, Doubletalk: The Story of the First Strategic Arms Limitation Talks (New York: Doubleday, 1980), pp. 35, 37.

5. Richard J. Barnet, The Giants: Russia and America (New York: Simon and Schuster, 1977), p. 102.

6. Strobe Talbott, Engame: The Inside Story of SALT II (New York: Harper and Row, 1980), p. 20.

7. Representative Gerald R. Ford, Address before the VFW State Convention, Grand Rapids, Michigan (June 17, 1972).

8. A full text is in Michael B. Donley, ed., The SALT Handbook (Washington, D.C.: The Heritage Foundation, 1979), p. 15.

9. Thomas W. Wolfe, The SALT Experience (Cambridge, Mass.: Ballinger, 1979), p. 123.

10. Report of Secretary of Defense James R. Schlesinger to the Congress on the FY 1975 Budget and FY 1975-1979 Defense Program (March 4, 1974), p. 6.

11. New York Times, October 13, 1975.

12. Paul C. Warnke, "The SALT Process," Department of State News Release (January 19, 1978), p. 3.

13. David C. Jones, General, USAF, Chairman, Joint Chiefs of Staff, "SALT II: The Opinion of the Joint Chiefs of Staff," Vital Speeches of the Day, 45 (August 15, 1979), p. 655.

14. Cited in Talbott, Endgame, p. 290.

15. Los Angeles Times, January 24, 1982.

16. New York Times, May 1, 1982.

17. Carnegie Panel on U.S. Security and the Future of Arms Control, Challenges for U.S. National Security: Nuclear Strategy Issues of the 1980s, Third Report (Washington, D.C.: Carnegie Endowment for International Peace, 1982), p. 68.

18. John H. Barton and Lawrence D. Weiler, eds., International Arms Control: Issues and Agreements (Stanford, Ca.: Stanford University Press. 1976), p. 226.

19. Robert J. Bresler and Robert C. Gray, "The Bargaining Chip and SALT," Political Science Quarterly, 92 (1977/78), p. 86.

20. Cited in Talbott, Endgame, p. 98 and Wolfe, SALT Experience, p. 261.

21. Jane M. O. Sharp, "Restructuring the SALT Dialogue," International Security, 6 (Winter 1981/82), p. 149.

CHAPTER 15

SALT PROCESS: PAST AND FUTURE

Ralph Earle II
Earle and Greene and Company
Washington, D. C. 20009

INTRODUCTION

I am pleased that so many of you are attending this second short course, thereby demonstrating your interest in teaching or studying the issues of the arms race in general and the nuclear arms race in particular.

President Reagan is not one of my principal sources of quotation, but on this issue he did say one thing last November that is worth repeating tonight:

> The prevention of conflict and the reduction of weapons are the most important public issues of our time. Yet, on no other issue are there more misconceptions and misunderstandings.

So let us hope that today some of those misconceptions and misunderstandings have been removed and that the way has been paved to continue the public education on this vital subject.

In preparing my remarks, I took quite literally the stated goal of this conference -- "to supply information" -- and that I will try to do. But I think it is too much to expect of one who spent nearly eight years in the pursuit of a nuclear arms control agreement to avoid voicing some opinions as I go along. I hope you scientists will bear with me for those departures from strict "information" or data.

HISTORY OF SALT/START NEGOTIATIONS

First, let me begin with some history of the SALT/START negotiations.

President Johnson, following his talks with Kosygin at Glassboro concluded that something had to be done about the nuclear arms race, and, after a year of study, was on the verge of announcing the beginning of SALT (Strategic Arms Limitation Talks) with the Soviet Union in the summer of 1968. But, literally only days before the planned announcement, the Soviets sent massive forces into Czechoslovakia and the arrangement was terminated.

That was followed closely by Mr. Nixon's election, the transition, and a re-study of the problem by the new administration. Happily, this new consideration also reached the conclusion that there should be such talks, and, in July 1969, the U.S. SALT Delegation, headed by Ambassador Gerard Smith, was sent to Belgium to consult with our NATO allies preparatory to the first meeting with the Soviets. I mention that meeting in Brussels, not only because I was present and had my first direct exposure to SALT,

0094-243X/83/1040262-08 $3.00 Copyright 1983 American Institute of Physics

but because it was the first of a continuing and frequent series of such consultations with our North Atlantic partners, which continues to this day.

The first bilateral meeting with the Soviets took place in Helsinki in November, 1969, and thereafter sessions lasting approximately two and a half months alternated between Helsinki and Vienna for the next two and half years. Initially, the sides attempted to achieve comprehensive limitations on both offensive and defensive strategic arms, but about half way through the negotiations it was agreed that that might be trying for too much too soon and that priority should be given to defensive weapons without total neglect of the offensive side of the coin. As a result of these negotiations, Nixon and Brezhnev, in May of 1972, signed two agreements, the Anti-Ballistic Missile (ABM) Treaty and the so-called Interim Agreement on Offensive Arms.

THE SALT AGREEMENTS

Most of you are familiar with these agreements, so I will be brief.

The ABM Treaty limited each side to two anti-missile defensive sites (later reduced to one in 1974) with no more than 100 launchers at any one site, it placed limitations on but did not ban further research and development of ABM systems, and it prohibited transfer of these systems to third countries. In other words, the two sides agreed that in the future (the Treaty being of "indefinite duration") each would be and would remain vulnerable to the missiles of the other side and thereby codified a deterrent of and for each. More about this later.

The Interim Agreement was to last five years and in effect was a freeze on the launchers of intercontinental ballistic missiles (ICBMs) and of submarine-launched ballistic missiles (SLBMs) of both sides. At this time, the Soviets had more of both, but this numerical advantage in launchers was balanced by the fact that the U.S. had begun its program of developing multiple independently-targetable re-entry vehicles (MIRVs) and thus had more deliverable warheads. There were no limitations placed on the third category of strategic weapons, heavy bombers; there the U.S. also had a numerical edge, but the Soviets were far ahead in air defenses.

These two SALT I agreements came into effect in October 1972, and the next month the SALT II discussions began, this time in only one city, Geneva. They lasted nearly seven years and focussed entirely on what we called "central systems" or SNDVs (strategic nuclear delivery vehicles): ICBM launchers, SLBM launchers and heavy bombers. The SALT II Treaty, signed at Vienna in June 1979, is a complex agreement and cannot be summarized briefly. A few major points: it calls for "equal aggregates" — initially a total of no more than 2400 SNDVs on each side, that number to be scaled down to 2250 (which would have required the Soviets to destroy nearly 300 weapons, the U.S. none); it limits MIRV launchers and heavy bombers with long-range cruise missiles; it sub-limits MIRV launchers and MIRV ICBM launchers; it puts some restraints on the number of reentry vehicles; it permits each side to produce only one new type of ICBM over the life of the Treaty (through 1985); it has many provisions which contribute to improved verification of compliance with its terms; and it has paragraphs and paragraphs of "boiler plate" -- definitions and so forth. The numbers of weapons permitted were indeed high, but ceilings were imposed from which a further scaling down could be foreseen.

As you all know, a number of events took place which resulted in the Treaty not coming to a vote on the Senate floor (although it had been favorably voted out by the Senate Foreign Relations Committee): the "discovery" of the Soviet brigade in Cuba, the taking of hostages in Teheran, and, finally, the Soviet invasion of Afghanistan.

Ronald Reagan campaigned against SALT II as "fatally flawed" and continues to describe it as such. However, after taking office he adopted the somewhat paradoxical position that the U.S. would live up to the terms of this "fatally flawed" agreement so long as the Soviets did the same. And both sides continue to do so.

After a considerable pause, the Reagan Administration and the Soviet Union returned to negotiations in Geneva, this time in two separate forms: the Strategic Arms Reductions talks (START) and the Intermediate range Nuclear Forces (INF) talks. The former is a successor to SALT and apparently has made little progress. The latter, INF, together with U.S. plan to deploy 572 missiles in Europe, has been and continues to be the subject of intense media coverage and public debate, both here and in Western Europe. Where either will lead is impossible to say at this time. SALT II was complex and difficult. The current negotiations are also complex and difficult, even more so.

THE PROCESS OF NEGOTIATIONS

At this point, let me say a few words about the form or procedure of the negotiations which are so often mentioned in the news media but which are so seldom described.

When Benjamin Franklin and his colleagues went off to Paris to try to negotiate an end to the Revolutionary War, his instructions were, in effect, "Bring back a good treaty", and then he was on his own. A far cry from today's world of huge bureaucracies and instant communication. At SALT, and now at START and INF, the American delegation is composed of a chief or chairman and senior represen- tatives of the State Department, the Department of Defense, the Joint Chiefs of Staff, the Arms Control and Disarmament Agency and the intelligence community, and each of these is supported by staff assistants -- a total of about 45 men and women on site. And, to the extent that the Soviet bureaucracy is similar to ours, the Soviet delegation is similarly staffed.

A comparable SALT, START or INF organization exists in Washington, with the President and the National Security Council at the top of the pyramid, and committees and sub-committees at various rank levels below them addressing the issues. As a result, the instructions sent to Geneva and their implementation in the negotiations by the delegation are the product of intense scrutiny by all interested agencies. This in turn results in a slow, sometimes snail-like, process. Every word the Soviets say is reported to Washington and analyzed, debated and considered there as well as by the people in Geneva. And we assume the same takes place in Moscow. But the stakes are very high, and the care and caution are understandable.

THE PRESENT AND FUTURE OF NEGOTIATIONS

Now I would like to turn to the present and the future — and offer some comments.

First, I think it important to bear in mind that strategic arms negotiations (and their product — agreements) have to be considered from two perspectives: military and political.

From a military point of view, the arms control policy and implementation is a part of overall national security considerations. A simple guideline would be that the negotiators should not give something away until and unless a judgment had been made at the highest level that such an agreement would not jeopardize our national security. An obvious example: the U.S. should not have agreed to limit its ABM systems without the same commitment from the Soviet Union (and it did not).

The political aspect is far more complex and far more difficult to deal with. Shortly before Jimmy Carter was to be inaugurated as President, he met with the Joint Chiefs of Staff and, inter alia, asked them how many missiles did we need to deter the Soviets — in other words, could we do with fewer, perhaps many fewer, than we then possessed. I believe that the correct answer, from a purely military point of view, would have been in the affirmative. But, when word of this meeting with the Chiefs leaked to the press, Carter was severely criticized by many for even asking the question. Some of the criticism came from the uninformed, and some of it came from political opponents who chose to see it as a sign of weakness or pacifism. But some criticism came from more thoughtful people, their reasoning being, and not without substance, that if in fact his question led to unilateral reductions which placed us in a numerically inferior position to the Soviet Union, even though unilaterally our deterrent would be maintained, it could or would be seen by the rest of the world as an American acceptance of Soviet superiority. Put another way, if the Soviet Union could destroy the U.S. twenty times over and we could destroy them only twice over, then the Soviets would be considered ten times as strong as the U.S.; this in turn could result in increased Soviet political and economic leverage all over the world. So political considerations of this sort must be taken into account.

With respect to the current negotiations and those that, hopefully, will take place in the future, there are certain fundamentals that should be kept in mind — they are fundamental but frequently neglected or ignored.

First, any strategic arms agreement should be tested with respect to strategic stability. Strategic stability, by my definition, is a situation in which neither side is capable of attacking the other without assurance of receiving unacceptable damage in return, and neither side perceives either that it can make a successful -- that is, unanswered -- attack or that the other side can make such an attack. And here the perception is just as important as the reality. Does the particular agreement under scrutiny maintain or, preferably, enhance strategic stability? That is one fundamental.

A second, and much disregarded, fundamental is that consideration must be given to the asymmetries between the sides. These asymmetries are largely or entirely immutable and therefore are relevant to any negotiation between the super powers.

One is geography. We have long coastlines with total access to non-ice-bound passages. The Soviet Union, although a huge land mass, has far fewer ports, and most of those lie behind so-called "choke points" -- for instance, their submarines must transit close to Norway, Great Britain, Iceland, Japan, and so forth -- ours are in the open sea the moment they leave, for instance, Charleston, South Carolina, Guam, or Bangor, Washington. This is a major reason why the Soviets have emphasized land-based missiles and we have put more effort into our submarine force. At the same time, we have a far larger percentage of our population on or near the coast which makes us more vulnerable to short flight-time SLBMs.

Second, the two powers have very different alliances. The Soviet allies are to one degree or another subject nations and without strategic capability of their own. On the other hand, our allies are free, with great economic potential, and two of them, Great Britain and France, are nuclear powers with their own missile-bearing submarines and, in the case of France, with their own land-based missiles.

There is also the matter of third countries. The Soviets are faced with a major threat over a border of thousands of miles long from one billion Chinese -- who also have nuclear weapons. As some one noted, the Soviet Union is the only nation in the world surrounded by unfriendly Communist countries. We have no such strategic problem.

A third, and also often-neglected, fundamental is that we should in our negotiations with the Soviets, or any one else for that matter, try to keep in mind their perceptions of the current situation, their perceptions of history, and their resultant attitudes and beliefs -- these all affect the relationship. For instance, we see, understandably, the Soviet Union as an aggressor (e.g. Afghanistan), a trouble maker (e.g. Angola) and a nation seeking the spread of its own brand of Communism (e.g. Cuba).

But how do they see themselves -- and us? They see the Soviet Union in a world in which five nations have nuclear weapons and four of them have those weapons aimed at Moscow. They see what they call "U.S. adventurism" in Korea, Vietnam, Lebanon, and so forth, and they see us as the first and only nation to use the atomic bomb.

Let me make it clear -- I am not defending the Soviet Union or condemning the U.S. -- I am simply saying that we can deal better with our adversary if we understand him and his perceptions.

A final rule for our future negotiations -- let us not let the best be the enemy of the good. Do not ask for too much too soon. Let me give you some examples of what I mean.

In March, 1977, Secretary Vance travelled to Moscow to put forward the so-called "comprehensive proposal" which called for dramatic reductions, especially in land-based weapons. From the American point of view, it was a position which made good sense and would have contributed greatly to an enhancement of strategic stability. But the Soviets saw it very differently -- as an effort to weaken their strength -- land-based ICBSs -- and in effect make them re-structure their forces at great cost and with no assurance that they would be equally secure after such restructuring. As a result, they rejected that proposal almost immediately. It was not a major setback for SALT, but it does demonstrate my point.

Subsequently, in considering the resultant SALT II Treaty, critics complained that we "let the Soviets have 308 heavy ICBM launchers" and the U.S. could have none. There was an implication in this criticism that the U.S. had somehow made a gift of these launchers to the Soviet Union. In fact, they had had them for a decade and had tacitly agreed at Vladivostok in 1974, four and a half years before, that they would stop pressing for limits on the U.S. forward-based systems in Europe in return for maintenance of these launchers. And the very critics who claimed that the Treaty should not be ratified because of this alleged flaw were in fact advocating a no-treaty world in which the Soviets could have 3,008 or 30,008 heavy launchers.

And on the other side of the coin, those who support a freeze at this time may be asking for too much. Personally, I am very sympathetic to the freeze movement as it reflects a massive public support for doing something about the nuclear arms race. But is a freeze really attainable? I would assume that most if not all of you want any future agreement with the Soviets to be verifiable. But how long would it take, for instance, to negotiate a truly verifiable ban on the production of nuclear weapons? Such a goal is obviously desirable in the long run, but would its pursuit not divert us from getting something almost as good in the near term?

AN ANTI-NUCLEAR UMBRELLA

I would now like to take a few minutes and talk some more about strategic stability and defensive strategic weapons, and in particular President Reagan's suggestion for an anti-nuclear umbrella for us and our allies. This appears to be an issue that will be before us for the foreseeable future (or at least two more years).

My basic problem with the proposal is that apparently its author fails to comprehend the difference between a nuclear and a non-nuclear world. I began preparing these remarks before the President's speech of March 23rd, so I did not sense that I should have to remind people of one more fundamental -- that is, strategic arms negotiations deal with nuclear weapons, and nuclear weapons are different from other weapons.

As you scientists know better than I, no weapons system is ever 100% effective. However, in the past that has not been vitally important as long as a particular system had a high degree of reliability.

Suppose, for instance, in 1940 Great Britain had been able to deploy an air defense weapon which was 95% effective against German bombers. Winston Churchill would have been very happy if he had been confident that only 5 of every 100 enemy bombers would reach London -- a 1000-plane raid would have meant the destruction of only 50 blocks in London, and the Nazi bomber command would have been decimated in a few days.

Let us also suppose that we go ahead and spend the billions of dollars necessary for Mr. Reagan's plan and that in the year 2000 we are able to deploy an anti-missile defense which is 99% effective (and that in the interim the Soviets do absolutely nothing, either with respect to anti-satellite weapons or penetration capability). If the Soviets were to attack with their existing ICBMs and SLBMs (a total of about 8,000 reentry vehicles), and only 1% got through, we would not lose 80 blocks of New York City -- we would lose New York and 79 other cities. A nuclear world is indeed different.

And what do we do to defend against the Soviet bombers, their long-range cruise missiles, and their depressed trajectory SLBMs launched close to our coasts?

Mr. Reagan also proposed that this new defense be designed and deployed so that our allies are equally protected. But they can be attacked, as can we, by a huge array of endoatmospheric weapons, and, for the allies, these include artillery rounds. What is the 100% defense against them? And, as I implied, the Soviets just might do something to counter our umbrella.

Even assuming that 100% reliability and effectiveness could be achieved against this mixed assortment of weapons -- ballistic missiles, bombers, cruise missiles and artillery -- I do not believe we will be better off -- in fact, the situation will be worse than ever. Let me explain.

Still assuming that the Soviets did nothing to improve their forces between now and 2000, there would come that moment nearly 20 years from now when we are about to deploy the impenetrable umbrella -- say, midnight, December 31, 2000. We have retained our offensive forces as a deterrent (also about 8,000 re-entry vehicles) -- necessary until the umbrella is in place -- and we are one minute away from launching the hundreds of satellites necessary for the defense. What is the Soviet perception? In one minute the U.S. will be invulnerable to Soviet attack and still have the offensive capacity to destroy the Soviet Union many times over. Obviously, crisis stability has vanished, and the Soviet Union, to protect itself from an intolerable situation, would see itself required to launch a pre-emptive attack. What would be the position of the President of the U.S. if the situation were reversed?

WHAT SHALL WE DO?

This is a dire picture, and hopefully it will not be realized. So I will put aside Darth Vader and Star Wars and return to reality.

What should we do?

I believe we should strive for a series of strategic arms control agreements, each one building on the ones before it. We have strategic stability now. These agreements should codify it and then enhance it through a cycle of mutual and verifiable reductions and increasing limitations on the development and deployment of new weapons. We should not seek to gain an edge over the Soviet Union -- an edge which we cannot achieve either through negotiation or through an arms build-up. We should limit and reduce the numbers of launchers and heavy bombers, we should limit and reduce the numbers of reentry vehicles, bombs and cruise missiles, and we should encourage and amplify the dialogue between the two superpowers on these subjects. We should seek negotiable incentives not to replace the more destabilizing weapons as they become obsolete. We should demonstrate to the world that not only do we want fewer weapons and greater stability but that we are doing something about that desire.

And we should pray.

270

APPENDICES

APPENDIX A

BIBLIOGRAPHY ON PHYSICS AND THE NUCLEAR ARMS RACE

Dietrich Schroeer
Department of Physics and Astronomy
University of North Carolina
Chapel Hill, North Carolina 27514

John Dowling
Department of Physics
School of Arts and Science, Mansfield State College
Mansfield, Pennsylvania 26933

This collection of article and book references is assembled to help physicists understand and teach about the nuclear arms race. Both technical and political aspects are covered. Articles or books marked with an asterisk(*) are included in the REPRINT VOLUME ON "PNAR-1: PHYSICS AND THE NUCLEAR ARMS RACE", to be published by the American Association of Physics Teachers in June of 1983.

I. INTRODUCTION

Teaching and study of the nuclear arms race requires resources that are hard to find. Physics and the arms race couple on three levels. There is real physics in the arms race. This physics in turn leads to important changes in some technologies. Finally, these physics-inspired technologies modify the politics of the arms race. It is these interconnections which make the arms race so interesting, and also make it so difficult to locate proper resource material.

This resource letter includes references dealing with each of these linkages. We have chosen to give a feeling for the resources that are readily available. Literature accessible only to experts in the field has been left out; a separate resource letter on Rand reports might well be worthwhile. Whenever an area has become a field in itself, we have tended to list the classics that provide a good introductory overview. Whenever a choice had to be made, we selected the well-written resource over more expert but also more obscure items. In general we chose those readings that one ought to have seen; nothing is so embarrassing as loosing an argument because one has not seen "the" book, only to discover later that it really did not contradict one's position. Video tapes are listed if they contribute significant technical insights.

This is a resource letter on "the nuclear arms race." Primary emphasis is on resources that focus on nuclear weapons, their effects, their usage, and their control. But the topic of nuclear defenses is also covered, as are the alternatives to strategic nuclear deterrence of conventional strategic bombing, of chemical-biological warfare, and of tactical nuclear warfighting capabilities.

A technical understanding of the nuclear arms race rquires a

0094-243X/83/1040271-25 $3.00 Copyright 1983 American Institute of Physics

feeling for the orders of magnitude of effects. Why is the A bomb as much as a million times more effective than TNT, and the H bomb yet another factor of a thousand more so? What does the term "effective" mean? How much does the nuclear power industry increase the probability of nuclear proliferation? What are likely to be the next developments in the nuclear arms race? The reference dealing with the science and technology of the nuclear arms race range from qualitative discussions through simple calculations to outlines of more sophisticated approaches. The emphasis is on technical literacy.

Both the technological imperative and social driving forces are involved in converting scientific possibilities into major new weapons system. Some of the references consider these in a journalistic format, but some are thorough political science studies.

As new technical developments lead to new weapons, so do new weapons lead to new military, political, and social strategies. Some of the references will consider these strategies, and how they might be predicted and controlled.

1.1. "Physics and modern warfare: The awkward silence," E. L. Woollett, Am. J. Phys. <u>48</u>, 104-111 (1980). Why should we talk about it?

1.2. "War, Peace, Science, and Technology in the Atomic Age--A Physics Course for the General Student," R. A. Uritam, Am. J. Phys. <u>40</u>, 1324-7 (1972). Description of a physics course exploring the interaction among science, technology, war and military affairs.

1.3. "Science and Society Test for Physicists: The Arms Race," D. Hafemeister, Am. J. Phys. <u>41</u>, 1191-96 (1973). Too many physicists flunk such a test.* See also his "Science and Society Test VIII: The Arms Race Revisited", Am. J. Phys., <u>51</u>, 215-225 (1983).

II. GENERAL REFERENCES

2.1. THE DYNAMICS OF THE NUCLEAR BALANCE, A. Legault and G. Lindsey (Cornell University, Ithaca, NY, revised edition, 1976). This comes closest to being a textbook for a physics-based arms race course. If only it had a few more calculations.

2.2. THE EFFECTS OF NUCLEAR WEAPONS, edited by S. Glasstone and P. J. Dolan (U.S. Government Printing Office, Washington, DC, 3rd ed., 1977). The "bible" on nuclear weapons effects, sophisticated yet understandable.

2.3. NUCLEAR WEAPONS DATA BOOK, (Ballinger Publ. Co., 1983).

2.4 FUNDAMENTALS OF STRATEGIC WEAPONS; OFFENSE AND DEFENSE SYSTEMS, James N. Constant (Martinus Nijhoff Publishers, Boston, 1981). A treasury of technical information about strategic weapons systems; with equations, numbers, summary tables, and extensive references not only about weapons systems per se, but also about the principles underlying them. Very expensive, but very useful for someone with professional interest.

2.5. WEAPONS SYSTEMS FUNDAMENTALS; VOL. I: ELEMENTS OF WEAPONS
SYSTEMS, VOL. II: ANALYSIS OF WEAPONS, VOL. III: SYNTHESIS OF
SYSTEMS, published under the direction of Commander, Naval
Ordnance Systems Command (NAVORDOP-3000) first revision
(Department of the Navy, Ordnance Systems Command, 1971).
Qualitative, and at times quantitative, description of naval
weapons systems, including missiles (and their trajectories),
etc. See also PRINCIPLES OF GUIDED MISSILES AND NUCLEAR
WEAPONS, Naval Training Command (U.S. Government Printing
Office, Washington, DC, revised edition, 1972).

2.6. ARMS CONTROL: READINGS FROM SCIENTIFIC AMERICAN, edited by H.
F. York (Freeman, San Francisco, 1973). This collection
provides the couplings between science, technology, and the
nuclear arms race on a qualitative but relatively sophisticated level. It is updated in PROGRESS IN ARMS CONTROL?
READINGS FROM SCIENTIFIC AMERICAN, edited by B. M. Russett and
B. G. Blair (Freeman, San Francisco, 1979).

B. General References

2.7. JANE'S WEAPONS SYSTEMS, R. T. Petty and H. R. Archer (Jane's
Yearbooks, London, latest of the annual editions). Authoritative descriptions of strategic and tactical weapons systems,
sometimes quantitative, political connections are at times
spelled out.

2.8. WORLD ARMAMENTS AND DISARMAMENT, SIPRI YEARBOOK, Stockholm
International Peace Research Institute (annually; 1982 edition
published by Oelgeschlager, Gunn & Hain, Inc., Cambridge,
MA). This annual report keeps track of the world's
armaments. SIPRI publishes various other regular and
occasional books on a variety of arms race topics.

2.9. THE MILITARY BALANCE (International Institute for Strategic
Studies, London, annually). A yearly quantitative assessment
of the military power and defense expenditures of countries
throughout the world. The IISS also publishes THE STRATEGIC
SURVEY each year, in which it reviews the past year's
security-related events and trends, including arms limitation
efforts. And its ADELPHI monographs analyze current and
future problems of international security affairs.

2.10. ANNUAL REPORT OF THE DEPARTMENT OF DEFENSE, by the Secretary
of Defense; and UNITED STATES MILITARY POSTURE, by the
Chairman of the Joint Chiefs of Staff (U.S. Government
Printing Office, Washington, DC, annually).

2.11. ARMS CONTROL IMPACT STATEMENTS, annual report by the U. S.
Arms Control and Disarmament Agency to the Committees on
Foreign Affairs and Foreign Relations of the House of
Representatives and Senate, respectively (U.S. Government
Printing Office, Washington, DC annually).

C. Journals

2.12. AVIATION WEEK AND SPACE TECHNOLOGY. This is the trade journal for aviation technology. It is frequently authoritative; but is also often overly enthusiastic about future technological capabilities, with not enough critical analysis to separate the facts from the speculations.

2.13. BULLETIN OF THE ATOMIC SCIENTISTS. Strongly in favor of nuclear arms control, it frequently addresses important arms race issues that have scientific and technological components. And sometime it gives interesting numbers.

2.14. FOREIGN POLICY. This journal of the Carnegie Endowment for International Peace publishes informative articles on arms control, proliferation, military strategy, and defense policy.

2.15. INTERNATIONAL SECURITY. Published by the Program for Science and International Affairs of Harvard University; it looks at the relations of science and technology to international security problems.

2.16. SCIENCE. The news columns of this journal of the American Association for the Advancement of Science frequently covers contemporary arms race issues. Sometimes it has quite advanced technical articles on arms issues.

2.17. SCIENTIFIC AMERICAN. Has arms race articles with good scientific and technical content. Unfortunately, it is qualitative, and the references are usually not very helpful. It is particularly useful as an assigned reading resource for teaching about the arms race to nonscientists (see Ref. 2.5 for collections of articles from this journal).

2.18. STRATEGIC REVIEW. Published by the U.S. Strategic Institute of Washington, it deals with problems of national security in a nuclear age.

2.19. PUBLIC INTEREST REPORT. The newsletter of the Federationn of American Scientists, obtainable from the FAS, 307 Massachusetts Ave., NE, Washington, DC 20002. Contains good summaries of current arms race controversies.

2.20. ARMS CONTROL TODAY. The newsletter of the Arms Control Association, obtainable from the ACA at 11 Dupont Circle, NW, Washington, DC 20036. Prodisarmament review of arms race issues.

2.21. THE DEFENSE MONITOR. The newsletter of the Center for Defense Information, obtainable from the CDI at 122 Maryland Ave., NE, Washington, DC 20002. A review of arms race issues.

2.22. WAR AND PEACE FILM GUIDE, J. Dowling (World Without War Council, Chicago, 1980). Describes over fifty films on the arms race and nuclear war.

III.   NUCLEAR ARMS

This section of references deals with the capability for massive destruction. The references cover the development of strategic bombing doctrines in World War II, and the building of A-bombs and H-bombs. The strategic bombing effort of World War II is a possible guide for the efficiency of nuclear strategic bombing. Nuclear fission weapons are considered from various perspectives: how they work, how are they made, and what effects they have. Historical and political references that address whether the discovery of nuclear fission was inevitable, whether the development of nuclear weapons was controllable, and whether their use was required. Some references describe how fusion works, and how an H-bomb is constructed. But others look at the efforts to guide the development of this even larger weapon, to submit it to political control from the beginning. Finally, some references explore the large-scale destruction that would result from nuclear war.

A. <u>Conventional Strategic War</u>

3.1. NUCLEAR ENERGY: ITS PHYSICS AND ITS SOCIAL CHALLENGE, D. R. Inglis (Addison-Wesley, Reading, MA, 1973). This text deals with the various forms of energy. See particularly pp. 11-52.

3.2. PHYSICS OF SHOCK WAVES AND HIGH-TEMPERATURE HYDRODYNAMIC PHENOMENA, Y. B. Zel'dovich and Y. P. Raizer (Academic Press, New York, 1967). An excellent advanced text on high-temperature gas dynamics and shock waves with deep physical insights.

3.3. STRATEGIC BOMBING IN WORLD WAR II: OVERALL REPORT, U. S. Strategic Bombing Survey (U.S. Government Printing Office, Washington, DC, 1945). Summarizes in quantitative terms the physical impact of conventional strategic bombing during the second world war. See also STRATEGIC BOMBING IN WORLD WAR TWO: THE STORY OF THE U.S. STRATEGIC BOMBING SURVEY, AND UNITED STATES STRATEGIC BOMBING SURVEY: A COLLECTION OF THE 31 MOST IMPORTANT REPORTS IN 10 VOLUMES, by D. MacIsaac (Garland, New York, 1976).

3.4. THE SOCIAL IMPACT OF BOMB DESTRUCTION, F. C. Ikle (University of Oklahoma, Norman, OK, 1958). Outlines both the physical and social impact of strategic conventional bombing in World War II, with emphasis on the recovery rates.

3.5. THE DESTRUCTION OF DRESDEN, D. Irving (Ballentine, New York, 1963). A description of the consequences on a human level of the strategic bombing of this city. The Dresden firestorm may have resembled the effects that might be expected from nuclear weapons. However, DRESDEN IM LUFTKRIEG, by G. Bergander (Boehlau, Koeln, 1977) suggests that the actual number of fatalities was much smaller than indicated by Irving. See also A TORCH TO THE ENEMY: THE FIRE RAID ON TOKYO, M. Caidin (Ballentine, New York, 1960).

3.6. SCIENCE AND GOVERNMENT, C. P. Snow (The New American Library, New York, 1962). The famous account of the feud during World War II between Sir Henry Tizard and F. A. Lindemann concerning the efficacy of strategic bombing.

B. <u>The Fission Bomb</u>

1. Technological aspects

3.7. NUCLEAR ENERGY: AN INTRODUCTION TO THE CONCEPTS, SYSTEMS, AND APPLICATIONS OF NUCLEAR PROCESSES, R. L. Murray (Pergamon, New

York, 2nd ed., 1980). Introductory nuclear physics.

3.8. INTRODUCTION TO NUCLEAR ENGINEERING, R. L. Murray (Prentice-Hall, Englewood Cliffs, NJ, 2nd ed., 1961). The early chapters nicely review nuclear physics, neutrons, fissions and the chain reaction, and the separation of isotopes. Presents good order-of-magnitude calculations, as of the energy required for isotopic separation by gaseous diffusion.

3.9. ELEMENTARY INTRODUCTION TO NUCLEAR REACTOR PHYSICS, S. E. Liverhant (Wiley, New York, 1960). This old book is excellent in its description and order-of-magnitude calculations of nuclear physics and the exploitation of the fission process. It is neither too theoretical in its physics nor too narrowly focussed on reactors.

3.10. SOURCEBOOK ON ATOMIC ENERGY, S. Glasstone (Van Nostrand-Reinhold, New York, 3rd ed., 1967). Contains semiquantitative reviews of nuclear fission and the utilization of nuclear energy.

3.11. ATOMIC ENERGY FOR MILITARY PURPOSES, H. D. Smyth (U.S. Government Printing Office, Washington, DC, 1946). A summary of the wartime Manhattan Project A-bomb technological developments.

3.12. "REVIEW OF NUCLEAR WEAPONS EFFECTS," H. L. Brode, Ann. Rev. Nucl. Sci. 18, 153-202 (1968). Descriptive, theoretical, and calculational presentations of a nuclear explosion and its physical consequences; high level of sophistication. See also Refs. 2.2 and 3.2.*

3.13. "NUCLEAR EXPLOSIONS," A. A. Broyles, Am. J. Phys. 50, 586-594 (1982).

3.14. "NUCLEAR WEAPONS AND POWER-REACTOR PLUTONIUM," A. B. Lovins, Nature 283, 817-23 (1980). A sophisticated and often quantitative discussion of the effects of Pu-240 on criticality and on premature-ignition problems.*

2. Historical aspects

3.15. BRIGHTER THAN A THOUSAND SUNS, R. Jungk (Harcourt and Brace, New York, 1958). A broad survey of the A-bomb and H-bomb history; biased, but nonetheless excellent, if only because so well written. See also THE GREATEST POWER ON EARTH: THE INTERNATIONAL RACE FOR NUCLEAR SUPREMACY, R. Clark (Harper and Row, New York, 1980).

3.16. MANHATTAN PROJECT, THE UNTOLD STORY OF THE MAKING OF THE ATOMIC BOMB, S. Groueff (Little and Brown, Boston, 1967). A very nice history of the Manhattan Project, offering insights into the triumphs and the tribulations.

3.17. A HISTORY OF THE U.S. ATOMIC ENERGY COMMISSION; THE NEW WORLD: 1939-1946, R. G. Hewlett and O. E. Anderson (Pennsylvania State University, University Park, PA, 1962). The official history of the Manhattan Project. More bland than Refs. 3.15 and 3.16, but also more thorough and detailed, giving a much more accurate impression of the enormous industrial effort involved.

3.18. THE GERMAN ATOMIC BOMB: THE HISTORY OF NUCLEAR RESEARCH IN NAZI GERMANY, D. Irving (Simon and Schuster, New York, 1968). Was Nazi Germany really in an A-bomb race with the U.S.? See also ALSOS, by S. A. Goudsmith (Schuman, New York, 1947).

3.19. BRITAIN AND ATOMIC ENERGY: 1939-1945, M. M. Gowing (St. Martin's, New York, 1964). A report of the British start on nuclear weapons in World War II, cooperation with the U.S., and the breakdown of this relationship by 1945.

3.20. "Nuclear Weapons History: Japan's Wartime Bomb Projects Revealed," D. Shapley, Science 199, 152-7 (1978); see also letters to the Editor of Science 199, 728, 1286-90 (1978) and 200, 486 (1978). The Japanese did think about a possible A-bomb project during World War II.

3.21. HIROSHIMA: THE DECISION TO USE THE A-BOMB, E. Fogelman (Scribner's, New York, 1964). Statements by American and Japanese scientific, military, and political leaders about the decision to use the A-bomb during World War II.

3.22. HIROSHIMA DIARY: THE JOURNAL OF A JAPANESE PHYSICIAN; AUGUST 6-SEPTEMBER 30, 1945, M. Hachiya (University of North Carolina, Chapel Hill, 1955). Describes what it was like to cope with the medical aftereffects of the first nuclear bomb.

3.23. HIROSHIMA AND NAGASAKI: THE PHYSICAL, MEDICAL, AND SOCIAL EFFECTS OF THE ATOMIC BOMBINGS, The Committee for the Compilation of Materials on the Damage Caused by the Atomic Bombs in Hiroshima and Nagaski (Basic, New York, 1981). See also MEDICAL EFFECTS OF THE ATOMIC BOMB IN JAPAN, edited by A. W. Oughterson and W. Shields (McGraw-Hill, New York, 1965).

3.24. DEATH IN LIFE: SURVIVORS OF HIROSHIMA, R. J. Lifton (Random House, New York, 1968). A review of the psychological effects on the survivors at Hiroshima.

3.25. "Hiroshima-Nagasaki: August 1945," a film produced by E. Barnouw and P. Ronder (Museum of Modern Art, Circulating Film Program, 11 W. 53rd St., New York, NY 10019; 15 min, 16 mm, B/W, $275 purchase, $25 rental). A film of the aftereffects of the A-bombs at Hiroshima and Nagasaki, taken by Japanese photographers. Sometimes gruesome, emotionally moving.

3.26. "Truman and the Atomic Bomb," produced by Learning Corporation of America (Learning Corp. of America, 1350Ave. of Americas, New York, NY 10019; 1969, 16 mm, B/W, 15 min, $125 purchase, $15 rental). An excellent film dealing with Truman's decision to drop the atomic bomb on Japan.

C. The Fusion Bomb

1. Technical aspects

3.27. "The H-bomb Secret," H. Morland, The Progressive 43 (11), 14-23 (November 1979); and "Errata," ibid, 36 (December 1979);* and THE SECRET THAT EXPLODED (Random House, New York, 1981); see also the letter about the H-bomb design by H. Hansen in the Chicago Tribune of September 18, 1979, pp. 12-13; and BORN

SECRET: THE H BOMB, THE PROGRESSIVE CASE AND NATIONAL SECURITY, A. de Volpi, G. E. Marsh, T. A. Postol, and G. S. Stanford (Pergamon, New York, 1981). The design information revealed here is very interesting. But the critical nature of the errata forces one to be skeptical about the design. There is much discussion about which parts, if any, of the secret were really "secret." See also Ref. 2.2.

√ 3.28. THE PHYSICAL PRINCIPLES OF THERMONUCLEAR EXPLOSIVE DEVICES, F. Winterberg (The Fusion Energy Foundation, New York, 1981). Advanced discussion of designs.

3.29. "The Nature of Radioactive Fall-out and Its Effects on Man," Hearings before the Special Subcommittee on Radiation of the Joint Committee on Atomic Energy, Congress of the United States, 85th Congress, First Session, May 27, 28 29 and June 3, 1957 (U.S. Government Printing Office, Washington, DC, Parts I, II, and III [Index], 1957). A potpourri of information.

3.30. ICHIBAN: RADIATION DOSIMETRY FOR THE SURVIVORS OF THE BOMBING OF HIROSHIMA AND NAGASAKI, J. A. Auxier (National Technical Information Service, Dept. of Commerce, Springfield, VA, TID-27080, 1977).

3.31. "Effects on Populations of Exposure to Low Levels of Ionizing Radiation," Committee on the Biological Effects of Ionizing Radiations (National Academy of Sciences, Washington, DC, 1972 and 1980). These are thorough reviews of all the known effects of radiation on people, expressed ultimately as some functional extrapolations down to very low dose levels. These reports, known as BEIR II and BEIR III, are at the same time both authoritative and controversial. See, for example, Bull. At. Sci. $\underline{29}$ (3), 47-49 (March 1973); and Science $\underline{212}$, 900-3 (1981); and "Genetic Effects of the Atomic Bomb: A Reappraisal," W. J. Schull, M. Otake, and J. V. Neel, Science $\underline{213}$, 1220-7 (1981).

3.32. "Cancer and Low-Level Ionizing Radiation," K. Z. Morgan, Bull. At. Sci. $\underline{34}$ (7), 30-41 (September 1978). A summary of the current understanding of the effects of low-level exposure to radiation.

3.33. "The War Game," produced by P. Watkins (British Film Institute Films Inc., 1144 Wilmette Ave., Wilmette, IL 60091, 1968; 16 mm or 3/4-in. videocassette, B/W, 49 min, $575 or $435 purchase, or $40 rental). The effects of a nuclear attack on Great Britain are shown in horrifying detail based on information supplied by experts in nuclear defense, economics, and medicine.

2. Historical aspects

3.34. ATOMIC SHIELD, 1947/1952: A HISTORY OF THE UNITED STATES ATOMIC ENERGY COMMISSION, VOL. II, R. G. Hewlett and F. Duncan (The Pennsylvania State University, University Park, PA, 1969). The authoriative biography of the Atomic Energy Commission, a continuation of Ref. 3.17.

3.35. THE ADVISORS: OPPENHEIMER, TELLER AND THE SUPERBOMB, H. F. York (Freeman, San Francisco, 1975). See also "The Debate over the Hydrogen Bomb," H. F. York, Sci. Am. 233 (4), 106-113 (October 1975). York analyzes the 1949 advice by the AEC General Advisory Committee under Oppenheimer that there should not be crash programs to develop the Super, the H bomb.

D. Full-Scale Nuclear War

1. Technical aspects

3.36. FALLOUT: A STUDY OF SUPERBOMBS, STRONTIUM-90 AND SURVIVAL, J. M. Fowler (Basic, New York, 1960). A nice survey of the facts about and effects of fallout.

3.37. "The Prompt and Delayed Effects of Nuclear Weapons," K. N. Lewis, Sci. Am. 241 (1), 35-47 (July 1979).

3.38. ENVIRONMENTAL EFFECTS OF NUCLEAR WEAPONS, R. U. Ayres (Hudson Institute, Harmon-on-Hudson, NY, 1965). Vol. I present particularly detailed calculations of fallout and other problems.

3.39. See also "Physics, Chemistry, and Meterology of Fallout," R. Bjoernersted and K. Edvarson, Ann. Rev. Nucl. Sci. 13, 505-34 (1963); "Movement of Fallout Radionuclides Through the Biosphere and Man," C. L. Comar, Ann. Rev. Nucl. Sci. 15, 175-206 (1965); "Effects of Radiation on Man," A. C. Upton, Ann. Rev. Nucl. Sci. 18, 495-528 (1968), and "Global Consequences of Nuclear Weaponry," J. C. Mark, Rev. Nucl. Sci. 26, 51-87 (1976).

3.40. LONG-TERM WORLD-WIDE EFFECTS OF MULTIPLE-NUCLEAR WEAPONS DETONATIONS, Committee to Study the Long-Term Worldwide Effects of Multiple Nuclear-Weapons Detonations (National Academy of Sciences, Washington, DC, 1975).* A review of possible global effects from 10^4 megatons of nuclear explosions, including fallout, ozone depletion, etc. See also review in Science 190, 248-50 (1975).

3.41. THE EFFECTS OF NUCLEAR WAR, Office of Technology Assessment (Allanheld and Osmun, Montclair, NJ, 1980 [1979]). A review of these effects, including some introductory tutorials.

3.42. WEAPONS OF MASS DESTRUCTION, A. H. Westing (Crane and Russak, New York, for SIPRI, 1977). Nice summaries, as in tables, of nuclear weapons effects. See also NUCLEAR RADIATION WARFARE, J. Rotblat (Oelgeschlager, Gunn & Hain, Inc., Cambridge, MA, for SIPRI, 1981).

3.43. NUCLEAR DISASTER, T. Stonier (World Publishing, Cleveland, 1964). Physical, biological, economic, social, and psychological effects of a nuclear attack.

3.44. LAST AID: THE MEDICAL DIMINSION OF WAR, E. Chivian, S. Chivian, R. J. Lifton, J. E. Mack, Eds., (W. H. Freeman & Co., San Francisco, 1982). A physician's perspective.

3.45. NUCLEAR WAR: THE AFTERMATH (based on a special issue of Ambio 11 (2-3), 1982), J. Peterson, D. Hinrichsen and B. Dampier,

Eds. (Pergamon Press, New York, 1983). The consequences of a specific full-scale war scenario are analyzed in detail.

3.46. LIFE AFTER NUCLEAR WAR, A. M. Katz (Ballinger Publ. Co., Cambridge, MA, 1982). An attempt to look at social consequences of nuclear war.

3.47. DESTRUCTION OF NUCLEAR ENERGY FACILITIES IN WAR: THE PROBLEM AND THE IMPLICATIONS: B. Ramberg (Lexington Books [Heath], Lexington, MA, 1980). See also "Catastrophic Releases of Radioactivity," S. A. Fetter and K. Tsipis, Sci. Am. 244 (4), 41-7 (April 1981).

3.48. ON THERMONUCLEAR WAR, H. Kahn (Princeton University, Princeton, NJ, 2nd ed., 1961). A truly classical examination of the consequences of, and recovery from nuclear war; covers deterrence and other objectives of nuclear arms, and relevant lessons from other wars.

2. Political aspects

3.49. NUCLEAR FORCES FOR MEDIUM POWERS; PART I: TARGETS AND WEAPONS SYSTEMS, G. Kemp (The International Institute for Strategic Studies, London, Adelphi Paper 106, 1974). Calculations of the strategic forces required to produce specific levels of destruction, a sample targeting of the Soviet Union.

3.50. HOW MUCH IS ENOUGH? SHAPING THE DEFENSE PROGRAM, 1961-1969, A. C. Enthoven and K. W. Smith (Harper and Row, New York, 1971). How to quantitatively analyze defense programs, and control technological progress.

3.51. THE BALANCE OF TERROR: A GUIDE TO THE ARMS RACE, E. M. Bottome (Beacon, Boston, 1971). A good review of the development of the strategy of massive retaliation.

3.52. THE FATE OF THE EARTH, J. Schell (A. Knopf, New york, 1982). The ethical perspective on fullscale nuclear war.

IV.   NUCLEAR BALANCE

In Section IV we list references about the various delivery systems for nuclear warheads, focussing primarily on the strategic triad of intercontinental bombers, intercontinental ballistic missiles and nuclear-missile submarines. Beyond the technology of the deterrence triad, the references look at the characteristics of the deterrent balance. Several alternative ways of evaluating the nuclear balance are described. The goals of deterrence are discussed, with emphasis on differentiating between the qualities and quantities of weapons systems.

A. Intercontinential bombers

1. Technical aspects

4.1. INTRODUCTION TO FLIGHT, J. D. Anderson (McGraw-Hill, New York, 1978). Very nice advanced textbook for engineers; covers basic aerodynamics, aircraft performance, stability and control; with calculations for "real" aircraft.

4.2. RADAR DETECTION AND TRACKING SYSTEMS, A. A. Hovanessian (Artech House, Dedham, MA, 1978 [1973]). Describes basic radar principles, probabilities of detection, new technologies like synthetic radar arrays, and specific applications like terrain avoidance. It can be understood by senior-level physics students.

4.3. INTRODUCTION TO RADAR SYSTEMS, Merril I. Skolnik (McGraw-Hill, New York, 2nd Ed., 1980).

4.4. RADAR PRINCIPLES FOR THE NON-SPECIALIST, J. C. Toomay (Lifetime Learning Publ., Belmont, CA, 1983).

4.5. APPLIED ECM, L. B. van Brunt (E. W. Engineering, Dunn Loring, VA, 1978). Technically advanced three-volume text on electronic countermeasures against, for example, radar detection.

4.6. "Defense Against Bomber Attack," R. D. English and D. I. Bolef, Sci. Am. <u>229</u> (27), 11-19 (August 1973). A description of the airborne warning and control system aircraft and its costs.

2. Policy aspects

4.7. MODERNIZING THE STRATEGIC BOMBER FORCE: WHY AND HOW, A. H. Quanbeck and A. L. Wood (The Brookings Institution, Washington, DC, 1976). A good quantitative analysis and comparison of the various alternatives to the deployment of the B-1 bomber.* It is criticized by SLOW TO TAKE OFFENSE: BOMBERS, CRUISE MISSILES AND PRUDENT DETERRENCE, F. P. Hoeber (Center for Strategic and International Studies, Georgetown University, 2nd ed., 1980).

4.8. THE TFX DECISION: McNAMARA AND THE MILITARY, R. J. Art (Little and Brown, New York, 1968). A case study of the weapons acquisition process and the extent to which it can be controlled.

4.9. "Dr. Strangelove or: How I Learned To Stop Worrying and Love the Bomb," produced and directed by S. Kubrick (Swank Motion Pictures, Inc., 393 Front St., Hempstead, NY 11550, 1964; 16mm, B/W, 93 min., $150 rental only). A biting and at times hilariously funny satire on the military mind; raises questions about the bomber portion of the Triad.

B. Intercontinental Ballistic Missiles

1. Technical aspects

4.10. GUIDED MISSILE ENGINEERING, A. E. Puckett and S. Ramo (McGraw-Hill, New York, 1959). In spite of its venerable age, this is a good tutorial on aerodynamics, stability in flight, rocket motors, information theory for signal transmission, etc.

4.11. PRINCIPLES OF SPACEFLIGHT PROPULSION, E. M. Goodger (Pergamon, New York, 1970). Reviews chemical and nuclear energies, multistage rockets, etc. See also ROCKET PROPULSION ELEMENTS: AN INTRODUCTION TO THE ENGINEERING OF ROCKETS, G. P. Sutton and D. M. Ross (Wiley, New York, 4th ed., 1976).

4.12. GUIDED WEAPONS CONTROL SYSTEMS, P. Garnell and D. J. East (Pergamon, New York, 1977). For advanced engineers.

4.13. See pp. 84-103 of Ref. 5.6 for the flight of ballistic missiles.

4.14. "Ballistic Missile Guidance," D. G. Hoag in IMPACT OF NEW TECHNOLOGIES ON THE ARMS RACE, edited by B. T. Feld, T. Greenwood, G. W. Rathjens, and W. Weinberg (MIT, Cambridge, MA, 1971), pp. 19-106. An excellent, even if old, article.

4.15. "The Accuracy of Strategic Missiles," K. Tsipis, Sci. Am. 233 (1), 14-23 (July 1975). The technology of superaccurate missiles with inertial guidance.

4.16. "Land Based Intercontinental Ballistic Missiles," B. T. Feld and K. Tsipis, Sci. Am. 241 (5), 50-61 (November 1979). Are ICBM's vulnerable, is the MX missile an appropriate response to such a vulnerability?

4.17. MX-MISSILE BASING, Office of Technology Assessment (U.S. Government Printing Office, Washington, DC, 1981). A review of the MX missile and its various possible basing modes.

4.18. MX: PRESCRIPTION FOR DISASTER, H. Scoville (MIT Press, Cambridge, MA, 1981).

4.19. NUCLEAR EXPLOSION EFFECTS ON MISSILE SILOS, K. Tsipis (Center for International Studies, MIT, Cambridge, MA, 1978).

4.20. "All you Ever Wanted to Know About MIRV and ICBM Calculations But Were not Cleared to Ask," L. E. Davis and W. R. Schilling, J. Conflict Res. 17, 207-42 (June 1973). Has the equations and does calculations about missile-silo vulnerabilities in terms of the K parameter.

4.21. "Physics and Calculus of Countercity and Counterforce Nuclear Attacks," K. Tsipis, Science 187, 393-7 (1975). More calculations about the K silo-kill parameter. See also Letters to the Editor of Science 190, 1117-9 (1975); 206, 510 (1979).*

2. ICBM, MIRV and the ICBM

4.22. DEVELOPING THE ICBM: A STUDY IN BUREAUCRATIC POLITICS, E. Beard (Columbia University, New York, 1976). A sophisticated case study of developing a new weapons system.

4.23. "Multiple-Warhead Missiles," H. F. York, Sci. Am. 229 (5), 18-27 (November 1973). Description of the MIRV technology.

4.24. MIRV AND THE ARMS RACE, R. L. Tammen (Praeger, New York, 1973). Analyzes the action-reaction component of this development. See also MAKING THE MIRV: A STUDY OF DEFENSE DECISION MAKING, T. Greenwood (Ballinger, Cambridge, MA, 1975).

C. SLBM Submarines

1. Technical aspects

4.25. MISSILE BENEATH THE SEA: THE STORY OF POLARIS, J. J. DiCerto (St. Martin's, new York, 1967). A good, even if qualitative description of the technology involved in the submarine and the missile.

4.26. SONAR AND UNDERWATER SOUND, A. W. Cox (Lexington Books [Heath], Lexington, MA, 1974). A good professional book on the subject.

4.27. TACTICAL AND STRATEGIC ANTISUBMARINE WARFARE, K. Tsipis (MIT, Cambridge, MA, 1974). Describes the technology and strategy of the underwater Cold War. See also THE FUTURE OF THE SEA-BASED DETERRENT, edited by K. Tsipis, A. H. Cahn and B. T. Feld (MIT, Cambridge, MA, 1973).

4.28. "Antisubmarine Warfare and National Security," R. L. Garwin, Sci. Am. 227 (1), 14-25 (July 1972). Qualitative, but technical description of ASW. See also "Missile Submarines and National Security," H. Scoville, Jr., Sci. Am. 226 (6), 15-27 (June 1972).

4.29. "Advances in Antisubmarine Warfare," J. S. Wit, Sci. Am. 244 (2), 31-41 (February 1981). Discusses improvements in ASW and their implications.*

4.30. THE SECRET WAR FOR THE OCEAN DEPTHS, T. S. Burns (Rawson, New york, 1978). A very readable book about the nuclear submarine forces, and U.S. and U.S.S.R. attitudes toward control of the oceans.

2. Political aspects

4.31. NUCLEAR NAVY: 1946-1962, R. G. Hewlett and F. Duncan (University of Chicago, Chicago, 1974). Detailed history of U. S. nuclear submarines and of the Polaris missile.

4.32. THE POLARIS SYSTEM DEVELOPMENT: BUREAUCRATIC AND PROGRAMMATIC SUCCESS IN GOVERNMENT, H. M. Sapolski (Harvard University, Cambridge, MA, 1972). Examines institutional reasons for the success of the Polaris Program.

4.33. SECURING THE SEAS: THE SOVIET NAVAL CHALLENGE AND WESTERN ALLIANCE OPTIONS, P. H. Nitze, L. Sullivan, Jr., and the Atlantic Working Group on Securing the Seas (Westview, Boulder, CO, 1979). A good overall survey of the naval balance.

3. ELF communications

4.34. "Long-Range Communications at Extremely Low Frequencies," S. L. Bernstein, et al., Proc. IEEE 62, 292-312 (March 1974). Good technical review of ELF communication and its problems.

4.35. "Project Sanguine," J. R. Wait, Science 178, 272-5 (1972); and "The U. S. Navy's Project Saguine—Will it Become a Billion-Dollar Boondoggle?". G. Nelson, Congr. Rev. 117 (12), E15378-86 (May 17, 1971). Contains some quite sophisticated material, as well as political comments. See also Science 192, 1213-5 (1976); 197, 964-8 1977), and THE ELF ODYSSEY: NATIONAL SECURITY VERSUS ENVIRONMENTAL PROTECTION, L. L. Klessig and V. L. Strite (Westview, Boulder, CO, 1980). A general review.

D. Nuclear Deterrence and Stability

1. Quantitative aspects

4.36. STABILITY AND STRATEGIC NUCLEAR ARMS, H. Afheldt and P. Sonntag (New York Institute for International Order, World Law Fund Ocasional paper 1, New York, 1971). Quite sophisticated mathematical analysis of the nuclear balance in a two-player matrix, defining quantitatively such concepts as international stability.

4.37. "Assuring Strategic Stability in an Era of Detente," P. H. Nitze, Foreign Affairs 54 (2), 207-232 (January 1976). Outlines the alternatives to the strategy of deterrence; and argues tht for limited nuclear exchanges the U. S. arsenal is inadequate. Widely cited.

4.38. THE USE OF FORCE, edited by R. J. Art and K. N. Waltz (Little and Brown, Boston, 1979). Contrasts stability and balance, discusses Kissinger and deterrence, and ABM, etc.

4.39. U.S./U.S.S.R. STRATEGIC FORCE—ASYMMETRICAL DEVELOPMENTS: A NET AMERICAN ASSESSMENT, J. R. Thomson (University Press of America, Washington, DC, 1977). A comprehensive review of the whole military balance, both in technical and political terms.

2. Political aspects

4.40. THE ROLE OF NUCLEAR FORCES IN CURRENT SOVIET STRATEGY, L. Goure, F. D. Kohler, and M. L. Harvey (Center for Advanced International Studies, University of Miami, Coral Gables, FL, 1974). Argues, on the basis of extensive quotations from Soviet literature, that the U.S.S.R. has not accepted MAD as a strategy.

4.41. THE DIPLOMACY OF DETENTE: THE KISSINGER ERA, Coral Bell (St. Martin's, New York, 1977). Very nice history of the arrival of MAD, deterrence and detente, as U.S. national policies.

4.42. THE SOVIET AMERICAN ARMS RACE, C. S. Gray (Lexington Book [Heath], Lexington, MA, 1976). Comparison of the forces, and how the arms race is progressing.

4.43. THE TECHNOLOGICAL LEVEL OF SOVIET INDUSTRY, R. Amann, J. Cooper, and R. W. Davis (Yale University, New Haven, CT, 1977); particularly Chapter 8 on "Computer Technology," pp. 377-406. See also INDUSTRIAL INNOVATION IN THE SOVIET UNION, R. Aman and J. Cooper (Yale University Press, New Haven, CT, 1982); and "Computer Sales to U.S.S.R.: Critics Look for Quid pro Quo," N. Wade, Science 183, 499-501 (1974); "American-Soviet Relations: The Cancelled Computer," N. Wade, Science 201, 422-6 (1978); and "New Chips Shed Light on Soviet Electronics," D. Shapley, Science 204, 283-4 (1979). Discussions of the Soviet-American computer race. For a historical explanation of the Soviet lag, see "Cybernetics," L. R. Graham in SCIENCE AND IDEOLOGY IN SOVIET SOCIETY, edited by G. Fischer (Atherton, New York, 1967), pp. 83-106.

V. ALTERNATIVES TO NUCLEAR DETERRENCE

This section of reference explores some alternatives to the docterine of mutual assured destruction (MAD). (i) Successful defense against a nuclear attack could eliminate MAD. A civil defense program represents a passive defense, while an antiballistic missile system represents an active defense. (ii) Chemical or biological warfare could replace nuclear weaponry as the guarantor of deterrence. (iii) Small-scale conventional wars continue to be fought under the umbrellas of nuclear deterrence. How far can such wars go before they spill over into the arena of deterrence? Can tactical nuclear weapons be used without an escalation up to the strategic nuclear level?

A. Civil (passive) Defense

5.1. SURVIVAL AND THE BOMB: METHODS OF CIVIL DEFENSE, edited by E. P. Wigner (Indiana University, Bloomington, IN, 1969). A collection of essays summarizing some of the more pertinent aspects of civil defense, edited by a long-time advocate of a strong civil defense program. Covers shelters, decontamination, economic recovery, psychological effects, objectives.

5.2. "Limited Nuclear War," S. D. Drell and F. von Hippel, Sci. Am. $\underline{235}$ (5), 27-37 (November 1976). How useful might civil defense be in a limited nuclear war? See also "Civil Defense in Limited War--a Debate," A. A. Broyles and E. P. Wigner, and S. D. Drell, Phys. Today $\underline{29}$ (4), 44-57 (April 1976); and letters to the Editor of Phys. Today $\underline{29}$ (12), 11-15 (December 1976).

5.3. WAR SURVIVAL IN SOVIET STRATEGY: USSR CIVIL DEFENSE, L. Goure (Center for Advanced International Studies, University of Miami, Coral Gables, FL, 1976). Professor Goure has read vast numbers of Soviet sources which he interprets as demonstrating a strong and growing Soviet civil defense effort. He argues that the Soviet Union continues to believe it can win a nuclear war with the U.S.

5.4. "In the Event of an Emergency," produced by WGBH for NOVA (WGBH Educational Foundation, 125 Western Ave., Boston, MA, 02134, 3/4 in. videocassette, color, 59 min.). A review, with interviews, of the pros and cons of civil defense.

B. ABM (active) Defense

1. Technical aspects

5.5. "Anti-Ballistic-Missile Systems," R. L. Garwin and H. A. Bethe, Sci. Am. $\underline{218}$ (3), 13-23 (March 1968). Good qualitative technical information, focussing on penetration aids. See also letters to the Editor of Sci. Am. $\underline{218}$ (5), 6-8 (May 1968).

5.6. INTRODUCTION OF DEFENSE RADAR SYSTEMS ENGINEERING. J. Constant (Spartan, New York, 1972). Very advanced but excellent text relating to the Safeguard ABM system. Chapters on target characteristics, missile flight paths, array antennas, beta blackout, etc.

5.7. RADAR SENSOR ENGINEERING, AND RADAR TARGETS, C. Y. Bachman (Lexington Books, Lexington MA, 1982).

5.8. BALLISTIC MISSILE DEFENSE AND DECEPTIVE BASING: A NEW CALCULUS FOR THE DEFENSE OF ICBMs, R. C. Starsman (National Defense University Press, Washington DC, 1980). A pamphlet with lots of calculations.

5.9. U. S. STRATEGIC-NUCLEAR POLICY AND BALLISTIC MISSILE DEFENSE: THE 1980s AND BEYOND, Schneider, et. al. (Institute for Foreign Policy Analysis, 1980).

2. Political aspects

5.10. ABM: AN EVALUATION OF THE DECISION TO DEPLOY AN ANTIBALLISTIC MISSILE SYSTEM, edited by A. Chayes and J. B. Wiesner (Harper and Row, New York, 1969); and WHY ABM: POLICY ISSUES IN THE MISSILE DEFENSE CONTROVERSY, edited by J. J. Holst and W. Schneider, Jr. (Pergamon, New York, 1969). The two sides of the ABM debate.*

5.11. THE MISSILE DEFENSE CONTROVERSY: STRATEGY, TECHNOLOGY AND POLITICS, 1955-1972, E. J. Yanarella (University Press of Kentucky, Lexington, KY, 1977). An excellent review of the political development of the ABM; asks questions about the technological imperative.

5.12. "American Scientists and the ABM: A Case Study of Controversy," A. H. Cahn in SCIENTISTS AND PUBLIC AFFAIRS, edited by A. H. Teich (MIT, Cambridge, MA, 1974), pp. 41-120. A discussion of the role played by scientific experts in the ABM debate. See also "Lobbying Against the ABM, 1967-1970," T. A. Halstead, Bull. At. Sci. 27 (4), 23-28 (April 1971). A controversy raged over whether scientists in the debate followed acceptable procedures. "Guidelines for the Practice of Operations Research," T. E. Caywood et al., Operations Res. 19 (5), 1123-1257 (September 1971); "The Obligations of Scientists as Councillors: Guidelines for the Practice of Operations Research," T. E. Caywood, Minerva 10 (1), 105-157 (January 1972); "Can Investigations Improve Scientific Advice: The Case of the ABM?," P. C. Doty, Minerva 10 (2), 280-294 (April 1972); and "The Scientists as Advisors: Operations Research Experience," H. E. Miser, Minerva 11 (1), 95-108 (January 1973).

3. Particle-beam weapons

5.13. "Particle Beam Weapons—A Technical Assessment," G. Bekefi, B. T. Feld, J. Parmentola, and K. Tsipis, Nature 284, 219-225 (1980). A technical review of the science and technology of this weaponry. See also PARTICLE BEAM WEAPONS, G. Bekefi, B. T. Feld, J. Parmentola, and K. Tsipis, Report 4 of the Program in Science and Technology for International Security, MIT Dept. of Physics, Cambridge, MA, 1978 and "Particle-Beam

Weapons," J. Parmentola and K. Tsipis, Sci. Am. <u>240</u>, (4), 54–65 (April 1979).

5.14. P. J. Klass and C. A. Robinson, Jr. have had several series of articles on laser and particle-beam weapons in AVIATION WEEK AND SPACE TECHNOLOGY. In these articles they summarize the evidence for the state of the art both in the U. S. and the Soviet Union. Their interpretation of the data are controversial. See pp. 34–39 (August 18, 1975); pp. 50–56 (September 1, 1975); pp. 53–59 (September 8, 1975); pp. 14–16 (August 7, 1978; pp. 38–47 (August 21, 1978); pp. 56–60 (August 28, 1978); pp. 16–23 (May 2, 1977); pp. 14–22 (October 2, 1978); pp. 32–66 (July 28, 1980); pp. 44–68 (August 4, 1980); pp. 52–71 (May 25, 1981).

5.15. DIRECTED-ENERGY WEAPONS; A JURIDICIAL ANALYSIS, E. A. Fessler (Praeger, New York, 1979). Accepts the reports of Ref. 5.14 and discusses their implications in terms of international space laws.

5.16. OUTER SPACE: A NEW DIMENSION OF THE ARMS RACE, B. Jasani, Ed. (Oelgeschlager, Gunn and Hain, Cambridge, MA, 1980).

5.17. "The Real War in Space," produced by the BBC (Time-Life Films, 100 Eisenhower Dr., Paramus, NJ 07652, 1980; 3/4 in. videocassette, color, 52 min., $200 purchase). Competition with the USSR on ABM and killer satellites.

C. Chemical and Biological Warfare

5.18. CHEMICAL WEAPONS AND CHEMICAL ARMS CONTROL, edited by M. Meselson (Carnegie Endowment for International Peace, New York, 1978). A 1977 conference discussing the pros and cons of CBW defense planning, and its relation to NATO. See also "Chemical Warfare and Chemical Disarmament," M. Meselson and J. P. Robinson, Sci. Am. <u>242</u> (4), 38–47 (April 1980).

5.19. CHEMICAL AND BIOLOGICAL WARFARE: A STUDY OF THE HISTORICAL, TECHNICAL, MILITARY, LEGAL AND POLITICAL ASPECTS OF CBW AND POSSIBLE DISARMAMENT MEASURES, six volumes of the Swedish International Peace Research Institute (Humanities Press for SIPRI, New York, 1971).

D. Tactical Nuclear War

1. The neutron bomb

5.20. "Enhanced-Radiation Weapons," F. M. Kaplan, Sci. Am. 238 (5), 44–51 (May 1978). Technical details about the neutron bomb.

5.21. THE NEUTRON BOMB: POLITICAL, TECHNOLOGICAL AND MILITARY ISSUES, S. T. Cohen (Institute for Foreign Policy Analysis, Cambridge, MA, 19780. Covers all social aspects of the neutron bomb.

5.22. TACTICAL NUCLEAR WEAPONS: EUROPEAN PERSPECTIVES, F. Barnaby on behalf of the Swedish International Peace Research Institute (Crane, Russak, New York, 1978). Defines, describes, and lists tactical nuclear weapons and their delivery systems in Europe.

2. Tactical nuclear war in Europe

5.23. DEFENDING THE CENTRAL FRONT: THE BALANCE OF FORCES, R. L. Fischer (International Institute for Strategic Studies, London, Adelphi Paper 127, 1976). A careful account of the balance.

5.24. "Precision-guided Weapons," P. F. Walker, Sci. Am. 245 (2), 36-45 (August 1981). Tactical nuclear weapons would most likely be used by NATO forces in Europe if conventional defenses are unsuccessful. Precision-guided munitions might help the conventional defenses.

5.25. THE SUPERWARRIORS: THE FANTASTIC WORLD OF PENTAGON SUPERWEAPONS, J. W. Canan (Weybright and Tally, New york, 1975); THE ELECTRONIC BATTLEFIELD, P. Dickson (Indiana University, Bloomington, IN, 1976); and PRECISION GUIDED WEAPONS, J. Digby (International Institute for Strategic Studies, London, Adelphi Paper 118, 1975). A pair of journalistic exposes and a review of the newest high technology for military purposes.

5.26. NEW TECHNOLOGY AND MILITARY POWER: GENERAL PURPOSE MILITARY FORCES FOR THE 1980s AND BEYOND, S. J. Deitchman (Westview, Boulder, CO, 1979). A very nice summary of the impact of technology on future forces.

5.27. INTRODUCTION TO BATTLEFIELD WEAPONS SYSTEMS AND TECHNOLOGY, R. G. Lee (Pergamon [Brassey's Publ. Ltd], New York, 1981).

5.28. INTERNATIONAL WEAPONS DEVELOPMENTS, R.U.S.I.-Brassey's (Pergamon [Brassey's Publ. Ltd], New York, 4th Ed., 1980).

5.29. "Intermediate-Range Nuclear Weapons," K. N. Lewis, Sci. Am. 243 (6), 63-73 (December 1980). Discusses the gray area between strategic-nuclear and conventional warfare.

5.30. TACTICAL NUCLEAR WEAPONS: AN EXAMINATION OF THE ISSUES, W. R. Van Cleave and S. T. Cohen (Crane, Russak, New York, 1978). A review.

5.31. U.S. NUCLEAR WEAPONS IN EUROPE, J. Record (Brookings Institution, Washington, DC, 1974). A good summary of the history and disposition of tactical nuclear weapons in Europe; interestingly the neutron bomb is never mentioned.

5.32. SOVIET MILITARY STRATEGY IN EUROPE, J. D. Douglas, Jr. (Pergamon, New York, 1980). Interprets original Soviet documents as arguing for a winable nuclear war. See also SOVIET STRATEGY FOR NUCLEAR WAR, J. D. Douglas, Jr., and A. M Hoeber (Hoover Institute, Stanford, CA, 1979).

5.33. "The Nuclear 'Balance of Terror' in Europe," H. F. York, Ambio 4 (5/6), 203-8 (1975). Effects of a nuclear war in Europe, as through fallout.*

5.34. NUCLEAR NIGHTMARES: AN INVESTIGATION INTO POSSIBLE WARS, N. Calder (Viking, New York, 1979). A look at four possible causes of nuclear war: escalation, proliferation, accident, and pre-emptive first strike. Also a BBC-TV production by P. Batty, directed by N. Calder (Corinth Films, 410 E. 62nd St., New York, NY 10021, 1980, $175 rental).

5.35. THE THIRD WORLD WAR; A FUTURE HISTORY, J. Hackett, et al.

(Berkley [Macmillan], New York, 1978). How World War III might break out in Europe, how it might progress, and how it might end.

5.36. EMP RADIATION AND PROTECTIVE TECHNIQUES, L. W. Ricketts, J. E. Bridges, and J. Milette (Wiley, New York, 1976); and FUNDAMENTALS OF NUCLEAR HARDENING OF ELECTRONIC EQUIPMENT, L. W. Ricketts (Wiley, New York, 1972). Advanced courses on nuclear effects on electronic equipment. See also "Nuclear Pulse (I): Awakening to the Chaos Factor," "Nuclear Pulse (II): Ensuring Delivery of the Doomsday Signal," and "Nuclear Pulse (III): Playing a Wild Card," W. J. Broad, Science 212, 1009-1012, 1116-1120, and 1248-1251 (1981). An examination of the possibility of fighting a limited nuclear war when such phenomena as electromagnetic pulses from nuclear explosions may destroy many communication channels.

5.37. "The Bull's-Eye War," produced by the BBC (Time-Life Films, 100 Eisenhower Dr., Paramus, NJ 07652, 1980; 3/4 in. videocassette, color, 52 min., $200 purchase). Precision-guided and smart bombs are leading to an era of electronic wars.

VI. ARMS CONTROL AND DISARMAMENT

The references in this section look at arms control and disarmament from various perspectives. It has frequently been argued that control of the arms race is made difficult by the technological imperative, by the intrinsic momentum of technological progress. The first set of references looks at that proposition, and at computers and the cruise missile which exemplify it. The second set of references examines one major consequence of this imperative: nuclear proliferation may be seen as an inevitable consequence of the worldwide spreading nuclear technology. The third and fourth sets of references deal with two arms-control efforts, with the nuclear Test Ban Treaty and the SALT agreements. The fifth set of references considers nuclear disarmament in general.

A. The Technological Imperative

6.1. ARMS, DEFENSE POLICY, AND ARMS CONTROL, edited by F. A. Long and G. W. Rathjens (Norton, New York, 1976). See particularly "Armaments and Arms Control: Exploring the Determinants of Military Weapons," G. T. Allison and F. A. Morris, pp. 99-129; "The Military Innovation System and the Qualitative Arms Race," H. Brooks, pp. 75-97; "Organizational and Political Dimensions of the Strategic Posture: The Problems of Reform," J. Steinbruner and B. Carter, pp. 131-154.

6.2. ANALYZING SOVIET STRATEGIC ARMS DECISIONS, K. F. Spielmann (Westview, Boulder, CO, 1978). Is the technological imperative operative in the Soviet Union? See also SOVIET DECISION-MAKING FOR DEFENSE: A CRITIQUE OF U.S. PERSPECTIVES

ON THE ARMS RACE, M. P. Gallagher and K. F. Spielmann, Jr. (Praeger, new York, 1972).

6.3. ARMS CONTROL AND TECHNOLOGICAL INNOVATION, edited by D. Carlton and C. Schaerf (Wiley, New York, 1976). Can political insitutions control the arms race in the face of accelerating technological change?

6.4. RACE TO OBLIVION: A PARTICIPANTS VIEW OF THE ARMS RACE, H. F. York (Simon and Schuster, New York, 1970). York worries that technologies have frequently a life of their own.

6.5. SCIENCE of 18 March 1977, and SCIENTIFIC AMERICAN of September 1977. These issues are both totally devoted to discussions of microelectronics, and of the implications for technology of minaturized electronic circuitry.

6.6. "Cruise Missiles," K. Tsipis, Sci. Am. 236 (2), 20-29 (February 1977); see also Letters to the Editor of Sci. Am. 236 (5), 8-10 (May 1977); 236 (6), 10 (June 1977); and 237 (2), 8, 13, 16 (February 1977). A good qualitative discussion of cruise missile technology.

6.7. THE CRUISE MISSILE: BARGAINING CHIP OR DEFENSE BARGAIN? R. L. Pfaltzgraff, Jr. (Institute for Foreign Policy Analysis, Cambridge, MA, 1977). A thorough review of the technical, military, and political aspects of the cruise missile development. See also CRUISE MISSILES: TECHNOLOGY, STRATEGY, POLITICS, R. K. Betts, Ed. (Brookings, Washington, DC, 1981).

6.8. THE ORIGINS OF THE STRATEGIC CRUISE MISSILE, R. Huiskens, (Praeger Publ., New York, 1981).

B. Nuclear Proliferation

1. Historical aspects

6.9. ATOMIC ENERGY IN THE SOVIET UNION, A. Kramish (Stanford University, Stanford, CA, 1959; I. V. KHURCHATOV: A SOCIALIST-REALIST BIOGRAPHY OF THE NUCLEAR SCIENTIST, I. N. Golovin (Selbstverlag, Bloomington, IN, 1969, translated by W. H. Dougherty from the Russian); FROM SCIENTIFIC SEARCH TO ATOMIC INDUSTRY: MODERN PROBLEMS OF ATOMIC SCIENCE AND TECHNOLOGY IN THE U.S.S.R., A. M. Petroyants (Interstate, Danville, IL, 1975). Histories of the Soviet A-bomb project, e.g., how much information from the Manhattan Project did it have?

6.10. INDEPENDENCE AND DETERRENCE: BRITAIN AND ATOMIC ENERGY, 1945-1952, VOL. I: POLICY MAKING, VOL. II: POLICY EXECUTION, M. Gowing, assisted by L. Arnold (St. Martin's, New York, 1975). A continuation of Ref. 3.19.

6.11. SCIENTISTS IN POWER, S. R. Weart (Harvard University, Cambridge, MA, 1979). About the development in France of nuclear power before, during, and after World War II, with a focus on Frederick Joliot-Curie.

6.12. GERMANY AND THE POLITICS OF NUCLEAR WEAPONS, C. M. Kelleher (Columbia University, New York, 1975). Detailed description of nuclear policy for a country that could have, but so far has not developed its own nuclear weapons.

6.13. JAPAN'S NUCLEAR OPTION: POLITICAL, TECHNICAL AND STRATEGIC FACTS, J. E. Endicott (Praeger, New York, 1975). Under what circumstances might Japan exercise a nuclear option?

6.14. THE NUCLEAR AXIS: SECRET COLLABORATION BETWEEN WEST GERMANY AND SOUTH AFRICA, Z. Cervenka and B. Rogers (Times, New York, 1978). An expose.

6.15. NUCLEAR PROLIFERATION AND SAFEGUARDS, Office of Technology Assessment (Praeger, New York, 1977). A complete, even if in spots superficial survey of all the proliferation issues.

6.16. NUCLEAR PROLIFERATION: THE SPENT FUEL PROBLEM, edited by F. C. Williams and D. A. Deese (Pergamon, New York, 1979). Good summary of the proliferation status of various countries and regions of the world as in Eastern Europe, the Indian Ocean Basin, Latin America, Asia, and the Middle East.

2. Proliferation technologies

6.17. INTRODUCTION TO NUCLEAR ENGINEERING, J. R. LaMarsh (Addison-Wesley, Reading, MA, 1977). A good engineering test on the level of physics seniors.

6.18. "Science and Society Test V: Nuclear Nonproliferation," D. W. Hafemeister, Am. J. Phys. 48 (2), 112-120 (1980). Gives numerical estimates on questions affecting the nonproliferation policy of the U.S.*

6.19. PROLIFERATION, PLUTONIUM AND POLICY: INSTITUTIONAL AND TECHNOLOGICAL IMPEDIMENTS TO NUCLEAR WEAPONS PROPAGATION, A. De. Volpi (Pergamon, New York, 1979). Has very good and extensive discussion both of the technical aspects of critical yields, denaturing, etc., and of the political aspects of institutional safeguards. See also Ref. 3.14.

6.20. URANIUM ENRICHMENT, edited by S. Villani (Springer, New York, 1979). Exhaustive and highly technical reviews of cascade theory, gaseous diffusion, centrifugation, separation nozzles, laser methods of isotope separation, plasma separation effects. See also "World Uranium Resources," K. W. Deffeyes and J. D. MacGregor, Sci. Am. 242 (1), 66-76 (January 1980); "The Gas Centrifuge," D. R. Olander, Sci. Am. 239 (2), 37-43 (August 1978); and "Laser Separation of Isotopes," R. N. Zare, Sci. Am. 236 (2), 86-98 (February 1977).

3. Nuclear proliferation

6.21. "A Ban on the Production of Fissionable Material for Weapons," W. Epstein, Sci. Am. 243 (1), 31-9 (July 1980). A review of the connection between civilian nuclear power and nuclear weapons proliferation.

6.22. NUCLEAR PROLIFERATION: MOTIVATIONS, CAPABILITIES, AND STRATEGIES FOR CONTROL, T. Greenwood, H. A. Feiveson, and T. B. Taylor (McGraw-Hill, New York, 1977). A good review.

6.23. "Superphenix: A Full-Scale Breeder Reactor," G. A. Vendryes, Sci. Am. 236 (3), 26-35 (March 1977). The breeder implies a Pu economy.

6.24. "Natural-Uranium Heavy-Water Reactors," H. C. McIntyre, Sci.

Am. <u>233</u> (4), 17–27 (October 1975). These reactors are more suitable for nuclear proliferation than are light-water reactors.

6.25. "THE REPROCESSING OF NUCLEAR FUELS," W. P. Beddington, Sci. Am. <u>235</u> (6), 30–41 (December 1976).

6.26. "Nuclear Power, Nuclear Weapons and International Security," D. J. Rose and R. K. Lester, Sci. Am. <u>238</u> (4), 45–57 (April 1978).

6.27. "Diversion by Non-governmental Organizations," T. B. Taylor in INTERNATIONAL SAFEGUARDS AND THE NUCLEAR INDUSTRY, edited by M. Willrich (Johns Hopkins University, Baltimore, MD, 1973), pp. 176-198. Contains the infamous list of references to help designers of A bombs.

6.28. THE CURVE OF BINDING ENERGY, J. McPhee (Farrar, Strauss and Giroux, New York, 1974). partially a biography of Ted Taylor, this very readable book leaves the impression that designing and building an A bomb is not so difficult.

6.29. "The Plutonium Connection," WGBH for NOVA (Time-Life Films, 100 Eisenhower Dr., Paramus, NJ 07652, 1975, 16 mm or 3/4 in. videocassette, color, 59 min., $200 video purchase). Deals with the possibility of diverting nuclear weapons-grade material from the nuclear fuel cycle to build A-bombs.

C. Nuclear Test Ban Treaty

1. Technical aspects

6.30. "Detection of Nuclear Explosions," R. Latter, R. F. Herbst, and K. M. Watson, Ann. Rev. Nucl. Sci. <u>11</u>, 371-418 (1961). Has, for example, equations for the decoupling phenomenon, for electromagnetic pulses from explosions, for x-ray fluxes and photon yields from nuclear tests in space.

6.31. "Nuclear Test Detection," a special issue of the Proc. IEEE <u>53</u> (#12) (December 1965). Contains articles on seismometers, seismic arrays, test detection, and identification, much on Vela satellites, on electronic pulses and on air fluorescence from atmospheric nuclear tests.

6.32. NUCLEAR EXPLOSIONS AND EARTHQUAKES: THE PARTED VEIL, B. A. Bolt (Freeman, San Francisco, 1976). An excellent book on earthquakes and underground nuclear explosions, not only on the science, but also on the history and politics of nuclear-test-ban efforts.

6.33. MONITORING NUCLEAR EXPLOSIONS, O. Dahlman and H. Israelson (Elsevier, New York, 1977). More detailed than Ref. 6.32, but less easy to follow.

6.34. NUCLEAR EXPLOSION SEISMOLOGY, H. C. Rodean (National Technical Information Service, Springfield, VA, TID-25572, 1971). Very sophisticated, yet understandable review of the generation of seismic signals by underground nuclear explosions, including the phenomenon of decoupling to minimize signals.

6.35. THE CONSTRUCTIVE USE OF NUCLEAR EXPLOSIVES, E. Teller, W. K. Talley, G. H. Higgins, and G. W. Johnson (McGraw-Hill, New

York, 1968). Describes the process of nuclear explosions, plowshare tests, and their computer simulation. Excellent text for first-year graduate students. See also EDUCATION FOR PEACEFUL USES OF NUCLEAR EXPLOSIVES, edited by L. E. Weaver (University of Arizona, Tucson, 1970). A conference report enthusiastically reporting on "plowshare" peaceful uses of nuclear explosives.

2. Political aspects

6.36. BLOWING ON THE WIND; THE NUCLEAR TEST BAN DEBATE: 1954-1960, R. A. Divine (Oxford University, New York, (1978). A very nice political history of the early efforts to reach a comprehensive nuclear test ban treaty.
6.37. AMERICAN SCIENTISTS AND NUCLEAR WEAPONS POLICY, R. Gilpin (Princeton University, Princeton, NJ, 1962). A classic discussion of the role scientists can play in scientific-political debates.
6.38. DIPLOMATS, SCIENTISTS AND POLITICIANS: THE U. S. AND THE NUCLEAR TEST BAN NEGOTIATIONS, H. K. Jacobson and E. Stein (University of Michigan, Ann Arbor, 1966).

D. Arms Limitations and SALT

1. Verification

6.39. RECONNAISSANCE, SURVEILLANCE AND ARMS CONTROL, T. Greenwood (International Institute of Strategic Studies, London, Adelphi Paper 88, 1972); see also his article in Sci. Am. 228 (2), 14-25 (February 1973);* SECRET SENTRIES IN SPACE, P. J. Klass (Random House, New York, 1971); and "Military Satellites," B. Jasani and "U. S. Air and Space Reconnaisance Programmes," H. F. York, in WORLD ARMAMENTS AND DISARMAMENT: SIPRI YEARBOOK 1975 (MIT, Cambridge, MA, 1975), pp. 103-179 and 180-187, respectively. These old references give indications of the extent and quality of spy-satellite reconnaissance efforts.
6.40. TELEMETRY SYSTEMS, L. E. Foster (Wiley, New York, 1965). An old text with some parts on telemetry systems that are still quite useful.
6.41. "Side-looking Airborne Radar," H. Jensen, L. C. Graham, L. J. Porcello, and E. N. Keith, Sci. Am. 237 (4), 84-95 (October 1977). Radar observation is independent of the weather or time of day; its capabilities are improving rapidly.
6.42. "The Verification of the SALT II Agreement," L. Aspin, Sci. Am. 240 (2), 38-45 (February 1979); and Letters to the Editor of Sci. Am. 241 (1), 8-9 (July 1979). Does the U. S. have adequate "national technical means" to monitor a SALT treaty?
6.43. VERIFICATION AND SALT: THE CHALLENGE OF STRATEGIC DECEPTION, W. C. Potter, Ed. (Westview Press, Boulder, CO, 1981).
6.44. SATELLITE RECONNAISSANCE: THE ROLE OF INFORMAL ARMS CONTROL NEGOTIATIONS, G. M. Steinberg (Praeger Publ., New York, 1982).

2. Politics

6.45. COLD DAWN: THE STORY OF SALT, J. Newhouse (Holt, Rinehart and Winston, New York, 1973). Classic description of the process whereby the SALT I agreements were reached. See also DOUBLETALK: THE STORY OF THE FIRST STRATEGIC ARMS LIMITATION TALKS, G. Smith (Doubleday, Garden City, NY, 1980).

6.46. THE SALT EXPERIENCE, T. W. Wolfe (Ballinger, Cambridge, MA, 1979); and END GAME: THE INSIDE STORY OF SALT II, S. Talbott (Harper and Row, New York, 1979). Very good journalistic tales about the development of the SALT II agreements.

6.47. GETTING TO YES: NEGOTIATING AGREEMENT WITHOUT GIVING IN, R. Fischer and W. Ury (Houghton-Mifflin, Boston, MA, 1981).

E. Disarmament

6.48. HOW TO THINK ABOUT ARMS CONTROL AND DISARMAMENT, J. E. Dougherty (Crane, Russak, New York, 1973). A review not only of the causes of war, but also of categories of arms control measures.

6.49. PEACE AGAINST WAR: THE ECOLOGY OF INTERNATIONAL VIOLENCE, F. A. Beer (W. H. Freeman and Co., San Francisco, 1981).

6.50. THE DEFENSE INDUSTRY, J. S. Gansler (MIT, Cambridge, MA, 1980). The operations of the defense industry; e.g., economic impacts, cost escalations, structure of the industry, foreign sales, etc.

6.51. THE ESTIMATION OF SOVIET DEFENSE EXPENDITURES; 1955–75: AN UNCONVENTIONAL APPROACH, W. T. LEE (PRAEGER, NEW YORK, 1977). Argues that Soviet military expenditures are larger than is generally accepted.

6.52. FROM GUNS TO BUTTER: TECHNOLOGY ORGANIZATIONS AND REDUCED MILITARY SPENDING IN WESTERN EUROPE, B. Udis (Ballinger [Lippincott], Cambridge, MA, 1978). What will happen to economics when military spending is reduced?

6.53. NATIONAL DEFENSE, J. M. Fallows (Random House, New York, 1981).

6.54. "Lewis F. Richardson's Mathematical Theory of War," A. Rapoport, J. Conflict Res. $\underline{1}$ (3), 249–99 (1957); see also Rapoport's FIGHTS, GAMES AND DEBATES (University of Michigan, Ann Arbor, 1960) and "The Use and Misuse of Game Theory," Sci. Am. $\underline{207}$ (6), 108–118 (December 1962). Outlines mathematical theories of war, and presents some of Richardson's results.

6.55. GAME THEORY, M. D. Davis (Basic, New York, 1970). A very nicely written text covering one-, two-, and n-person games and utility theory.

6.56. MATHEMATICAL MODELS OF ARMS CONTROL AND DISARMAMENT, T. L. Saaty (Wiley, New York, 1968). Focusses mathematical modelling and game theory specifically on arms control and disarmament.

6.57. INTERNATIONAL ARMS CONTROL: ISSUES AND AGREEMENTS, edited by J. H. Barton and L. D. Weiler (Stanford University, Stanford, CA, 1976). Particularly the chapters on the economics of arms control, the European example in regional arms control, and control of convenient arms.

6.58. THE GAME OF DISARMAMENT, A. Myrdal (Pantheon, New York, 1976); see the review of Sci. Am. 236 (5), 139-40 (May 1977). Swedish advocacy of disarmament. See also her article "The International Control of Disarmament," Sci. Am. 231 (4), 21-33 (October 1974).

6.59. POLITICS OF ARMS CONTROL, D. L. Clark (Free Press, Glencoe, IL, 1979). An assessment of the U. S. Arms Control and Disarmament Agency.

6.60. THE PRICE OF DEFENSE, The Boston Study Group (Times, New York, 1979). Details a new strategy for military spending, how to reduce it while maintaining sufficient military capability. See also "A New Strategy for Military Spending," P. Morrison and P. F. Walker, Sci. Am. 239 (4), 48-61 (October 1978); and Letters to the Editor, Sci. Am. 240 (1), 8-10 (January 1979). Proposals for major cuts in armaments, and analysis of the consequences of such cuts.

6.61. WORLD MILITARY AND SOCIAL EXPENDITURES: 1982, R. L. Sivard, (World Priorities, Leesburg, VA, 1982).

6.62. REPORT FROM IRON MOUNTAIN ON THE POSSIBILITY AND DESIRABILITY OF PEACE, L. C. Lewin (Dell, New York, 1967). A study of the problems that would confront the U.S. if a condition of "permanent" peace should arrive. This satire is a very proper ending for this reference list.

VII. ACKNOWLEDGMENT

We would like to acknowledge financial support of some of this bibliographic work by the UNC University Research Council.

APPENDIX B
A CHRONOLOGY OF THE NUCLEAR ARMS RACE
D.W. HAFEMEISTER
PHYSICS DEPARTMENT
CALIFORNIA POLYTECHNIC UNIVERSITY

1919
--June; Rutherford creates oxygen from nitrogen. "Talk softly please. I have been engaged in experiments which suggest that the atom can be artificially disintegrated. If it is true, it is of far greater importance than a war." ($He^4 + N^{14} \rightarrow H^1 + O^{17}$)

1920
--Rutherford speculates on the existence of the neutron at the Royal Society.

1931
--November; Urey discovers deuterium.

1932
--February; Chadwick discovers the neutron; ($He^4 + Be^9 \rightarrow 3He^4 + n^1$)

1933
--January 28; Hitler becomes Chancellor of Germany.
--March 4; Roosevelt becomes President of the U.S.
--April; Born, Courant, Franck and many other scientists were compelled to leave the University of Gottingen because of their "Jewish physics;" The "Aryan physics" of Stark and Lenard was not widely accepted by physicsts.
--October; Szilard recollects that "it occurred to me in October, 1933 that a chain reaction might be set up if an element could be found that would emit two neutrons when it swallowed one neutron." This idea was classified as a British patent in 1934 before fission was discovered.

1934
--Artificial radioactivity discovered by Curie/Joliet (with alpha particles) and by Fermi (with neutrons).

1938
--December; Fermi receives the Nobel prize for the discovery of the transuranic elements (actually fission of uranium) and departs for the "new world."
--December 22; Hahn, Strassman, Meitner, and Frisch conclude that the indentifiction of barium implies that the uranium nucleus has been been fissioned by neutrons.

1939
--January to May; Many experiments on uranium fission (a brief list is at the end of this appendix);
--April 29; Conference in Berlin to consider the German

0094-243X/83/1040296-11 $3.00 Copyright 1983 American Institute of Physic

uranium burner and bomb.
--August 2; Szilard and Teller obtain a letter from
Einstein on the possibility of a uranium weapon;
Roosevelt receives the letter on October 11, 1939 from
Sachs.
--September 1; Hitler invades Poland.

1940
--June 3; Hartek fails to observe neutron multiplication
with his reactor in Hamburg (185 kg of uranium oxide, 15
tons of CO_2 ice).

1941
--January; Using the natural uranium (300 kg) reactor in
the "virus house" in Berlin, the Germans reject graphite
as a moderator since neutrons did not diffuse adequately
through the graphite. This mistake caused by impurities
in the graphite forced the German program to rely only on
the heavy water from Vermork, Norway.
--July; British "Maud" Committee reports that a weapon
could be made with 10 kg of U-235; U.S. National Academy
of Sciences endorses bomb program.
--August; Hamburg group begins construction of the
ultracentrifuge to obtain U-235; some centrifuges explode
in April, 1942, but they did obtain 7% enrichment levels
by March, 1943.
--October; Bohr and Heisenberg indirectly discuss
termination of uranium research.
--December 6; Roosevelt directs substantial financial and
technical resources to construct the uranium weapon.
--December 7; Japan attacks Pearl Harbor.

1942
--May; Heisenberg and Dopel observe the first
multiplication of neutrons (13%) in Leipzig using 570 kg
of uranium and 140 kg of heavy ice.
--December 2; First nuclear chain reaction at Chicago's
Stagg Field by Fermi.

1943
--February 28; Vermork heavy water factory destroyed by
the allies; the attack of November 19, 1942 failed.
--March 15; Oppenheimer moves to Los Alamos.
--March; Seaborg suggests Pu weapons might be jeopardized
by Pu^{240} if it is a spontaneous neutron emitter.
--Resumption of Soviet nuclear experiments.

1944
--August 26; Bohr presents his memorandum on
international control of nuclear weapons to Roosevelt
(and Churchill).
--November; First batch of spent fuel obtained from
Hanford reactors.

--November; Goudsmit's Alsos mission obtains documents in Strasborg from the German bomb project which implied that their rate of progress had diminished.

1945
--January; First Pu reprocessing production run at Hanford.
--January 20; First U-235 seperated at the K25 gaseous diffusion plant, Oak Ridge.
--April 25; U.N. Charter signed by 50 nations in San Francisco.
--May 4; End of war in Europe.
--June 11; The Franck Report on the demonstration of the bomb and its international control was sent to the Secretary of War.
--July 16; U.S. explodes first atomic bomb, Trinity, at Alamogordo, N.M. Electronics equipment shielded by Fermi to avoid EMP pulse.
--August 6, 9; Atomic bombs dropped by U.S. on Hiroshima ("thin man," uranium, 9000 pounds, 10 feet by 28 inches in diameter) and Nagasaki ("fat man," plutonium, 11 feet by 5 feet in diameter).
--August 15; End of war in the Pacific.

1946
--June 14; Baruch presents the Acheson-Lilenthal plan to internationalize the atom to the U.N. "We are here to make a choice between the quick and the dead."
--June 30; First subsurface detonation by U.S. at Bakini.
--July; Demonstrations in Times Square, New York, against nuclear testing.
--December 31; AEC takes over nuclear weapons program from the army.

1948
--April, May; U.S. atomic tests, Eniwetok Atoll.
--November 4, 1948; U.N. General Assembly adopts, 40 to 6, U.S. plan for international atomic control; U.S.S.R opposed.

1949
--April 4; NATO established.
--August 29; First Soviet atomic detonation, in the Ustyurt desert.
--October 30; General Advisory Committee on the AEC recommends that the more powerful atomic bombs should be built rather than the hydrogen bomb.

1950
--January 27; Fuchs confessed that he transmitted atomic secrets to the Soviets.
--January 31; Truman announces the descion to proceed with the H bomb.

--March; Worldwide peace offensive to "ban the bomb" (the Stockholm Appeal) signed by more than 500 million people.
--June 25, 1950 - July, 1953; North Korean army crosses the 38th parallel.

1952
--January; U.N. Security Council establishes Disarmament Committee.
--June; Lawrence-Livermore Laborary established.
--October 3; First British atomic detonation, Monte Bello Islands, Australia.
--October 31; U.S. explodes first fusion device, Mike, of 10.4 Mt at Eniwetok (liquid deuterium, not deliverable).

1953
--March; U.S. above-ground tests start in Nevada.
--May; Representative Springfellow of Utah asks AEC to stop tests at the Nevada Test Site because of public alarm over fallout.
--August 12; First Soviet fusion device exploded on a tower in Siberia (relatively low yield, probably not deliverable, but used LiD).
--Fall; India and Australia propose to U.N. a total ban on nuclear weapons.
--December 8; "Atoms for Peace" speech by Eisenhower at U.N.

1954
--January 21; USS Nautilus, first atomic-powered submarine, launched.
--March; The "Bravo" event; Marshall Islanders affected by fallout from large U.S. tests.
--April; British Parliment petitions Churchill, Eisenhower, and Malenkov to meet on the control of nuclear weapons.
--April 12 to May 6; Oppenheimer hearings that resulted in his denial of access to atomic secrets.
--August 30; Atomic Energy Act of 1954 to emphasize peaceful uses of atomic energy.

1955
--January-December; Reports of increasing fallout.
--May 6; W. Germany joins NATO.
--May 14; Warsaw Pact Organization established.
--August 8-20; First International Conference on Peaceful uses of atomic energy, Geneva.
--November 23; First relatively high yield deliverable Soviet H bomb.

1956
--U.S. National National Academy of Sciences panel finds genetic effects of fallout from nuclear testing to be slight compared with natural radiation background;

Stevenson makes fallout an issue in the U.S. Presidential campaign.

1957
--May 15; First British H bomb exploded at Chirstmas Island.
--July 6-11; First Pugwash Conference advocates test ban; Soviet scientists attend.
--August; AEC inaugurates Plowshare Program for peaceful uses of nuclear explosions.
--September 19; First underground test, Ranier, 1.7 kt.
--October 1; IAEA inaugurated in Vienna.
--October 4; First artificial Earth satellite, Sputnik I, put into orbit by the U.S.S.R.
--December 2; Shippingport reactor reached full power of 60 MW.

1958
--January 1; Euratom (European Atomic Energy Community) established.
--January 31; Explorer I, the First U.S. satellite.
--November, 1958 to September, 1961; U.S., U.K., and U.S.S.R. agree to a moratorium on atmospheric tests.

1959
--June 12; Results on Panel on Seismic Improvement made public; Geneva system assessed effective only above 20 kt; bold research program in seismology recommended.
--September 2; Vela Uniform seismic project established with Department of Defense.
--November 24; U.S. and U.S.S.R. sign a memorandum of cooperation for the utilization of atomic energy.

1960
--February 13; First nuclear test by France, Sahara desert.
--May 1; U.S. spy plane (U-2) shot down over the U.S.S.R.
--July 26; U.S.S.R. sugggets 3 on-site inspections/year as part of a test ban.
--November 15; First Polaris missile launched from a sub.

1961
--January 17; Eisenhower's farewell address.
--February 1; U.S. launched Minuteman I.
--March 31; Pravda suggests use of H bombs to obtain fresh water from Soviet glaciers.
--April 12; Gagarin becomes the first cosmonaut in orbit.
--May 29; U.S. and U.K. agree to draft CTB treaty with 12 on-site inspections/year.
--August; Installation by U.S. of first seismic stations of a network in 60 countries.
--September 1; U.S.S.R. resumes nuclear tests, including a 58 Mt explosion on October 31.

--September 15; U.S. resumes nuclear tests.

1962
--February 20; Glenn becomes the first U.S. astronaut in orbit.
--July 6; Sedan excavation experiment by Plowshare; 10.4 Mt of earth displaced.
--July 8; Electromagnetic pulse (EMP) from the high altitude (400 km above Johnston Island) "Fishbowl" test (1.4 Mt) destroys 300 streetlights on Oahu, Hawaii (1200 km away).
--September 3-7; Tenth Pugwash Conference proposed "black box" for complete test ban (CTB) verification.
--October; U.S. reduces requirement to 8-10 inspections/year.
--October 23 to November 20; Cuban missile crisis and blockade.
--December 19; Khrushchev accepts 2-3 inspections/year and 3 unmanned seismographic stations (black boxes) in the U.S.S.R.

1963
--January; Secret test-ban talks between U.S. and U.S.S.R.
--February 19; U.S. accepts 7 inspections/year provided any mysterious event can be challenged.
--April; Khrushchev withdraws offer of 3 inspections/year.
--June 20; US/USSR sign "hot line" agreement.
--August 5; Limited Test Ban Treaty signed in Moscow by U.S., U.S.S.R. and U.K.; bans nuclear explosions in atmosphere, in space, and underwater.

1964
--August 2; "Gulf of Tonkin" resolution allows U.S. troops to be sent to Vietnam.
--October 16; China (PRC) expodes first atom bomb.
--October 22; U.S. muffling experiment Salmon in a salt dome in Mississippi.

1965
--June; Large Seismic Array (LASA) opened in Montana for nuclear detection.

1966
--January 17; U.S. B-52 bomber crashes near Palomares, Spain with 4 unarmed H bombs.
--September 24; First French H bomb, Tuamoto Islands.

1967
--January 27; Outer Space and Celestial Bodies Treaty bans nuclear weapons being placed in orbit.
--February 14; Treaty for Prohibition of Nuclear Weapons

in Latin America signed in Mexico City (Tlatelolco); all
the Latin nations must ratify for the treaty to entr into
force.
--June 17; First Chinese H bomb exploded in the
atmosphere.
--June 23-25; Johnson and Kosygin hold talks in
Glassboro, N.J.
--December 10; Gasbuggy Plowshare explosion for gas
production, New Mexico.

1968
--July 1; The Non-Proliferation Treaty on nuclear weapons
(NPT) opened for signature.
--August; Soviet GALOSH ABM system deployed around
Moscow.

1969
--July 20; U.S. Appollo 11 landed on moon.
--November 3; Committee on Disarmament reports to U.N. on
comprehensive test ban; proposes worldwide exchange of
seismological data.
--November to December; Preliminary SALT talks in
Helsinki.

1970
--January; Nixon administration cuts Plowshare budget.
--March 2-6; Talks by 21 nations at the IAAEA on peaceful
uses of nuclear explosions.
--March 5; NPT entered into force; 50 nations ratified
the NPT by 1970, over 100 by 1980.
--November 30; Atlantic-Pacific Interoceanic Canal Study
Commission rejets use of nuclear explosives.

1971
--March 30: U.S. deploys Poseidon SLBM.
--June 15; Izvestia mentions three Russian nuclear
charges exploded in oil fields to increase productivity.

1972
--May 26; SALT I Treaties limiting ABM and offensive
missiles signed by Nixon and Brezhnev in Moscow.
--June; U.N. Environmental Conference in Stockholm votes
48 to 2 to halt all testing of nuclear weapons; France
and China against, U.S. and U.K. abstain, U.S.S.R absent.
--October; First U.S. detection of flight test of Soviet
SS-18 ICBM.

1973
--June 22; World Court issues injunction to France not to
carry out Mururoa tests.
--October 17, 1973 to March 17, 1974; OPEC embargos
petroleum to U.S. and the Nethererlands.

1974
--May 18; India sets off a low-yield device (10-15 kt)
under Rajasthan desert, expanding the nuclear club beyond
the "big five" of World War II.
--November 24; U.S./U.S.S.R. agree to limit the number of
strategic launchers (2400) and MIRV launchers (1320).

1975
--January 19; AEC reorganized into the NRC (regulatory)
and ERDA (developmental, later DOE on August 4, 1977).

1976
--January 29; The order for the South Korean reprocessing
plant from France was cancelled. (Pakistan plant
cancelled in 1978)
--October 28; Ford postpones reprocessing of spent fuel
to obtain commercial Pu.

1977
--April 7; Carter postpones indefinitely reprocessing of
commercial spent fuel, slows progress on the Clinch River
Breeder Reactor, and calls for an International Fuel
Cycle Evaluation (INFCE) to investiate ways to make the
nuclear cycle more resistant to proliferation.
--October 19, 1977 to February 26, 1980; Over 40 nations
prepare a report at INFCE.

1978
--January 11, The Guidelines of the Nuclear Suppliers
(London meetings) were forwarded to the IAEA.
--March 10, The Nuclear Nonproliferation Act of 1978
(NNPA) was signed into law.
--April 4; Camp David Accord signed by Egypt and Isreal
in Wshington.
--April 27; Carter issues an Executive Order to export
low-enriched uranium fuel to India; this was the first
override of the NRC as allowed for in the NNPA; the
Congress did not veto this action.
--July 13; Euratom ended the the embargo of nuclear fuel
from the U.S. (required by NNPA) and agreed to discuss
renegotation of the US-Euratom agreement on the issue of
retranfer and reprocessing rights.

1979
--April, 6; The U.S. cut off economic and military aid to
Pakistan after concluding that Pakistan was building an
enrichment plant to produce weapons grade uranium.
--June 8; First Trident SLBM launched.
--June 18; The SALT II Treaty was signed in Vienna by
Brezhnev and Carter.
--December 26; U.S.S.R. invades Afganistan; SALT II
treaty removed from consideration by the Senate.

1980
--July 15; Presidential Office of Science and Technology
Policy reported that the light signals recorded by a VELA
satellite over the South Atlantic on September 22, 1979
was probably not from a nuclear explosion.
--December; Tomahawk submarine lauched cruise missile
flies out to sea and returns 600 miles to land; later it
obtains an accuracy of about 5 meters on a flight from
Pt. Mugu to Nevada.

1981
--April 12-14; Space shuttle Columbia flies 36 orbits.
--June 7, Isreal destroys Iraq's Osirak reactor.
--November 13; Senate ratifies Protocol I of Tlatelolco
Treaty; U.S. cannot store, deploy, or use nuclear weapons
in Puerto Rico, Virgin Islands, or Guantanomo Naval Base.

1982
--June 29; START talks begin in Geneva.
--1982; U.S.S.R. places about 300 SS-20 missiles on its
western border.
--November; Reagan adopts the dense pack basing mode for
the MX missile.

ACKNOWLEDGEMENT AND REFERENCES

In preparing this list of chronological data, we
would like to acknowledge the lists and suggestions of
others; in particular we would like to thank Bruce Bolt,
Prentice Dean, Warren Donnelly, and Spencer Weart. Some
references that we have found to be useful are as
follows:

1. B. Bolt; Nuclear Explosions and Earthquakes, Freeman,
San Francisco, 1976.
2. P. Dean, Energy History Chronology from World War II
to the Present, U.S. Dept. of Energy, DOE/ES-0002,
Washington, D.C., 1982.

3. W. Donnelly and D. Kramer, Nuclear Proliferation
Factbook, Congressional Research Service, Library of
Congress, 1980.

4. B. Goldschmidt, The Atomic Complex, Am. Nuc. Soc., La
Grange Park, IL, 1982.

5. R. Hewlett and O. Anderson, A History of the U.S.
Atomic Energy Commission: The New World: 1939-46, Penn.
St. Univ. Press, University Park, PA, 1962. R. Hewlett
and F. Duncan, Atomic Shield: 1947-52, Penn. St. Univ.
Press, University Park, PA, 1969.

6. D. Irving, The German Atom Bomb, Simon and Schuster, NY, 1967.

7. D. Kevles, The Physicsts, Vintage, N.Y., 1979.

8. H. Smyth, "Atomic Energy for Military Purposes," Rev. Mod. Phys. 17, 351 (1945).

9. H. York, Race to Oblivion, Simon and Schuster, NY, 1970.

ADDITIONAL REFERENCES ON URANIUM FISSION (1939):
(Dates will be in the order of submitted/published)

November 3, 1938/January 15, 1939; Nier (Phys. Rev. 55, 150 (1939)); "A mass spectrographic determination ... found U-238/U-235 = 139 $\pm$ 1 percent."

February 16/March 1; Fermi, et al (Phys. Rev. 55, 511 (1939)); "If we assume that the energy release in the fission is approximately 200 Mev, and that the two fragments may have somewhat different masses, then fragments with energies up to 120 or 130 Mev might be expected in some cases."

March 8/March 18; Joliot, et al (Nature 143, 470 (1939)); "...more than one neutron must be produced. This seems to be the case..."

March 10/April 1; Roberts, et al (Phys. Rev. 55, 664 (1939)); "direct neutron emission is responsible for the delayed neutrons)."

March 16/April 15; Fermi, et al (Phys. Rev. 55, 797 (1939)); "...it would correspond to a yield of about two neutrons..."

March 16/April 15; Szilard and Zinn (Phys. Rev. 55, 799 (1939)); "...we find the number of neutrons emitted per fission to be about two."

April 6/June 1; W. Furry, R. Clark, and L. Onsager (Phys. Rev. 55, 1083 (1939)); "On the Theory of Isotope Seperation by Thermal Diffusion."

April 7/April 22; Joliot, et al (Nature 143, 680 (1939)); the number of neutrons emitted per fission = ν = 3.5 $\pm$ 0.7

May 20/June 3; Joliot, et al (Nature 143, 939 (1939));

"...neutrons possessing an energy of at least 11 Mev are
liberated in uranium irradiated with thermal neutrons."

May 31/August 1; J. Beams and C. Skarstrom (Phys. Rev.
56, 266 (1939)); "The Concentration of Isotopes by the
Evaporative Centrifuge Method."

July 3/August 1; H. Anderson, E. Fermi, and L. Szilard
(Phys. Rev. 56, 284 (1939)); "Neutron Production and
Absorption in Uranium... It has been found that there is
an abundant emission of neutrons from uranium under the
action of slow neutrons."

June 28/September 1 (the day Hitler invaded Poland); N.
Bohr and J. Wheeler (Phys. Rev. 56, 426 (1939)); "The
Mechanism of Nuclear Fission;" includes theory of fission
by fast neutrons.

July 10/September 1; J. Oppenheimer and H. Snyder (Phys.
Rev. 56, 455 (1939)); "On Continual Gravitational
Contraction... When all thermonuclear sources are
exhausted a sufficiently heavy star will collapse."

SIPRI: LIST OF FIRST NUCLEAR EXPLOSIONS

Fission devices				Thermonuclear devices		
Country	Year of first explosion	Fissile material	Source of fissile material	Year of first explosion	Fissile material	Source of fissile material
USA	1945	Pu-239	Reactor	1952	U-235	Gaseous diffusion
USSR	1949	Pu-239	Reactor	1952	U-235	Gaseous diffusion
UK	1952	Pu-239	Reactor	1957	U-235	Gaseous diffusion
France	1960	Pu-239	Reactor	1968	U-235?	Gaseous diffusion
China	1964	U-235	Gaseous diffusion	1967	U-235	Gaseous diffusion
India	1974	Pu-239	Reactor	-	-	-

Source: Stockholm International Peace Research Institute, Nuclear
energy and other nuclear weapons proliferation. London:
Taylor and Francis Ltd., 1979, p.2.

APPENDIX C: ARMS CONTROL AGREEMENTS AND LAWS

This section will contain the key provisions* for the following arms control agreements and laws:

1. The Limited Test Ban Treaty (1963).
2. The Non-Proliferation Treaty (1970).
3. The Anti-Ballistic Missile Systems Treaty (1972).
4. The Nuclear Non-Proliferation Act of 1978.
5. Strategic Arms Limitation Talks (SALT II, 1979).
6. Other Treaties (1928-1978).

The parties to the limited test ban treaty undertake "not to carry out any nuclear weapon test explosion, or any other nuclear explosion," in the atmosphere, under water, or in outer space, or in any other environment if the explosion would cause radioactive debris to be present outside the borders of the state conducting the explosion. As explained by Acting Secretary of State Ball in a subsequent report to President Kennedy, "The phrase 'any other nuclear explosion' includes explosions for peaceful purposes. Such explosions are prohibited by the treaty because of the difficulty of differentiating between weapon test explosions and peaceful explosions without additional controls."

The treaty is of unlimited duration, with provisions for amendment or withdrawl. Article III opens the treaty to all states, and most of the countries of the world have now signed it (108 nations are parties, 15 nations have signed but not ratified it). The treaty has not been signed by France or by the People's Republic of China.

2.   THE NON-PROLIFERATION TREATY (1970)

Approximately 119 nations have ratified the Non-Proliferation Treaty (NPT) since it went into force in 1970. The basic provisions of the treaty are as follows:

Article I: Each non-nuclear-weapon State Party to the Treaty undertakes not to transfer to any recipient whatsoever nuclear weapons, other nuclear explosive devices, or to assist any non-nuclear-weapon State to manufacture or àcquire nuclear weapons. The nuclear-weapon States are the United States, Russia, Great Britain, France, and China; the latter two have not ratified the NPT.

---

* Arms Control and Disarmament Agreements, U.S. Arms Control and Disarmament Agency, Washington, D.C., 1980.

Article II: Each non-nuclear-weapon State Party to the Treaty undertakes not to receive nuclear weapons or other nuclear explosive devices, or to manufacture or otherwise acquire nuclear weapons or other nuclear explosive devices.

Article III: Each non-nuclear-weapon State Party to the Treaty undertakes to accept safeguards, as set forth in an agreement with the International Atomic Energy Agency. The safeguards required by this article shall be applied to all source or special fissionable material in all peaceful nuclear activities within the territory of such State, under its jurisdiction, or carried out under its control anywhere. The safeguards required by this article shall be implemented in a manner designed to comply with article IV of this Treaty, and to avoid hampering the economic or technological development of the Parties.

Article IV: Nothing in this Treaty shall be interpreted as affecting the inalienable right of all the Parties to the Treaty to develop research, production and use of nuclear energy for peaceful purposes without discrimination. Parties to the Treaty in a position to do so shall also cooperate in contributing alone or together with other States to further develop the applications of nuclear energy for peaceful purposes, especially in the territories of the non-nuclear-weapon States.

Article VI: Each of the Parties to the Treaty undertakes to pursue negotiations in good faith on effective measures relating to cessation of the nuclear arms race at an early date and to nuclear disarmament, and on a treaty on general and complete disarmament under strict and effective international control.

Article VII: Every five years a conference can be held in order to review the operation of the NPT.

3. TREATY TO LIMIT ANTI-BALLISTIC MISSILE SYSTEMS (1972)

In the ABM Treaty the United States and the Soviet Union agree that each may have only two ABM deployment areas; this was subsequently reduced to one area. The ABM Treaty is for an unlimited duration.

Precise quantitative and qualitative limits are imposed on the ABM systems that may be deployed. At each site there may be no more than 100 interceptor missiles and 100 launchers. Agreement on the number and characteristics of radars to be permitted had required extensive and complex technical negotiations, and the provisions governing these important components of ABM systems are spelled out in very specific detail in the treaty and further clarified in the "Agreed Statements" accompanying it.

Both parties agreed to limit qualitative improvement of their ABM technology, e.g. not to develop, test, or deploy ABM launchers capable of launching more than one interceptor missile at a time or modify existing launchers to give them this capability, and systems for rapid reload of launchers are similarly barred. These provisions, the Agreed Statements clarify, also ban interceptor missiles with more than one independently guided warhead.

4. THE NUCLEAR NON-PROLIFERATION ACT OF 1978 (NNPA)

This act established the current criteria for exports of nuclear reactors and fuel from the U.S. as well as the criteria for retransferring and reprocessing spent fuel rods. These criteria are summarized below (Department of State, 1978):

I. IMMEDIATELY APPLICABLE EXPORT LICENSING CRITERIA

In addition to the requirement that an export not be inimical to U.S. common defense and security, specific criteria to be applied to U.S. supply are established for NRC export licensing of nuclear fuel and reactors:
-- IAEA safeguards;
-- No explosive use;
-- Maintaining adequate physical security;
-- U.S. consent for retransfers*;
-- U.S. consent for reprocessing*;
-- Application of above criteria to materials or equipment produced through any transferred sensitive nuclear technology.

II. FULL-SCOPE SAFEGUARDS

-- IAEA safeguards on all peaceful nuclear activities in non-nuclear-weapons states will be required, after 24 months following enactment, as a condition of continued U.S. export. This export licensing criterion may be waived by the President under exceptional circumstances, subject, however, to Congressional veto by concurrent resolution.

III. CRITERIA FOR NEW AND AMENDED AGREEMENTS FOR COOPERATION

Beyond the immediate export criteria, the following additional provisions must be included in new and amended agreements:
-- IAEA safeguards of indefinite duration on U.S. supply and, as a continuing condition of U.S. supply on all peaceful nuclear activities in non-nuclear-weapons states;
-- Return to the U.S. of nuclear materials if a recipient non-nuclear-weapons state detonates a nuclear explosive device, terminates a safeguards agreement, or materially violates the cooperation agreement;
-- U.S. consent to the retransfer and/or reprocessing of spent fuel irradiated in U.S.-supplied reactors;

---

* Exports to groups of nations (Euratom) are exempted from the retransfer and reprocessing requirements for 24 months (with possible one-year extensions) if they agree to renegotiate.

-- U.S. approval of facilities for storage of weapons usable material;
-- Application of all of the above conditions to any nuclear materials produced through the use of transferred sensitive nuclear technology;
-- One or more of these requirements may be waived by the President if U.S. non-proliferation objectives or national security would otherwise be jeopardized. New or amended agreements must lie before Congress for 60 days and then come into effect unless the Congress passes a concurrent resolution disapproving the agreement.

IV. EVENTS REQUIRING TERMINATION OF U.S. NUCLEAR SUPPLY OR MILITARY AND ECONOMIC AID

-- Exports to any non-nuclear-weapons state will be terminated if that state:
  - detonates a nuclear explosive device;
  - terminates or abrogates IAEA safeguards;
  - is found by the President to have materially violated IAEA safeguards;
  - uses nuclear material in activities of direct significance to the acquisition of explosive devices.
-- Exports will be terminated to any state or group of states that:
  - violates its agreement for cooperation or other terms of supply;
  - assists a non-nuclear-weapons state in activities of direct significance to the acquisition of nuclear explosives;
  - enters into an agreement for transferring reprocessing capabilities except in connection with INFCE or a subsequent international arrangement to which the U.S. subscribes;
  - forbids export of military and economic aid to a non-nuclear-weapons state that explodes a nuclear device, or imports an enrichment or reprocessing plant on a national basis (Foreign Assistance Act).

V. SUBSEQUENT ARRANGEMENTS (UNDER EXISTING AND NEW AGREEMENTS) FOR REPROCESSING AND HANDLING OF SPENT FUEL

-- U.S. consent for reprocessing or transfer of separated plutonium may only be given if it will not result in a significant increase in the risk of proliferation. In making this evaluation, foremost consideration will be given to whether the reprocessing or retransfer will take place under conditions that ensure timely warning to the U.S. of any diversion well in advance of the time at which a non-nuclear-weapons state could transform that diverted material

into a nuclear explosive device. There is an exception for
reprocessing at facilities that processed power reactor fuel
prior to enactment of this legislation but the U.S. must
attempt to ensure the same standard is applied to reprocessing at these facilities as well;
-- Any proposed return of foreign spent fuel to the U.S. would
be subject to review by the Executive Branch and the Congress.

VI. LICENSING OF NUCLEAR COMPONENTS NOT SPECIFICALLY COVERED
IN THE LEGISLATION

-- The NRC is directed to define components which are of special relevance for nuclear explosives. These would be
licensed by NRC which would determine whether the export
was inimical to the common defense and security and the
license would also be subject to the following standards:
- IAEA safeguards;
- no explosive pledge;
- no retransfers without U.S. consent.

VII.   EXPORT LICENSING PROCEDURES WITH A VIEW TO
EXPEDITING THE EXPORT LICENSING PROCESS: THE ACT

-- Requires the NRC to adopt procedures which will be exclusive
basis for hearings before the Commission;
-- Permits the NRC to issue licenses for multiple shipments or
give expedited treatment to follow-on licenses if no material change in circumstances has occurred;
-- Allows the President to authorize an export after an NRC
negative decision, subject to Congressional review and veto;
-- Mandates the Executive Branch and the NRC normally to act on
licenses within 60 days; and
-- Limits Government-to-Government transfers (without NRC
licenses) to small quantities or to emergency situations in
order to achieve uniformity in all transactions.

VIII.   FUEL ASSURANCES INITIATIVES

-- The Secretary of Energy is directed to proceed with the construction and operation of expanded uranium enrichment capacity;
-- The President is directed to study the need for additional
U.S. enrichment capacity for foreign and domestic needs and
further to report to Congress on the desirability of inviting
foreign participation in new U.S. uranium enrichment facilities;
-- The President is also given a mandate to begin the negotiation of an International Nuclear Fuel Authority to provide
fuel services, and international spent-fuel repositories.

5. STRATEGIC ARMS LIMITATION TALKS (SALT)

The first series of SALT negotiations between the United States and the Soviet Union extended from November 1969 to May 1972; these SALT I talks resulted in the ABM treaty (Section 3 above) and the Interim Agreement which set a five year limit on the number of ICBM and SLBM launchers. In 1974, both sides agreed to limit the aggregate number of nuclear delivery vehicles (ICBMs, SLBMs and heavy bombers) to 2400 with no more than 1320 of these being MIRV systems. This agreement set one stage for the final round of talks on SALT II. The Treaty was signed in 1979, and it was intended to be in force until 1985; it has not been ratified by the U.S. Senate. The SALT II Treaty provides for:

-- an equal aggregate limit on the number of strategic nuclear delivery vehicles-ICBM and SLBM launchers, heavy bombers, and air-to-surface ballistic missiles (ASBMs). Initially, this ceiling will be 2,400, as agreed at Vladivostok. The ceiling will be lowered to 2,250 at the end of 1981;

-- an equal aggregate limit of 1,320 on the total number of launchers of MIRVed ballistic missiles and heavy bombers with long-range cruise missiles;

-- an equal aggregate limit of 1,200 on the total number of launchers of MIRVed ballistic missiles; and

-- an equal aggregate limit of 820 on launchers of MIRVed ICBMs.

In addition to these numerical limits, the agreement includes:

-- a ban on construction of additional fixed ICBM launchers, and on increases in the number of fixed heavy ICBM launchers;

-- a ban on heavy mobile ICBM launchers, and on launchers of heavy SLBMs and ASBMs;

-- a ban on flight-testing or deployment of new types of ICBMs, with an exception of one new type of light ICBM for each side;

-- a ban on increasing the numbers of warheads on existing types of ICBMs, and a limit of 10 warheads on the one new type of ICBM permitted to each Party, a limit of 14 warheads on SLBMs, and 10 warheads on ASBMs. The number of long-range (over 600 km) cruise missiles per heavy bomber is limited to an average of 28; and the number of long-range cruise missiles per heavy bomber of existing types is limited to 20;

-- ceilings on the launch weight and throw weight of strategic ballistic missiles and a ban on the conversion of light ICBM launchers to launchers of heavy ICBMs;

-- a ban on the Soviet SS-16 ICBM;

-- a ban on rapid reload ICBM systems;

-- a ban on certain new types of strategic offensive systems which are technologically feasible, but which have not yet been deployed. Such systems include long-range ballistic missiles on surface ships, and ballistic and cruise missile launchers on the seabeds;

-- an agreement to exchange data on a regular basis on the

numbers deployed for weapons systems constrained in the
agreement;
-- advance notification of certain ICBM test launches;
-- an agreement not to encryp telemetric data from missile tests;
-- an agreement to use counting rules which require that any
   missile which has been tested with MIRVs be counted as MIRVed,
   even if it contains a single warhead.

The protocol to the Treaty would have remained in force through
1981; it provided for the following temporary limitations:
-- a ban on the flight-testing of ICBMs from mobile launchers
   and the deployment of mobile ICBM launchers;
-- a ban on the testing and deployment of ASBMs;
-- a ban on the deployment of ground-launched and sea-launched
   cruise missiles having ranges greater than 600 km.

ALLOTMENT OF STRATEGIC NUCLEAR DELIVERY VEHICLES IN SALT II
Department of State, 1979

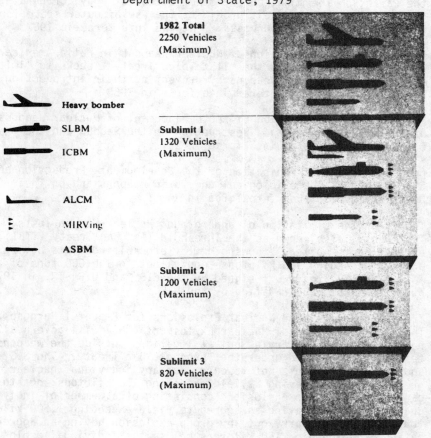

Heavy bomber

SLBM

ICBM

ALCM

MIRVing

ASBM

1982 Total
2250 Vehicles
(Maximum)

Sublimit 1
1320 Vehicles
(Maximum)

Sublimit 2
1200 Vehicles
(Maximum)

Sublimit 3
820 Vehicles
(Maximum)

6. OTHER TREATIES

A. General Protocol for the Prohibition of the Use in War of Asphyxiating, Poisonous or other Gases, and Bacteriological Methods of Warfare. Entered into force in 1928 and ratified by U.S. in 1975.

B. The Antarctica Treaty. The treaty internationalized and demilitarized the Antarctic continent and provided for its cooperative exploration and further use. Entered into force in 1961.

C. "Hot Line" Agreement. Entered into force in 1963 (and updated in 1971) by the U.S. and the U.S.S.R.

D. Outer Space Treaty. Article IV of the treaty restricts military activities in two ways: First, it forbids the placing of nuclear weapons (or other weapons of mass destruction) in orbit around the Earth, on the moon, on any other celestial body, or on a space station. Second, it prohibits the use of outer space for conventional military purposes. Entered into force in 1967.

E. Treaty for the Prohibition of Nuclear Weapons in Latin America. The Treaty, also called the "Tlatelolco Treaty", seeks to limit the spread of nuclear weapons by preventing their introduction into Latin America. Entered into force in 1968.

F. Treaty on the Prohibition of the Emplacement of Nuclear Weapons and other Weapons of Mass Destruction on the Seabed and the Ocean Floor. Entered into force in 1972.

G. Convention on the Prohibition of the Development, Production and Stockpiling of Bacteriological and Toxin Weapons and on Their Destruction. Entered into force in 1975.

H. Treaty on the Limitation of Underground Nuclear Weapon Tests. The Treaty, also known as the Threshold Test Ban Treaty (TTBT), establishes a nuclear "threshold", by prohibiting tests having a yield exceeding 150 kilotons (equivalent to 150,000 tons of TNT). The TTB Treaty was signed by the U.S. and U.S.S.R. in 1974, but it has not been ratified by the U.S.

I. Treaty on Underground Nuclear Explosions for Peaceful Purposes. The Treaty on Peaceful Nuclear Explosions (PNE) will govern all nuclear explosions carried out at locations outside the weapons test sites specified under the Threshold Ban Treaty. The U.S. and U.S.S.R agreed: not to carry out any individual nuclear group explosions having a yield exceeding 150 kilotons; not to carry out any group explosion (consisting of a number of individual explosions) having an aggregate yield exceeding 1,500 kilotons; and not to carry out any group explosion having an aggregate yield exceeding 150 kilotons unless the individual explosions in the group could be identified and measured by agreed verification procedures. The PNE treaty was signed in 1976, but it has not been ratified by the U.S.

J. Convention on the Prohibition of Military or Any Other Hostile Use of Environmental Modification Techniques. The Convention prohibits the use of any environmental or geophysical modification activity as a weapon of war. The Convention defines environmental modification techniques as changing - through the deliberate manipulation of natural processes - the dynamics, composition or structure of the earth, including its biota, lithosphere, hydrosphere, and atmosphere, or of outer space. Changes in weather or climate patterns, in ocean currents, or in the state of the ozone layer or ionosphere, or an upset in the ecological balance of a region are some of the effects which might result from the use of environmental modification techniques. The Convention entered into force in 1978.

APPENDIX D

PRINCIPAL FINDINGS OF THE OFFICE OF
TECHNOLOGY ASSESSMENT OF THE U.S. CONGRESS:

I.   THE EFFECTS OF NUCLEAR WAR (1979).

II. MX MISSILE BASING (1981).

III. NUCLEAR PROLIFERATION AND SAFEGURADS (1977).

I.   THE EFFECTS OF NUCLEAR WAR (1979).

1. The effects of a nuclear war that cannot be calculated are at least as important as those for which calculations are attempted. Moreover, even these limited calculations are subject to very large uncertainties.

Conservative military planners tend to base their calculations on factors that can be either controlled or predicted, and to make pessimistic assumptions where control or prediction are impossible. For example, planning for strategic nuclear warfare looks at the extent to which civilian targets will be destroyed by blast, and discounts the additional damage which may be caused by fires that the blast could ignite. This is not because fires are unlikely to cause damage, but because the extent of fire damage depends on factors such as weather and details of building construction that make it much more difficult to predict than blast damage. While it is proper for a military plan to provide for the destruction of key targets by the surest means even in unfavorable circumstances, the nonmilitary observer should remember that actual damage is likely to be greater than that reflected in the military calculations. This is particularly true for indirect effects such as deaths resulting from injuries and the unavailability of medical care, or for economic damage resulting from disruption and disorganization rather than from direct destruction.

For more than a decade, the declared policy of the United States has given prominence to a concept of "assured destruction:" the capabilities of U.S. nuclear weapons have been described in terms of the level of damage they can surely inflict even in the most unfavorable circumstances. It should be understood that in the event of an actual nuclear war, the destruction resulting from an all-out nuclear attack would probably be far greater. In addition to the tens of millions of deaths during the days and weeks after the attack, there would probably be further millions (perhaps further tens of millions) of deaths in the ensuing months or years. In addition to the enormous economic destruction caused by the actual nuclear explosions, there would be some years during which the residual economy would decline further, as stocks were consumed and machines wore out faster than recovered production could replace them. Nobody knows how to estimate the likelihood that industrial civilization might collapse in the areas attacked; additionally, the possibility of significant long-term ecological damage cannot be excluded.

2. The impact of even a "small" or "limited" nuclear attack would be enormous. Although predictions of the effects of such an attack are subject to the same uncertainties as predictions of the effects of an all-out attack, the possibilities can be bounded. OTA examined the impact of a small attack on economic targets (an attack on oil refineries limited to 10 missiles), and found that while economic recovery would be possible, the economic damage and social dislocation could be immense. A review of calculations of the effects on civilian populations and economies of major counterforce attacks found that while the consequences might be endurable (since they would be on a scale with wars and epidemics that nations have endured in the past), the number of deaths might be as high as 20 million. Moreover, the uncertainties are such that no government could predict with any confidence what the results of a limited attack or counterattack would be even if there was no further escalation.

3. It is therefore reasonable to suppose that the extreme uncertainties about the effects of a nuclear attack, as well as the certainty that the minimum consequences would be enormous, both play a role in the deterrent effect of nuclear weapons.

4. There are major differences between the United States and the Soviet Union that affect the nature of their vulnerability to nuclear attacks, despite the fact that both are large and diversified industrial countries. Differences between the two countries in terms of population distribution, closeness of population to other targets, vulnerability of agricultural systems, vulnerability of cities to fire, socioeconomic system, and political system create significant asymmetries in the potential effects of nuclear attacks. Differences in civil defense preparations and in the structure of the strategic arsenals compound these asymmetries. By and large, the Soviet Union is favored by having a bigger and better economy and (perhaps) a greater capacity for effective decentralization. The larger size of Soviet weapons also means that they are likely to kill more people while aiming at something else.

5. Although it is true that effective sheltering and/or evacuation could save lives, it is not clear that a civil defense program based on providing shelters or planning evacuation would necessarily be effective. To save lives, it is not only necessary to provide shelter in, or evacuation to, the right place (and only extreme measures of dispersion would overcome the problem that the location of safe places cannot be reliably predicted), it is also necessary to provide food, water, medical supplies, sanitation, security against other people, possibly filtered air, etc. After fallout diminishes, there must be enough supplies and enough organization to keep people alive while production is being restored. The effectiveness of civil defense measures depends, among other things, on the events leading up to the attack, the enemy's targeting policy, and sheer luck.

6. The situation in which the survivors of a nuclear attack find themselves will be quite unprecedented. The surviving nation would be weaker—economically, socially, and politically—than one would calculate by adding up the surviving economic assets and the numbers and skills of the surviving people. Natural resources would be destroyed; surviving equipment would be designed to use materials and skills that might no longer exist; and indeed some regions might be almost uninhabitable. Furthermore, prewar patterns of behavior would surely change, though in unpredictable ways. Finally, the entire society would suffer from the enormous psychological shock of having discovered the extent of its vulnerability.

7. From an economic point of view, and possibly from a political and social viewpoint as well, conditions after an attack would get worse before they started to get better. For a period of time, people could live off supplies (and, in a sense, off habits) left over from before the war. But shortages and uncertainties would get worse. The survivors would find themselves in a race to achieve viability (i.e., production at least equaling consumption plus depreciation) before stocks ran out completely. A failure to achieve viability, or even a slow recovery, would result in many additional deaths, and much additional economic, political, and social deterioration. This postwar damage could be as devastating as the damage from the actual nuclear explosion.

II. MX MISSILE BASING (1981).

1. There are five basing modes that appear feasible and offer reasonable prospects of providing survivability and meeting established performance criteria for ICBMs. They are: 1) MPS basing of the type now under development by the Air Force or in one of several variants. MPS basing involves hiding the missiles among a much larger number of shelters, so that the Soviets would have to target all the shelters in order to attack all the missiles. If there are more shelters than the Soviets could effectively target, then some of the missiles would survive. This approach was the choice of the Carter administration, and one variant of MPS is now under engineering development by the Air Force. 2) MPS basing defended by a low-altitude ABM system known as LoADS (Low Altitude Defense System); 3) reliance on launch under attack so that the missiles would be used before the Soviets could destroy them; 4) basing MX on small submarines; and 5) air-mobile basing in which missiles would be dropped from wide-bodied aircraft and launched while falling. As described below, each of these alternatives has serious risks and drawbacks, and it is believed that choosing which risks and drawbacks are most tolerable is a judgment which cannot be made on technical grounds alone.

2. No basing mode is likely to provide a substantial number of survivable MX missiles much before the end of this decade. While some basing modes would permit the first missiles to be operational as soon as 1986 or 1987, these missiles could not be considered more survivable than the existing Minuteman missiles until additional elements of the basing system were in place.

3. MPS basing would preserve the existing characteristics and improve the capabilities of land-based ICBMs, but has three principal drawbacks.

- **MX missiles based in MPS would provide better accuracy and endurance, and comparable responsiveness, time-on-target control, and retargeting capability, when compared to other feasible basing modes.**
- Survivability depends on what the Air Force calls "preservation of location uncertainty" (PLU), that is, preventing the Soviets from determining which shelters hold the actual missiles. PLU amounts to a new technology, and while it might well be carried out successfully, **confidence in PLU will be limited until prototypes have been successfully tested.** Even then lingering doubts might remain.
- MPS basing cannot ensure the survivability of the missiles unless the number of shelters is large enough relative to the size of the Soviet threat. **The "baseline" system of 200 MX missiles and 4,600 shelters would not be large enough if the Soviets choose to continue to increase their inventory of warheads.** If the trends shown in recent Soviet force modernization efforts continue into the future, an MPS deployment of about 350 missiles and 8,250 shelters would be needed by 1990 to provide survivability. Although the number of missiles and shelters needed depends on what the Soviets do, the leadtimes for construction are so long that decisions on size must be made before intelligence data on actual, as distinct from possible, Soviet programs are available.
- **MPS would severely impact the socioeconomic and physical characteristics of the deployment region.** At a minimum, the deployment area would suffer the impacts generally associated with the rapid population growth in rural communities; but larger urban areas would also be affected by economic uncertainties regarding the size of the MPS construction work force and its regional distribution. The physical impacts of MPS would be characterized by the impacts of major construction projects in arid regions; but because the grid pattern of MPS would mean that a very large area would be close to construction activities, it is possible that thousands

of square miles of rangeland could be rendered unproductive.

4. None of the variants of MPS would reduce the risks and uncertainties associated with PLU or significantly alter the number of the shelters required. However, split basing or the selection of a different deployment area would mitigate the regional impacts. The variants that OTA examined include changes from horizontal to vertical shelters, from "individual cluster" to "valley cluster" basing, and from Utah/Nevada basing to basing divided between Utah/Nevada and west Texas/New Mexico. A further variant would be to construct additional silos in the existing Minuteman basing areas to create a Minuteman/MPS system. This would be substantially cheaper than the proposed MX/MPS system, but would not be significantly quicker to construct.

5. A LoADS ABM system could effectively double the number of shelters in an MPS deployment provided two conditions were met. A LoADS system would have a high probability of shooting down the first Soviet warhead aimed at each MX missile, forcing the Soviets to attack each shelter with two warheads. The conditions for LoADS' effectiveness are: 1) PLU both for the MX and for the LoADS defense unit, and 2) survival and operation of the defense in the presence of nearby nuclear detonations. Since the LoADS defense unit would be concealed in a shelter and would be indistinguishable from the missiles and the decoys, LoADS deployment would compound the difficulties of PLU. These difficulties would be greater still if the LoADS addition were not planned at the time the MPS system was being designed. The LoADS defense unit would be required to endure nuclear effects of a severity unprecedented for so complex a piece of equipment.

A LoADS deployment would require the United States either to seek amendment of, or to withdraw from, the ABM Treaty reached at SALT I.

6. Basing MX missiles in silos and relying on launching the missiles before a Soviet attack could destroy them (launch under attack, or LUA) would be technically feasible, but it would create extreme requirements for availability of, and rapid decisionmaking by, National Command Authorities. A substantial upgrading of existing warning and communications systems would be required to ensure this capability against a determined Soviet attempt at disruption. Reliance on this capability would, however, impose extremely stringent requirements that the President be in communication with both the warning systems and the forces, and that an unprecedentedly weighty decision be made in a few minutes on the basis of information supplied by remote sensors. Finally, there would always be concern about whether the system was really immune to disruption or errors.

7. MX missiles based on small submarines would be highly survivable. Submarine-based MX would not be significantly less capable than land-based MX, but submarine-basing would involve a re-

orientation of U.S. strategic forces. An MX force based on small diesel-electric or nuclear submarines operating 1,000 to 1,500 miles from the U.S. coast could offer weapon effectiveness (i.e., accuracy, responsiveness, time-on-target control, and rapid retargeting) almost as good as land basing and would probably be adequate to carry out any strategic mission. A command, control, and communications (C³) system to support submarine basing would be different from that used for landbasing but would not necessarily be less capable. However, submarine basing of MX would change the relative importance of land- and submarine-based strategic forces. Although OTA could find no scientific basis for predicting such an occurrence, the possibility cannot be excluded that an unexpected Soviet capability in antisubmarine warfare which threatened the U.S. force of Poseidon and Trident submarines might also threaten a force of MX missiles on small submarines. The cost of providing 100 MX missiles on alert at all times on a small submarine force would be roughly comparable to the cost of the baseline MPS system, and would be less than the cost of an MPS system sized to meet a larger Soviet threat. A significant problem is that such a force of small submarines could not be constructed quickly; existing U.S. submarine construction programs are already behind schedule, and it would require a number of years for a shipyard which is not building submarines to learn how to do so. It is therefore unlikely that initial MX deployment on small submarines could take place before 1990. However, the first MX missiles deployed would be survivable even before the rest of the deployment was complete.

8. An air-mobile MX—carried on wide-bodied aircraft and launched in midair—would be survivable provided that the aircraft received timely warning and took off immediately. Its dependence on prompt response to timely warning of submarine-launched ballistic missile attack would give such a force a common failure mode with the bomber force. (Removing dependence on warning by means of continuous airborne alert would be prohibitively expensive; acquisition and 10 years of operation for such a force could cost $80 billion to $100 billion (fiscal year 1980 dollars).) On the other hand, an air mobile force could not be threatened by the Soviet ICBM force unless the Soviets deployed many more ICBM *missiles* than they now possess and used them to barrage the entire Central United States. The outcome of such an attack would be insensitive to Soviet improvement in the fractionation and accuracy of their ICBMs. An air mobile MX force could not endure long after an attack if the Soviets attacked every airfield on which such planes could land to refuel. In this case, the National Command Authorities would have to "use or lose" the MX missiles within 5 to 6 hours of a Soviet attack. Providing endurance by increasing the number of airfields at which the planes could refuel would be enormously expensive ($10 billion to $30 billion for up to 4,600 airfields), and growth of the Soviet threat to plausible levels for the 1990's would require so many airfields that they would essentially fill the continental United States. The aircraft would have to takeoff to launch their missiles, which could mean slow response time, longer warning for the Soviets of a U.S. strike, and the possibility that the Soviets would mistake dispersal during a crisis for preparation for a U.S. first strike. Warning, communications, and guidance systems for an airmobile force could be complex.

9. The problems associated with other basing modes studied by OTA appear more substantial. An ABM defense of MX missiles based in fixed silos against a large Soviet threat would require the use of a complex system based on frontier technology and potentially vulnerable to Soviet countermeasures. The technical risks appear too high to support a decision today to rely on such a system for MX basing. Basing MX on surface ships appears to offer no serious advantages and significantly less survivability than submarine basing. Basing MX in "superhardened" shelters (e.g., very deep underground) would likely involve a period of several days between a launch order and the actual launch of the missile. Rail mobile MX would involve problems of force management and vulnerability to peacetime accidents or sabotage. Road-mobile basing appears infeasible because of the size of the missile; off-road mobile basing appears to offer few advantages and several drawbacks compared to MPS.

10. In comparing MPS, MPS with LoADS, LUA, small submarine basing, and air mobile, it is found that:

- All offer reasonable prospects for feasibility and survivability. MPS depends for survivability on concealing its location (PLU) which creates a degree of technical risk, and which would be become still more difficult if LoADS is used to defend MPS.
- All are compatible with high weapon effectiveness for the MX missile, although MPS, MPS with LoADS, and LUA would provide slightly better accuracy than submarine basing or air mobile.
- MPS would endure in an operational condition for a long time if it survived; small submarines would endure for several months; air mobile might endure for only a few hours, depending on the nature of the Soviet attack; the endurance of LoADS would depend on the speed and effectiveness of surviving Soviet reconnaissance and retargeting capabilities; LUA would have no endurance at all.
- All are compatible with adequate C^3, but obtaining such C^3 for any of them would require time, effort, and money.
- MPS could complicate future arms control. MPS with LoADS would require amending or withdrawing from the ABM Treaty reached at SALT I. LUA, small submarines, and air mobile appear compatible with existing arms control concepts.
- MPS, or MPS with LoADS, would have an impact on both the socioeconomic and physical environment in the deployment region that would be so great as to be different in kind from the impacts of any of the other systems. LUA would have virtually no environmental impact. Impacts from submarine

basing and air mobile would be relatively small and limited to the areas of the operating bases.

- Assuming a requirement for 100 surviving MX missiles, costs of baseline MPS, submarine basing, and air mobile would be roughly comparable: costs of acquisition and 10 years of operations for nominal designs are estimated to be roughly $40 billion (fiscal year 1980 dollars). Rebasing Minuteman III in an MPS mode would cost 10 to 20 percent less. Growth in the Soviet threat would require increases in the costs of MPS systems, but not in the others. If the Soviet threat grew to a level OTA considers plausible for 1990, the United States could assure survivability of the MX/MPS either by adding LoADS (at an additional cost of $10 billion to $15 billion) or by expanding the number of shelters and MX missiles (at an additional cost of $15 billion to $20 billion). Continued growth of the Soviet threat into the 1990's would drive the cost of survivability as high as $80 billion. Costs of LUA would be the lowest: procurement of the MX missiles, modification of existing silos, and upgraded C³ and warning systems could be $20 billion cheaper than the alternatives.

- MPS could provide a small, *nonsurvivable* force by 1986 or 1987, and a large, survivable force by about 1990. MX deployment relying on launch under attack could begin in 1986, but completion of necessary upgrading of warning and C³ systems would require several years longer. Air mobile could be deployed near the end of the decade. MPS with LoADS could be available around 1990. Small submarines could be deployed beginning around 1990, and would be survivable immediately. Thus, none of the basing modes could close the so-called "window of vulnerability" before the end of the decade.

III. NUCLEAR PROLIFERATION AND SAFEGURADS (1977).

The Problem

Issue 1

Are More Countries Likely To Acquire Nuclear Weapons, and If So, Will This Proliferation Jeopardize U.S. and Global Interests?

Findings

The technical and economic barriers to proliferation are declining as accessibility to nuclear weapons material becomes more widespread. Consequently, the decision whether or not to acquire a nuclear weapons capability has become increasingly a political one. The choice will turn on whether a nation views the possession of such a capability as being, on balance, in its national interest.

That balance will be affected by certain global trends. The diffusion of global power and the erosion of bipolar alliance systems and great power security guarantees tend to increase the incentives to proliferation. On the other hand, a number of states have long had the capability to acquire nuclear weapons but have been persuaded by a variety of political considerations to refrain. These disincentives may also be persuasive in the future to the growing number of countries which find nuclear weapons within their capability. With internationally derived incentives and disincentives broadly offsetting one another, the decision on acquiring a nuclear weapons capability will tend to hinge on the particular circumstances of each Nth country and the policy pursued by present nuclear weapons states, especially the United States.

Press reports indicate that at least two states (Israel and South Africa) are at the verge of acquiring or have already acquired nuclear weapons. Several other countries are close to a weapons capability and a few may choose to attain it over the next few years.

As for the consequences of proliferation, it can be argued that proliferation will have a stabilizing effect on international politics due to the deterrent value of nuclear weapons. The alternative, and more persuasive possibility, is that further proliferation will jeopardize regional and global stability, increase the likelihood of nuclear war (local or general), exacerbate the threat of nuclear armed non-state terrorism, and greatly complicate U.S. relations with new (potential or actual) nuclear weapons states. The extent to which proliferation has a disequilibrating effect on international politics also impacts directly on American foreign policy, which has had the maintenance of global stability as its overriding objective in recent years. From this perspective, the threat to American interests derives not so much from the mere number of Nth countries but from the probability that proliferation will tend to be greatest in regions with the highest potential for international conflict, e.g., the Middle East, Southern Africa, and East Asia. (See chapters III and IV.)

Issue 2

What Will Be the Proliferation Impact of the International Spread of Plutonium Recycle Facilities?

Findings

Reprocessing provides the strongest link between commercial nuclear power and proliferation. Possession of such a facility gives a nation access to weapons material (plutonium) by slow covert diversion which would be difficult for safeguards to detect. An overt seizure of the plant or associated plutonium stockpiles following abrogation of safeguards commitments could, if preceded by a clandestine weapons development program, result in the fabrication of nuclear explosives within days. Furthermore, such a plant reduces a nation's susceptibility to international restraints (sanctions) by enhancing fuel cycle independence. Finally, plutonium recycle is the most likely source for both black market fissile material and direct theft by terrorists.

Most nations expect to have their nuclear fuel reprocessed despite these obvious complications for the task of preventing further proliferation, and several (none of them Nth countries) are constructing large reprocessing plants. There have been increasing doubts as to the economic feasibility of reprocessing in the United States, but other countries perceive reprocessing as being attractive. Their more limited energy resources make the energy of plutonium more valuable, and possibly less-stringent regulatory requirements may make the facilities less expensive. In addition, if nuclear energy is to be a long-term option, reprocessing will eventually have to be an integral part of the fuel cycle, although uranium resources may be adequate to last until about the year 2000 even without reprocessing. Hence, nonproliferation strategies that involve a total renouncement of reprocessing will be difficult and probably expensive to implement.

Reprocessing in the United States and other weapons states is not a direct proliferation issue (except for terrorists). Other supplier states such as West Germany and Japan are also unlikely to use their commercial facilities to procure weapons material. The less advanced countries that might misuse facilities have been precluded from importing them by supplier agreements (except for Brazil and Pakistan who already have contracts and are resisting pressure to cancel them). The technology is uncomplicated enough for some Nth countries to develop on a commercial basis, but this endeavor would almost certainly be commercially uneconomical if other energy sources are available. A double standard approach to reprocessing would further strain relations between suppliers and importers. Multinationally controlled facilities may be necessary to alleviate this tension if reprocessing does become widespread in the supplier states (see chapters III, V, and X).

Issue 3

How Will U.S. Decisions on Domestic Plutonium Recycle Affect Efforts To Curb the International Spread of Reprocessing?

Findings

Decisions on the future of reprocessing and plutonium recycle in the United States must be made in the near future because of the imminence of operation of the large plant at

Barnwell, S.C. Nonproliferation will clearly be best served if no one reprocesses. Other nations, however, have a stronger interest in reprocessing (as described in Issue 2) and will be unsympathetic to efforts to convince them to refrain. If the United States alone refrains, the nonproliferation effort could actually be damaged because the resulting unavailability of fuel cycle services would induce more nations to build their own facilities. If the United States does not refrain, however, the credibility of its efforts to dissuade others will be diminished. There is general agreement that Nth country possession of reprocessing plants would be inconsistent with efforts to contain proliferation. The key factors shaping positions on this issue are:

- The effect of a double standard, where supplier states build their own reprocessing plants but deny exports to other nations: Importing states have expressed resentment over discriminatory export policies, and this policy would be certain to annoy some. It is significant, however, that few Nth countries will have enough reactors in this century to make an indigenous reprocessing plant more economical than having the service provided by a supplier state.

- The ability of the United States to persuade other nations to forgo reprocessing: A U.S. decision to refrain would have slight impact on other suppliers unless accompanied by costly political and economic pressure. Their commitment to reprocessing and to early deployment of breeder reactors (which require reprocessing) is much stronger than that of the United States. Importing states, however, are more likely to be impressed by such a gesture (see chapters III, VII, and X).

Issue 4

Would Deployment of Fast Breeder Reactors (LMFBRs) Be Compatible With a Policy To Curtail Proliferation?

Findings

The LMFBR is the highest priority energy development program in most nuclear supplier states. It was chosen because, of all the long-term options for essentially inexhaustible energy, it may well be most economic. It is also in a relatively advanced state of development, and thus the most likely to be available for widespread deployment by the end of the century.

Proliferation, however, was not a major consideration in the elevation of the LMFBR to its present priority. Certain characteristics of the LMFBR system as presently envisaged will conflict with efforts to control proliferation. These are:

- National possession of a full LMFBR cycle would eliminate all technical barriers to acquiring weapons material. It would also provide virtual immunity to an international embargo on fuel shipments because the LMFBR produces more than enough plutonium to refuel itself.

- Even a national LMFBR tied into an international fuel cycle (e.g., fuel leasing or multinational fuel centers) increases the opportunity for proliferation. Many nations could eliminate their dependence on the international or foreign fuel services by constructing indigenous fuel fabrication and reprocessing plants or by processing the fresh fuel or partially spent fuel within the reactor.

- Some of the plutonium produced by the LMFBR is of extremely high quality for weapons.

- The LMFBR requires reprocessing, which creates opportunities for diversion of plutonium by nations or non-state adversaries.

An overall assessment of the desirability of the breeder must weigh its benefits as an energy source against its liabilities relative to proliferation, as well as other problems in comparison with alternative energy sources (see chapter VII, "Diversion From Commercial Power Systems").

Issue 5

Do Uranium Enrichment Facilities Have a High Potential for Proliferation?

Findings

Any enrichment plant can theoretically be used for the production of weapons material while simultaneously providing immunity from international nuclear fuel embargoes, but only one type of enrichment plant—the centrifuge type—increases opportunities for proliferation on the same scale as reprocessing plants. Diffusion plants are economical only on a very large scale, so this enrichment route is out of the question for all but the largest and most highly developed countries.

The nozzle method is currently under development in South Africa and Germany. It promises to cost less than diffusion and be fundamentally simpler. It does demand highly precise manufacturing techniques, and its operation requires about twice as much power as a diffusion plant of the same capacity. This makes it commercially impractical for nations lacking low-cost power such as hydroelectricity. Despite its simplicity, it does not appear to be a good choice for a small facility dedicated to weapons material production.

By contrast, centrifuge plants may be sufficiently economical in small sizes for many nations to find them commercially attractive. These plants could only be developed by technically advanced nations, but could be purchased and operated by less advanced nations.

If sold to less advanced nations, centrifuge plants would be exceptionally vulnerable to clandestine diversion. Moreover, as with a reprocessing plant, a centrifuge facility could be seized and used to produce weapons material in a short time.

Advanced enrichment techniques could not be developed except by technically advanced countries. Barring an unforeseen breakthrough, commercial laser isotope separation (LIS) facilities will probably not be feasible for even advanced countries until the late 1980's or early 1990's, and then only if a number of very difficult problems are solved. The United States, U.S.S.R., and France, among others, are actively developing LIS technology. The proliferation potential for LIS and other advanced technologies stems from the high enrichment achieved per stage. Thus, it may be possible to produce weapons material in a very few steps. In addition, LIS facilities may be economical on a very small scale, making them attractive purchases for nations with small nuclear programs. The United States, by guaranteeing enrichment services at a low fee or at cost might slow down the spread of advanced enrichment technologies (see chapter VII, "Dedicated Facilities").

Issue 6

How Feasible Would it be to Use Commercial Nuclear Reactors as a Source of Weapons Material?

Findings

- The power reactors presently available for export (LWR and CANDU) do not involve material that could be used directly for nuclear explosives. A nation would also have to have a reprocessing or enrichment facility to use its nuclear system as a source of weapons material.

- Spent fuel from either reactor type does contain plutonium, which could be recovered in a small indigenously developed reprocessing plant. This opportunity can be decreased by fuel leasing or buy-back arrangements which prevent the long-term storage of spent fuel by Nth countries. This would restrict the availability of spent fuel to that in the reactor, the use of which would probably result in the loss of the reactor as a power source.

- Reactors and short-term spent-fuel storage facilities can be effectively safeguarded. Consequently, diversion from them would have to take the form of overt nationalization (i.e., seizure).

- The additional expertise a nation acquires in operating its own reactor would be useful should it decide to develop weapons.

Abandoning nuclear power would reduce, but not eliminate, the possibility of further weapons proliferation. Countries could still construct facilities dedicated to the production of weapons material or, alternatively, they might be able to purchase or steal either material or a finished weapon (see chapter VII).

Issue 7

Could a Nation Acquire Nuclear Weapons Without Diverting Fissile Material From Its Commercial Nuclear Power Facilities?

Findings

None of the countries which now have nuclear weapons diverted fissile material from their power facilities. They all built facilities specifically dedicated to the production or reprocessing of nuclear weapons material.

The only dedicated facility option open to a nation which is not technologically advanced is a small, natural uranium-fueled plutonium production reactor, producing about 10 kg of weapons-grade plutonium per year (enough for one or two explosives), and a small reprocessing plant. The total capital costs of these facilities would be several tens of millions of dollars. Such a facility might escape detection, especially if the nation were not considered to be among the five or six most likely Nth countries.

A technologically advanced nation would be able to build a dedicated facility to support a large weapons program, but it is unlikely that the existence of such a facility could be kept secret (see chapter VII, "Dedicated Facilities").

Issue 8

How Plausible is the Direct Acquisition of Fissile Material or Weapons by Purchase or Theft?

Findings

If plutonium becomes a commonly traded commodity, minimal intermittent black market transactions seem plausible, simply because the large amounts of material that could be circulating would be difficult to safeguard perfectly. Theft of existing weapons would be more probable if proliferation continues and security in the new nuclear states is lax. (See chapter VI.)

Issue 9

How Critical is Nuclear Power to Future Global Energy Requirements?

Findings

Projections of growth in global nuclear energy use have been repeatedly revised downwards in recent years. The lowest projections presently available are the most plausible. Nevertheless, many governments, especially in Europe and Japan, still feel that nuclear energy will be crucial to their well-being as global oil and gas reserves are depleted. Many developing countries are also counting heavily on nuclear energy. Coal, another major alternative to oil and gas, is abundant in some countries but fraught with environmental hazards. The economics of other resources (e.g., solar) are more speculative. Hence, nuclear power is likely to be a significant factor for at least the next few decades (see chapter X).

Issue 10

How Difficult Would It Be for a Nation To Construct a Nuclear Weapon?

Findings

Many nations are capable of designing and constructing nuclear explosives which could be confidently expected, even without nuclear testing, to have predictable and reliable yields up to 10 to 20 kilotons TNT equivalent (using U^{235}, U^{233}, or weapons-grade plutonium) or in the kiloton range (using reactor-grade plutonium).

A national effort to achieve the above objective would require a group of more than a dozen well-trained and very competent persons with experience in several fields of science and engineering. They would need a high explosive field-test facility and the support of a modest, already established, scientific, technical, and organizational infrastructure. If the program is properly executed, the objective might be attained approximately 2 years after the start of the program, at a cost of a few tens of millions of dollars. This estimate does not include the time and money to obtain the fissile material or to establish the infrastructure assumed above.

The success or failure of a national effort will depend more on the strengths and weaknesses of the particular people involved in the effort than on specifics of the technological base of the country (see chapter VI, "Nuclear Fission Explosive Weapons" for further details.)

Issue 11

Is a Non-State Adversary Group Likely To Turn to Nuclear Means of Extortion or Violence?

Findings

There is no evidence that any non-state group has ever made any attempt to acquire weapons material for use in a nuclear explosive. The incidents that have occurred to date involving nuclear material or facilities have mostly been low-level incidents of vandalism or sabotage. However, the present record of nuclear incidents was assembled in an era when nuclear reactors were relatively few. The expansion of nuclear power, the advent of plutonium recycle, and trends towards increased violence could lead non-state adversaries to attempt large-scale nuclear threats or violence.

Non-state adversary groups have not yet gone to the limits of their ability to cause harm by non-nuclear means. Historical analysis of adversary tactics suggests reasons for this restraint. However, non-state adversaries, particularly terrorists or revolutionaries, may not behave in the future as they have in the past. The psychological impact of the threat or use of nuclear weapons would be enormous, and an adversary group may decide to attempt to exploit this leverage.

The entire subject of adversary actions involving massive threats or destruction has apparently just started to receive systematic study. When considering if non-state adversary groups will turn to massive extortion or violence, all routes to the same end—conventional explosives, other chemicals, nuclear and biological agents—should be considered. (See chapter V.)

Issue 12

How Difficult Would it be for a Non-State Adversary Group To Acquire Nuclear Material for a Nuclear Explosive Device?

Findings

It would be extremely difficult, verging on impossible, for a non-state adversary group to convert material diverted from LWR or CANDU fuel cycles to explosive's material, unless the spent fuel is commercially reprocessed to recover and recycle the plutonium.

In the LWR with plutonium recycle as presently planned, material suitable for nuclear explosives or easily convertible to weapons-useable material will be found at the reprocessing plant, in transit between the reprocessing plant and the fuel fabrication plant, and at the input area of the fuel fabrication plant. There are technologies and configurations (coprecipitation and colocation) under consideration that could eliminate most opportunities for the diversion of material easily converted to weapons material.

In the United States at present, the NRC is reportedly in the process of upgrading security at licensees handling plutonium or highly enriched uranium, requiring them to meet a threat of two or more insiders in collusion with several heavily armed attakers from the outside. Present safeguards and physical security may place undue reliance on one element of physical security—armed guards. It is not clear how well presently designed safeguards system can handle the problem of several insiders acting in collusion, or outsiders attacking with guile and deception rather than straightforward armed assault.

Some observers have also expressed doubts about the effectiveness of guard forces in handling diversion attempts, partly because of the questionable status of their exact legal powers. The subject of a Federal security force to protect plutonium and highly enriched uranium should be reopened, especially in view of the increased threat levels licensees are being required to meet.

Both ERDA and NRC have very promising safeguards programs in the development stage, but their ultimate effectiveness cannot be assessed at this time.

A vital point to note is that non-state adversaries are highly mobile, and capable of finding and attacking the weakest targets. No nation, however invulnerable its own facilities, can feel secure against non-state adversary nuclear threats and violence unless all facilities handling weapons-grade material worldwide are equally well protected. Physical security is generally left to the discretion of the individual nation, although supplier states are insisting on a minimum level as a condition for export. The International Atomic Energy Agency has no physical security enforcement powers, (see chapters V and VIII).

Issue 13

Could a Non-State Adversary Design and Construct Its Own Nuclear Explosive?

Findings

Given the weapons material and a fraction of a million dollars, a small group of people, none of whom have ever had access to the classified literature, could possibly design and build a crude nuclear explosive device. The group would have to include, at a minimum, a person capable of searching and understanding the technical literature in several fields, and a jack-of-all-trades technician. They would probably not be able to develop an accurate prediction of the yield of their device, and it could be a total failure because of either faulty design or faulty construction. If a member of the group is careless or incompetent, he might suffer serious or fatal injury. However, there is a clear possibility that a clever and competent group could design and construct a device which would produce a significant nuclear yield (see chapter VI "Nuclear Fission Explosive Weapons" for details).

Issue 14

What Are the Civil Liberties Implications of Safeguarding Nuclear Power, in Particular, Plutonium Recycle?

Findings

The civil liberties implications of safeguards turn on the scope of a security clearance program, the standards and procedures used in employee clearance, the scope and intrusiveness of domestic intelligence activities, and the nature of a recovery effort should a diversion occur.

There is disagreement among experts as to whether a safeguards program can be adequate for security without fundamentally infringing upon civil liberties. One position believes adequate safeguards will necessarily violate basic liberties for employees and political dissidents. A second position treats safeguards as an acceptable extension of existing clearance programs and blackmail threat responses in other fields of high security. A third position believes safeguards could be installed without doing serious damage to civil liberties, but only if a "least intrusive measures" approach is adopted and a zero-risk goal is rejected.

Although a safeguards system that would be extremely respectful of civil liberties can be designed, three potential dangers exist:

1. A gradual erosion of civil liberties as the safeguards system is "strengthened,"

2. A shunting aside of civil liberties during a recovery operation if weapons material were diverted and a convincing threat received; and

3. A public demand for Draconian safeguards in the future, even at the expense of civil liberties, if a diversion followed by a convincing threat or an actual act of destruction occurred.

Measures can be envisaged that would reduce the probability of the above three occurrences. Continued public monitoring of safeguards systems for civil liberties infractions, new technologies or configurations (e.g., coprecipitation or colocation), and response planning integrated at the local, State, regional, and Federal levels with authority clearly delineated could reduce the probability of civil liberties infractions in a strong safeguards system.

The Control

Issue 15

What is the Outlook for Control of Proliferation?

Findings

It is not too late to contain proliferation at a level which can be assimilated by the international political system. However, there are no single or all-purpose solutions; no short cuts. A viable nonproliferation policy will require the coordinated, planned use of a wide variety of measures: (a) political, economic, institutional, technological; (b) unilateral, bilateral, multilateral, international; and (c) executive and legislative.

Components of a nonproliferation policy would include: (a) Steps designed to tip the balance of political incentives and disincentives regarding the acquisition of weapons in favor of disincentives; (b) A comprehensive safeguards regime to prevent the diversion of nuclear material from civilian energy programs to weapons use; (c) Controls over exports, particularly with regard to enrichment and reprocessing capabilities, in conjunction with arrangements for the return of spent fuel to the supplier or any international repository; (d) A broad range of domestic and foreign policy supporting actions, including steps to upgrade physical security measures to prevent theft of nuclear materials, expansion of reactor-grade uranium production to obviate the need for reprocessing, and arms control negotiations; and (e) Steps to assure that other countries can meet their energy requirements without resorting to enrichment and/or reprocessing national facilities.

Moreover, because each Nth country is to some degree unique, policy must be tailored to fit particular national circumstances. This is especially true because of the potential for serious conflict between nonproliferation and other foreign policy objectives. The nature and severity of that conflict will vary from one Nth country to another, a fact which policy must take carefully into account. (Chapters III and IV.)

Issue 16

What Influence Can the United States Exert Upon Potential Weapons States?

Findings

In the long run two general rules apply: (a) Solutions to the proliferation problem will have to be found primarily, though not exclusively, through multilateral actions, and (b) The extent of U.S. influence will vary from country to country.

As American preeminence in the international market for nuclear fuel, facilities, and technology has been allowed to erode, the ability of the United States to unilaterally determine the ground rules of international nuclear cooperation has diminished. With the entrance of other suppliers into the market, importers have the option to turn to non-U.S. sources. If the United States were to remove itself from the global market entirely, other suppliers could quickly replace the withdrawn capacity. As a consequence American actions will tend to be most effective in a multilateral context—particularly in conjunction with other suppliers. The effectiveness of this approach has been demonstrated in the negotiations which led to the NPT, and more recently in the Suppliers' Conference.

There remains, however, significant scope for the unilateral assertion of U.S. influence—both in terms of positive inducements and negative sanctions. The recent successful U.S. effort inducing South Korea to abandon plans for purchasing a French reprocessing facility is an instance of the effective use of unilateral influence. Some of the more obvious levers available to Washington include:

- security guarantees;
- assistance to civilian nuclear energy programs;
- foreign economic aid (including U.S. influence in international lending institutions);
- military assistance programs;
- political pressures and diplomatic persuasion;
- mediation of international disputes with proliferation implications;
- controls on the export of sensitive nuclear technology;
- assistance concerning non-nuclear energy sources; and
- domestic policy initiatives (e.g., concerning reprocessing) which might enhance the credibility of U.S. efforts to persuade other countries to take similar steps.

The single most effective instrument of U.S. influence would be the capability to guarantee adequate low-enriched uranium exports to meet the needs of overseas users while, at the same time, providing for the collection and return of spent fuel.

An effective effort to assert U.S. influence will combine the carrot and the stick, with principal reliance on the former for the longer term. Such an effort will also take into account the wide variation in leverage available to Washington when dealing with one Nth country or another. Thus U.S. influence with nations dependent upon American military or economic assistance (e.g., South Korea) is very substantial but where such dependence is lacking (e.g., Argentina) U.S. influence declines.

Issue 17

What Influence Can the United States Exert Upon Other Supplier States?

Findings

Efforts by the United States inducing other supplier states to pursue policies supportive of nonproliferation will generally be most effective if they are formulated in a multilateral context and emphasize positive inducements. Possible measures include:

- political-diplomatic persuasion (e.g., the Suppliers' Conference),
- tie-in agreements guaranteeing U.S. enrichment services at nondiscriminatory prices to reactor customers of other suppliers,
- joint-venture enrichment and/or reprocessing facilities,
- market sharing agreements,
- multinational enrichment and/or reprocessing facilities,
- international fuel storage repositories, and
- a multilateral study of alternatives to reprocessing.

The problem of reprocessing is extremely difficult for two reasons. First, other supplier states (such as Germany) have already made a basic national decision in favor of reprocessing and the breeder. They regard this policy as a vital element in their efforts to assure adequate energy in the future. European breeder technology is the most advanced in the world. Second, other major suppliers are also America's principal allies and trading partners. The linkages of mutual interest and dependence are so extensive as to render most attempts to apply coercive pressures self-damaging. Consequently, U.S. efforts to obtain a global moratorium on reprocessing will encounter stiff European and Japanese resistance. The one area where agreement is demonstratively possible concerns control on exports of reprocessing facilities.

Issue 18

How Effective Are International Atomic Energy Agency Safeguards?

Findings

(a) Safeguards for reactors can be very effective. Nuclear material is contained in a relatively small number of discrete items, the fuel elements. Exact item accountability can be accomplished without great difficulty.

(b) Safeguards procedures for reprocessing plants, enrichment plants, and other fuel-cycle facilities which handle very large flows of nuclear material are in the experimental stage. It will be difficult to detect significant diversion of uranium or plutonium using nuclear material accountancy alone, even if the most advanced analytical techniques and accountancy methods are used. The task is further complicated by restrictions on IAEA inspection effort, inspector access, and the full use of IAEA surveillance devices.

(c) Containment and surveillance must play a key role in safeguards and must be regarded as more than supplementary to materials accountancy. Effective safeguards systems for enrichment and reprocessing plants will have to include the most advanced online monitoring and real-time accounting systems as well as highly reliable, instantaneously reporting, tamper-indicating surveillance equipment.

(d) A credible safeguards system provides a significant deterrent to diversion, by both increasing the chances of detection and establishing standards of legal behavior that buttress the position of political groups opposed to proliferation.

(e) No safeguards system can prevent an overt national seizure of a facility and its operation for weapons purposes.

Issue 19

Are Multinational Fuel-Cycle Facilities (MFCFs), on Balance, a Useful Approach for the Control of Proliferation?

Findings

The primary intent of MFCFs is to remove sensitive facilities (particularly enrichment and reprocessing) from national control. A part owner/operator of such a facility will find it much harder to tamper with equipment for purposes of diversion or to seize the plant outright even if on its own territory. It also offers economies of scale to nations with only a few reactors and improved security against nonstate adversary actions.

A great many political, economic, and institutional questions must be resolved before the concept can be considered viable. Member nations may not find acceptable sites in other members' territory. Another problem is the possibility that membership in a sensitive facility could provide sufficient access to the technology for members to recreate it indigenously. Thus MFCFs could spread the very problem they are intended to prevent.

Issue 20

Are Sanctions a Useful Instrument of Nonproliferation Policy Toward Nth Countries?

Findings

Provisions for modest sanctions (e.g., the cutoff of nuclear assistance) already are contained in U.S. and IAEA nuclear agreements with nonweapons states, and a variety of stronger sanctions can be postulated. To be most effective, sanctions should be applied jointly or multilaterally rather than unilaterally. Threats should be accompanied by inducements and rewards designed to relieve the pressures toward proliferation. A sophisticated approach will also combine automatic with more discretionary and flexible sanctions.

Depending upon the prospective proliferator, a significant degree of vulnerability to one or more of the available levers is likely to be present. In cases such as Taiwan, where the Nth country is dependent on the United States for security support as well as nuclear imports, the scope for the imposition of unilateral U.S. sanctions is substantial. In other cases, such as that of Brazil, resort to sanctions could probably prove futile. Sanctions could be more effective in all cases as an instrument to prevent proliferation than as a means to punish or "roll back" proliferation after it has occured. The most effective channel for imposition of unilateral sanctions will probably be the Suppliers' Conference. Because user states comprise a majority of IAEA membership, there are serious questions as to whether the agency could muster the political will to impose sanctions on a recipient—particularly if the circumstances surrounding the alleged violation are at all ambiguous.

Issue 21

Would an Arms-Reduction Agreement by Present Nuclear States Significantly Strengthen the International Norm Against Proliferation?

Findings

A meaningful multilateral arms reduction by the nuclear weapons states would demonstrate a commitment to the objective of nonproliferation and, in particular, to the Non-Proliferation Treaty. The extent of the impact of this demonstration is not clear, but the public stance of some of the non-weapons states indicates that it could be substantial. A

corollary benefit might be a reduction in the prestige attached to nuclear weapons.

Issue 22

To What Extent Can Improvements in Technology Help Contain or Limit Further Proliferation?

Findings

There is no technological fix that can eliminate the problem of proliferation, but concepts under development could, if successful, make diversion from commercial facilities much more difficult or even close to impossible.

One of the most promising medium-term approaches is the nonproliferating reactor. This concept is a fundamentally new approach both to reactor design and to nonproliferation. By incorporating nonproliferation requirements into the design of the reactor, the diversion routes which are present in current and projected power reactor systems could be largely eliminated. This approach deserves a thorough assessment and open-minded comparison with other alternatives to determine if it should be funded at an expanded scale.

Less radical changes are alternate fuel cycles (thorium) and modifications to present fuel cycles (e.g., coprecipitation and tandem cycles). New approaches are also being developed in safeguards technology. Integrating safeguards systems into facility designs would considerably strengthen safeguards effectiveness. Greater R&D emphasis on non-nuclear energy sources, especially those most appropriate for developing countries, could reduce the dependence on nuclear power and postpone or even eliminate the eventual need to move to more sensitive systems (such as fast breeders).

Issue 23

Can the Non-Proliferation Treaty (NPT) Play a Useful Role in Containing Proliferation?

Findings

The NPT has important weaknesses. It lacks universal adherence and a party can, under some conditions, legally withdraw with only 3 months notice. Nonparties to the treaty include a number of the strongest candidates for the acquisition of nuclear weapons. Moreover, the sanctions provided for in the treaty are not particularly impressive and there is a serious question whether even they could be enforced in the event of a violation.

Nevertheless, the NPT remains a key component of an effective nonproliferation policy. The fact that there have been no known violations of the treaty suggests that it acts as an important constraint upon Nth countries. It embodies a number of a basic international consensus that proliferation poses a serious threat to global well-being and should be contained. It also provides an agreed framework of mutual rights and obligations constituting a fundamental bargain between supplier and user states. As such, it sets forth a standard by which to measure and perhaps influence the behavior of states. For example, the NPT may provide some of the impetus behind current efforts by the superpowers to negotiate a new arms control agreement.

Two additional features of the NPT give it particular significance. First, by allowing the IAEA to impose safeguards on their domestic nuclear programs, the nonweapon parties to the treaty relinquish a significant measure of their sovereignty. This establishes an important principle upon which to build stronger international arrangements for controlling proliferation. Second, in addition to providing a statement of principles and objectives, the NPT encompasses an institutional mechanism (IAEA) for their implementation. The NPT is more than a treaty: it is an ongoing program. (Chapters III and VIII "International Control of Proliferation.")

Issue 24

What Issues Require Priority Attention, i.e., What Developments Threaten to Foreclose Future Options?

Findings

The following subject areas require immediate consideration by policymakers and legislators if the course of proliferation is not to be determined by default.

- Domestic (U.S.) reprocessing.
- U.S. enrichment capacity.
- Upgrading of supplier (export) controls.
- Sanctions and inducements to be applied to Nth countries.
- Research and development priorities (LMFBR vs. other breeders and non-nuclear sources).

APPENDIX E

DATA TABLES ON THE NUCLEAR BALANCE

DIETRICH SCHROEER
PHYSICS DEPARTMENT
UNIVERSITY OF NORTH CAROLINA

Table 1. The long-range and medium-range strategic bombing forces of the U.S.S.R. and the U.S.A. Unless otherwise indicated, numbers are from THE MILITARY BALANCE: 1982-1983 of the London International Institute for Strategic Studies, 1982.

Bombers	Range	Top Speed	Bomb Payload	Number	EMT Total[1]
			UNITED STATES		
B-52-D/F	9900 km	0.95 Mach[3]	30 tons; 4x1 Mt[4] +2 SRAMx200 kt[5]	75	351 EMT
B-52-G	12000 km	0.95 Mach[3]	35 tons; 4x1 Mt[4] +4 SRAMx200 kt[5]	151	311 EMT
B-52-G	12000 km	0.95 Mach[3]	35 tons; 4x1 Mt[4] +4 SRAMx200 kt[5] +12 CMx300 kt[6,7]	16[4]	172 EMT
B-52-H	16000 km	0.95 Mach[3]	35 tons; 4x1 Mt[4] +4 SRAMx0.2 Mt[5]	90	483 EMT
B-52	inactive storage			(187)	
FB-111A	4700 km	2.5 Mach[8]	19 tons; 2 SRAMx200 kt	(60)	
F-111/E/F (strike)	4700 km	2.5 Mach	14 tons; 2 SRAMx200 kt	(156)	
Tanker		0.95 Mach		(646)	
B-1	9800 km	1.6 Mach		(R&D)	
			U.S. TOTAL	332[9]	1817 EMT[9]
			SOVIET UNION		
Tu-95[10]	12800 km	0.78 Mach	20 tons;[11] 3x1 Mt[4,12]	105	315 EMT
Mya-4	11200 km	0.87 Mach[13]	10 tons;[11] 2x1 Mt[4]	45	90 EMT
Tu-16	4800 km	0.8 Mach	10 tons;[11]	(580)	
Backfire	8000 km	2.5 Mach[8,14]	9 tons;[11]	(180)	
Su-24 (strike)	4000 km	2.3 Mach[8]	4 tons;[11]	(550)	
Tankers				(45)	
			U.S.S.R. TOTAL	1509	405 EMT[9]
			UNITED KINGDOM		
Vulcan B2	6400 km	1.6 Mach	10.5 tons;	(48)[15]	
			U.K. TOTAL	0[9]	0 EMT[9]

0094-243X/83/1040336-12 $3.00 Copyright 1983 American Institute of Physics

NOTES FOR TABLE 1

[1] EMT (equivalent megatons) is equal to $Y^{2/3}$, with Y in megatons. For the 0.2-Mt SRAM the EMT=0.342 EMT, for the 0.3-Mt cruise missile the EMT=0.448 EMT.

[2] J. W. R. Taylor, Ed., JANE'S ALL THE WORLD'S AIRCRAFT: 1982-83, London: Jane's Publ. Co., 1982.

[3] Reference 2: At high altitude; the penetration speed at low altitude is 0.66 Mach (660 km/hr).

[4] Stockholm International Peace Research Institute, WORLD ARMAMENTS AND DISARMAMENT: SIPRI YEARBOOK 1982, Cambridge, MA: Oelgeschlager, Gunn & Hain, Inc., 1982. In general the data and numbers in this reference are similar to those in the IISS publication.

[5] Anthony H. Cordesman, "M-X and the Balance of Power: Reasserting America's Strength," Armed Forces Journal International 120 (4), 21-51 (December 1982), Table 5, p. 41. This table is slightly more detailed in listing the separate models; the totals are close to those in the IISS publication.

[6] Reference 2: Starting in 1982 173 B-52-Gs will be modified to carry 20 SRAM, six on each wing, plus eight inside the bomb bay; later in the 1980s this will be done to other B-52s.

[7] Reference 4 lists the yield of the cruise missile warhead as 200 kt.

[8] Reference 2: At high altitudes.

[9] Counting only truly intercontinental bombers; i.e. not counting the F-111 or the Vulcan B2, or the Tu-16 Badger, the Tu-26 Backfire, or the Su-24 Fencer.

[10] A turboprop aircraft, all others are jets.

[11] Reference 2 lists the speeds of the Tu-26 and the Su-24 respectively as 2 Mach and 2 or more Mach. It lists the payloads of the Tu-95, Mya-4, Tu-16, Tu-26 and Su-24 respectively as 13 tons, 5 tons, 5 tons, 12 tons, and 2 tons. These payloads are probably the typical operational payloads to achieve the maximum ranges listed, while the payloads given by the IISS may be the maximum payloads leading to a reduced range.

[12] Reference 2: A few carry the Kangeroo cruise missile.

[13] Reference 2: At an altitude of 11 km.

[14] Reference 2: At high altitudes. At low altitudes the maximum speed is 0.9 Mach.

[15] Prior to the Falkland island war, these were being phased out.

Table 2. ICBM missiles presently deployed and under development. All data are from THE MILITARY BALANCE 1982–1983 of the London International Institute for Strategic Studies, 1982, unless otherwise indicated.

Missile	Stages	Launch	Range	Accuracy	Payload
UNITED STATES					
Titan II[1]	2	hot	15,000 km	1300 m	4.2 tons
Minuteman II	3	hot	11,300 km	370 m	0.8 tons
Minuteman III	3	hot	13,000 km	280 m	1.2 tons
Minuteman IIIA	3	hot	13,000 km	220 m	1.2 tons
MX	3[2]	cold	11,000 km[3]	80 m	4.0 tons
SOVIET UNION					
SS-111 (MOD 1)	2	hot	10,500 km	1400 m	1.0 tons
(MOD 3)	2	hot	8,800 km	1100 m	1.3 tons
SS-13	3	hot	10,000 km	2000 m	0.5 tons
SS-17[1] (MOD 1)	2	cold	10,000 km	450 m	3 tons
(MOD 2)	2	cold	11,000 km	450 m	1.8 tons
SS-18[1] (MOD 1)	2	cold	12,000 km[4]	450 m	8.2 tons
(MOD 2)	2	cold	11,000 km[5]	450 m	8.3 tons
(MOD 3)	2	cold	10,500 km	350 m	8 tons
(MOD 4)	2	cold	9,000 km	300 m	8.3 tons
(MOD 5)	2	cold	(9,000 km)	(250 m)	(8 tons)
SS-19[1] (MOD 1)	2	hot	11,000 km	500 m	4 tons
(MOD 2)	2	hot	10,000 km	300 m	3.7 tons
(MOD 3)	2	hot	(10,000 km)	300 m	4 tons

[1] These rockets are liquid fueled.

[2] The post-boost bus has a rocket fuelled by storable liquid propellant.

[3] Office of Technology Assessment, MX MISSILE BASING, Washington, DC: U.S. Government Printing Office, 1981.

[4] R. T. Pretty, JANE'S WEAPONS SYSTEMS: 1982–83, London: Jane's Publ. Co., 1982; says 10,500 km.

[5] Reference 4 says 9,250 km.

[6] Equivalent megatonnage (EMT) is equal to $Y^{2/3}$, with the yield Y in megatons.

[7] Reference 4 says 200 kt.

[8] Anthony H. Cordesman, "M-X and the Balance of Power: Reasserting America's Strength," Armed Forces J. Int'l. <u>120</u> (4), 21–51 (December 1982); p. 41. This reference gives subtotals for the various modifications, and assumes 52 more SS-11s have been replaced by SS-19s.

[9] Out of service.

Table 2. continued

Missile		Warheads	# ICBMs	# RV	Mt	EMT
		UNITED STATES				
Titan II		1 @ 9 Mt	52	52	468	225
Minuteman II		1 @ 1½ Mt	450	450	675	590
Minuteman III		3 MIRV @ 0.170 Mt[7]	250	750	127	230
Minuteman IIIA		3 MIRV @ 0.335 Mt	300	900	302	434
MX		10 MIRV @ 0.335 Mt	(R&D)	0	0	0
		U.S. TOTAL	1052	2152	1572	1479
		SOVIET UNION				
SS-11	(MOD 1)	1 @ 1 Mt	518(-)[8]	518(-)	518	518
	(MOD 3)	3 MRV @ 0.1-0.3 Mt	some	some	(0)	(0)
SS-13		1 @ 0.75 Mt	60	60	45	50
SS-17	(MOD 1)	4 MIRV @ 0.75 Mt	120[8]	480	360	396
	(MOD 2)	1 @ 6 Mt	32[8]	32	192	106
SS-18	(MOD 1)	1 @ 20 Mt	58(-)[8]	58(-)	1160	427
	(MOD 2)	8 MIRV @ 0.9 Mt	175[8]	1400	1260	1305
	(MOD 3)	1 @ 20 Mt	some[8]	some	(0)	(0)
	(MOD 4)	10 MIRV @ 0.5 Mt	75[8]	750	375	472
	(MOD 5)	(10 MIRV @ 0.75 Mt)	(R&D)	0	0	0
SS-19	(MOD 1)	6 MIRV @ 0.55 Mt[7]	9	0	0	0
	(MOD 2)	1 @ 5 Mt	60[8]	60	300	175
	(MOD 3)	6 MIRV @ 0.55 Mt	300[8]	1800	990	1208
		U.S.S.R. TOTAL	1398	5158	5200	4657

Table 3. Listing of the submarine deterrent. All numbers are from Anthony Cordesman, "M-X and the Balance of Power: Reasserting America's Strength," Armed Forces J. Int'l. <u>120</u> (4), 21-51 (December 1982); unless otherwise indicated. These numbers are somewhat more detailed than those in Ref. 1; but basically the two sets of numbers are in agreement.

Submarines	Function	# of Missiles	# of Subs
UNITED STATES			
Polaris	SLBM Nuclear Subs	16 Polaris A-3	decom.'82
Poseidon	SLBM Nuclear Subs	16 Poseidon A-3	19
Poseidon	SLBM Nuclear Subs	16 Trident-I C-4	12
Trident	SLBM Nuclear Subs	24 Trident-I C-4	2
SSN-688	Nuclear Attack Subs	Harpoon & SUBROC	(18)[1]
Other Nuclear Subs	Attack Subs	SUBROC	(67)[1]
		U.S. TOTAL	33
SOVIET UNION			
Golf	SLBM Diesel Sub	6 SS-N-8	(1)[2]
Hotel II	SLBM Nuclear Subs	3 SS-N-5	(6)[2]
Hotel III	SLBM Nuclear Sub	6 SS-N-8	(1)[2]
Yankee I	SLBM Nuclear Subs	16 SS-N-6	25
Yankee II	SLBM Nuclear Sub	12 SS-N-17	1
Delta I	SLBM Nuclear Subs	12 SS-N-8	18
Delta II	SLBM Nuclear Subs	16 SS-N-8	4
Delta III	SLBM Nuclear Subs	16 SS-N-18	13
Typhoon	SLBM Nuclear Sub	20 SS-NX-20	1[3]
Other Nuclear Subs	Cruise-Missile Subs	24 SS-N-19, etc.	(49)[1]
Other Nuclear Subs	Attack Subs	SS-N-15/16, etc.	(56)[1]
		U.S.S.R. TOTAL	62
UNITED KINGDOM			
Resolution	SLBM Nuclear Sub	16 Polaris A-3	4
Other Nuclear Subs	Attack Subs		(11)[1]
		U.K. TOTAL	4
FRANCE			
Redoutable	SLBM Nuclear Sub	16 MSBS M-20	5
Rubis	Nuclear Attack Subs		(1)[4]
		French TOTAL	5

[1] International Institute for Strategic Studies, THE MILITARY BALANCE: 1982-1983, London: IISS, 1982.

[2] Missiles but not submarines are counted in SALT II.

[3] Stockholm International Peace Research Institute, WORLD ARMAMENTS AND DISARMAMENT: SIPRI YEARBOOK 1982, Cambridge, MA: Oelgeschlager, Gunn & Hain, Inc., 1982; considers this submarine to become operational in the mid 1980s.

[4] Henri Le Masson, LES SOUS-MARINS FRANCAIS, Paris: Editions de la Cite-Brest-Paris, 1980.

Table 4. Listing of nuclear-tipped SLBMs carried on submarines. All numbers, unless otherwise indicated, are from THE MILITARY BALANCE 1982-1983, by the London International Institute for Strategic Studies, 1982.

Missiles	Range	RV/SLBM	CEP	RV-'82	Total EMT[1]
		UNITED STATES			
Poseidon	4600 km	10 MIRV/50 kt	450 m	3040	413 EMT
Trident I	7400 km	8 MIRV/100 kt	450 m	1920	414 EMT
Trident II	11000 km	14 MaRV/150 kt[2]	?	0	0 EMT
		U.S. Total		4960	827 EMT
		SOVIET UNION			
SS-N-5	1400 km	1/1 Mt	2800 m	18	18 EMT
SS-N-6[3]	3000 km	1/1 Mt	900 m	400	400 EMT
SS-N-8[3,4]	7800 km	1/1 Mt	⩾900 m[5]	292	292 EMT
SS-NX-17	3900 km	1/1 Mt	1500 m	12	12 EMT
SS-N-18[3]	8300 km	7 MIRV/0.2 Mt	600 m	1456	498 EMT
SS-NX-20	8300 km	(12 MIRV/dev.)	?	(240)[6]	0 EMT
		U.S.S.R. Total		2178	1220 EMT
		GREAT BRITAIN			
Polaris	4600 km	3 MRV/170 kt	900 m	192	60 EMT
Chevaline	n.a.	3 MRV[2]	?	0	0 EMT
		U.K. Total		192	60 EMT
		FRANCE			
MSBS M-20	3000 km	1/1 Mt	?	80	80 EMT
M-4	4000 km	MRV	?	0	0 EMT
		French Total		80	80 EMT

[1] R. T. Pretty, Ed., JANE'S WEAPONS SYSTEMS: 1982-83, London: Jane's Publ. Co., 1982.

[2] Stockholm International Peace Research Institute, WORLD ARMAMENTS AND DISARMAMENT: SIPRI YEARBOOK 1982, Cambridge, MA: Oelgeschlager, Gunn & Hain, Inc., 1982.

[3] These missiles are liquid fueled.

[4] Mod 2 is listed in Ref. 1 as having a range of 9100 km with a payload of 4 tons and a single 800-kt warhead; that payload seems much too large to be reasonable.

[5] Reference 1 reports a claim that the accuracy of this missile might be 400 meters due to combining stellar with inertial guidance.

[6] Reference 2 considers this an experimental model, not becoming operational until 1985; other, such as the U.S. DoD's SOVIET MILITARY POWER, 1983, claim first deployment in 1983.

Table 5. Details of the strategic balance between the U.S. and the U.S.S.R. Damage area is taken as 30 mi^2 per equivalent megaton. The missile-silo blast resistances are taken as 2000 psi for all except the Titan II and SS-11 missiles, which are assumed to be hardened to 300 psi. The Minuteman IIIA may indeed be protected to that level; the older Soviet silos may be considerably less protected, some of the newer Soviet SS-17/18/19 silos may have higher levels of hardening. After Tables 1-4.

Missiles	# Warheads	Deterrence Capability Mt per Warhead	EMT per Warhead	Area Destroyed Each	Total
Titan II	52x1	9.0 Mt	4.3 EMT	130 mi^2	6,700 mi^2
Minuteman II	450x1	1.5 Mt	1.3 EMT	39 mi^2	17,700 mi^2
Minuteman III	250x3	0.17 Mt	0.31 EMT	9 mi^2	6,900 mi^2
Minuteman IIIA	300x3	.335 Mt	0.48 EMT	14 mi^2	13,000 mi^2
Poseidon C-3	304x10	0.05 Mt	0.14 EMT	4 mi^2	12,400 mi^2
Trident I	240x8	0.10 Mt	0.22 EMT	6 mi^2	12,400 mi^2
Bombers-bombs	332x4	1.0 Mt	1.0 EMT	30 mi^2	39,800 mi^2
—SRAM	1,178	0.2 Mt	0.34 EMT	10 mi^2	11,800 mi^2
—cruise	192	0.3 Mt	0.45 EMT	14 mi^2	2,600 mi^2
U.S. TOTAL	9,810	n.a.	n.a.	n.a.	123,300 mi^2
SS-11	518x1	1.0 Mt	1.0 EMT	30 mi^2	15,500 mi^2
SS-13	60x1	0.75 Mt	0.83 EMT	25 mi^2	1,500 mi^2
SS-17 (Mod 1)	120x4	0.75 Mt	0.83 EMT	25 mi^2	11,900 mi^2
(Mod 2)	32x1	6 Mt	3.3 EMT	99 mi^2	3,200 mi^2
SS-18 (Mod 1)	58x1	20 Mt	7.4 EMT	221 mi^2	12,800 mi^2
(Mod 2)	175x8	0.9 Mt	0.93 EMT	28 mi^2	39,200 mi^2
(Mod 4)	75x10	0.5 Mt	0.62 EMT	19 mi^2	14,200 mi^2
SS-19 (Mod 2)	60x1	5.0 Mt	2.9 EMT	88 mi^2	5,300 mi^2
(Mod 3)	300x6	0.55 Mt	0.67 EMT	20 mi^2	36,200 mi^2
SS-N-5	18x1	1 Mt	1 EMT	30 mi^2	500 mi^2
SS-N-6	400x1	1 Mt	1 EMT	30 mi^2	12,000 mi^2
SS-N-8	292x1	1 Mt	1 EMT	30 mi^2	8,800 mi^2
SS-N-17	12x1	1 Mt	1 EMT	30 mi^2	400 mi^2
SS-N-18	208x7	0.2 Mt	0.34 EMT	10 mi^2	14,900 mi^2
Bombers-bombs	405	1 Mt	1 EMT	30 mi^2	12,200 mi^2
U.S.S.R. TOTAL	7,741	n.a.	n.a.	n.a.	188,600 mi^2

Table 5. Continued

Missiles	Vulnerability (K required)			Counterforce Capability (K available)		
	Each-90%-Total	97%-Total		CEP	Each	Total
Titan II	30	1,600	2,700	1,300 m	9	500
Minuteman II	120	54,000	82,800	370 m	32	14,600
Minuteman III	120	30,000	46,000	280 m	7	5,400
Minuteman IIIA	120	36,000	55,200	220 m	33	29,700
Poseidon C-3	n.a.	n.a.	n.a.	450 m	2	6,700
Trident I	n.a.	n.a.	n.a.	450 m	4	6,700
Bombers-bombs	n.a.	n.a.	n.a.	n.a.	n.a.	n.a.
-SRAM	n.a.	n.a.	n.a.	n.a.	n.a.	n.a.
-cruise	n.a.	n.a.	n.a.	n.a.	n.a.	n.a.
U.S. TOTAL	n.a.	121,600	186,700	n.a.	n.a.	63,600
SS-11	30	15,500	26,900	900 m	2	880
SS-13	30	1,800	3,100	900 m	1	40
SS-17 (Mod 1)	120	14,400	22,100	450 m	14	6,480
(Mod 2)	120	3,800	5,900	450 m	54	1,700
SS-18 (Mod 1)	120	7,000	10,700	450 m	121	7,700
(Mod 2)	120	21,000	32,200	450 m	15	21,300
(Mod 4)	120	9,000	13,800	300 m	23	17,400
SS-19 (Mod 2)	120	7,200	11,000	300 m	108	6,500
(Mod 3)	120	36,000	55,200	300 m	25	44,500
SS-N-5	n.a.	n.a.	n.a.	2,800 m	0	0
SS-N-6	n.a.	n.a.	n.a.	900 m	4	1,640
SS-N-8	n.a.	n.a.	n.a.	900 m	4	1,200
SS-N-17	n.a.	n.a.	n.a.	1,500 m	2	20
SS-N-18	n.a.	n.a.	n.a.	600 m	9	13,400
Bombers-bombs	n.a.	n.a.	n.a.	n.a.	n.a.	n.a.
U.S.S.R. TOTAL	n.a.	n.a.	n.a.	n.a.	n.a.	122,700

Table 6. The strategic nuclear balance between the U.S. and the U.S.S.R.; a summary of Tables 1-4.

Delivery Systems	←-----Quantities				Qualities-----→	
	Number of Launchers	Throw weight			Number of Warheads	Median CEP
		tons	Y(Mt)	Y(EMT)		
UNITED STATES						
ICBM	1,052	1,238	1,572	1,479	2,152	280 m
Submarines	544	780	344	827	4,960	450 m
Bombers	332	11,245	1,621	1,817	2,698	n.a.
US TOTAL	1,928	13,263	3,537	4,123	9,810	n.a.
SOVIET UNION						
ICBM	1,398	4,938	5,200	4,657	5,158	450 m
Submarines	930	1,054	1,013	1,220	2,178	600 m
Bombers	150	2,550	405	405	405	n.a.
USSR TOTAL	2,478	8,542	6,618	6,282	7,741	n.a.

Table 7. The U.S.S.R./U.S. strategic balance expressed in terms of the area each can destroy, the counterforce available, and the counterforce each country would need for a 90% or 97% successful attack on the enemy's missile silos. A summary of Tables 1-6.

Delivery Systems	Area Destroyable	Silo Kill Factor K		
		Available	Reqd. for 90%	Reqd. for 97%
UNITED STATES				
ICBMs	44,300 mi^2	50,200	121,600	186,700
Submarines	24,800 mi^2	13,400	n.a.	n.a.
Bombers	54,200 mi^2	n.a.	n.a.	n.a.
US TOTAL	123,300 mi^2	63,600	121,600	186,700
SOVIET UNION				
ICBMs	139,800 mi^2	106,500	115,700	180,900
Submarines	36,600 mi^2	16,260	n.a.	n.a.
Bombers	12,200 mi^2	n.a.	n.a.	n.a.
USSR TOTAL	188,600 mi^2	122,760	115,700	180,900

Table 8. The balance of tactical nuclear delivery systems in Europe. From THE MILITARY BALANCE: 1982-83, London: International Institute for Strategic Studies, 1982; unless otherwise indicated.

	First Deploy	Range/Combat Radius[1]	Warheads Number	Warheads Each	Warheads Yield
NORTH ATLANTIC TREATY ORGANIZATION					
IRBM					
SSBS S-3 (FR)	1980	3,000 km	18	1	1 Mt
SRBM					
Pershing IA	1962	160-720 km	180	1	400 kt
Pershing II	1983[2]	1,500 km	108[2]	3	250 kt
Gr.-Launch Cruise	1983[3]	2,250 km	464[3]	1	300 kt?
SLBM[4]					
Polaris A-3 (GB)	1967	4,600 km	64	3[5]	200 kt
MSBS M-20 (FR)	1977	3,000 km	80	1	1 Mt
Land-based Aircraft					
Vulcan B-2 (GB)	1960	2,800 km	48	2	
F-111E/F	1967	1,900 km	156	2	
Mirage IVA (FR)	1964	1,600 km	34	1	
Bucaneer (GB)	1962	950 km	50	2	
F-104 (common)	1958	800 km	290	1	
F-4 (common)	1962	750 km	424	1	
F-16 (US,BE)	1979	900 km	68	1	
Jaguar (GB,FR)	1974	720 km	117	1	
Mirage IIIE (FR)	1964	600 km	30	1	
Carrier-based Aircraft					
A-6E	1963	1,000 km	20	2	
A-7E	1966	900 km	48	2	
Super Etendard (FR)	1980	560 km	16	2	15 kt

Table 8. Continued

	First Deploy	Range/Combat Radius[1]	Warheads Number	Each	Yield
WARSAW PACT ORGANIZATION					
IRBM					
SS-20	1977	5,000 km	210[6]	3	150 kt
SS-5 Skean	1961	4,100 km	16	1	1 Mt
MRBM					
SS-4 Sandal	1959	1,900 km	275	1	1 Mt
SLBM[f]					
SS-N-5 Serb	1964	1,400 km	57	1	Mt range
Battlefield					
SS-12 Scaleboard	1969	900 km	70	1	200 kt
Scud A/B	1965	300 km	450	1	kt range
Scud B/C	1965	300 km	143	1	kt range
SS-22	1978	1,000 km	100[7]	1	500 kt
SS-23	1980	350 km	10[7]	1	dual
Land-Based Aircraft					
Tu-26 Backfire	1974	4,025 km	100[8]	4	
Tu-16 Badger	1955	2,800 km	310	2	
Tu-22 Blinder	1962	3,100 km	125	2	
Su-24/19 Fencer	1974	1,699 km	550	2	
MiG-27 Flogger D	1971	720 km	550	1	
Su-17 Fitter C/D	1974	600 km	688	1	
Su-7 Fitter A	1959	400 km	265	1	
MiG-21 Fishbed J-N	1970	400 km	100	1	

[1] Range applies for missiles, combat radius is for unrefueled aircraft assuming high-altitude approach, low-altitude penetration, and an average payload.

[2] The Pershing II missile is scheduled to be first deployed in 1983, assuming its testing proceeds well, and it is accepted for stationing on their soil by NATO countries.

[3] The ground-launched cruise missile with a 2,400 km range is scheduled to be first deployed in Europe in 1983, if its stationing is accepted by NATO countries.

[4] The SLBM of France and Great Britain and the SS-N-5s of the U.S.S.R. are counted as theater nuclear weapons for NATO, since they are not counted in the SALT agreements.

[5] The British Polaris A-3 missiles carry 3 MRV warheads.

[6] Two-thirds of the 315 SS-20 missiles are thought to be targeted on Europe.

[7] These are the total nuclear-capable units available.

[8] This is the total thought at most to be assigned a nuclear strike role; only a fraction of these are likely to be used in that manner.

APPENDIX F

MISCELLANEOUS DATA ON THE ARMS RACE

TARGET RESOLUTION REQUIRED FOR INTERPRETATION TASKS

Target	Detection	General Identification	Precise Identification	Description
Bridge	20 Ft	15 Ft	5 Ft	3 Ft
Communications				
Radar	10 Ft	3 Ft	1 Ft	6 In
Radio	10 Ft	5 Ft	1 Ft	6 In
Supply Dump	5 Ft	2 Ft	1 Ft	1 In
Troop Units (Bivouac, Road)	70 Ft	7 Ft	4 Ft	1 Ft
Airfield Facilities	20 Ft	15 Ft	10 Ft	1 Ft
Rockets and Artillery	1 Ft	2 Ft	6 In	2 In
Aircraft	15 Ft	5 Ft	3 Ft	6 In
Command and Control Headquarters	10 Ft	5 Ft	3 Ft	6 In
Missile Sites (SSM,SAM)	10 Ft	5 Ft	2 Ft	1 Ft
Surface Ships	25 Ft	15 Ft	2 Ft	1 Ft
Nuclear Weapons Components	8 Ft	5 Ft	1 Ft	1 In
Vehicles	5 Ft	2 Ft	1 Ft	2 In
Land Minefields	30 Ft	20 Ft	3 Ft	1 In
Ports and Harbors	100 Ft	50 Ft	20 Ft	10 Ft
Coasts and Landing Beaches	100 Ft	15 Ft	10 Ft	5 Ft
Railroad Yards and Shops	100 Ft	50 Ft	20 Ft	5 Ft
Roads	30 Ft	20 Ft	6 Ft	2 Ft
Urban Area	200 Ft	100 Ft	10 Ft	10 Ft
Terrain	-	300 Ft	15 Ft	5 Ft
Surfaced Submarines	100 Ft	20 Ft	5 Ft	3 Ft

Source: U.S. Congress, Committee on Commerce, Science and Transportation, 1978 NASA Authorization, Washington, D.C.: Government Printing Office, 1977, pt. 3, p. 1597.

DEFINITION OF INTERPRETATION TASKS

Task	Requirement
Detection	Location of a class of units, object, or activity of military interest.
General Identification	Determination of general target type.
Precise Identification	Discrimination within target type of known types.
Description	Size/dimensions, configuration layout, components-construction, count of equipment, etc.

Source: U.S. Congress, Senate, Committee on Commerce, Science and Transportation, 1978 NASA Authorization, Washington, D.C.: Government Printing Office, 1977, pt. 3, p. 1597.

TARGET RESOLUTION REQUIRED FOR INTERPRETATION

Target Type	Detection (feet)	General Identification (feet)	Precise Identification (feet)	Description (feet)
ICBMs	10	5	2	1
SSBNs	100	20	5	3
Bombers	15	5	3	0.5
Surface Ships	25	15	2	1

VERIFICATION OF A NUCLEAR

INTELLIGENCE SYSTEMS: MONITORING TASKS:	Imaging Reconnaissance Satellites	Electronic Reconnaissance Satellites	Ocean Surveillance Satellites	Missile Warning Satellite	Nuclear Explosion Satellites "Vela Hotel" IONDS
I. Deployment Freeze					
A. Count fixed ICBM/IRBM launchers*	X	X			
B. Count mobile ICBM/IRBM/GLCM launchers*	X	X			
C. Count SLBM launchers*	X	X	X		
D. Count launchers for MIRVed missiles*	X				
E. Count strategic bombers (incl. ALCM)*	X	X			
F. Count other primary nuclear mission aircraft (e.g., FB-111, Backfire...)	X	X			
G. Count nuclear-armed ships/subs (incl. those with SLCMs, ASROCs, SUBROCs,...)	X	?	X		
H. Count nuclear artillery/battlefield missile units, weapon depots	X	?			
II. Delivery Vehicle Testing Freeze					
A. To monitor (prohibited) testing of new ICBMs/ SLBMs/IRBMs, monitor flight tests of existing missiles to detect:					
1. Changes in length, diameter, launch-weight and throw-weight (no greater than 5%)	X	X			
2. Number of stages/type of propellant (no change permitted)	X	X			
3. Number of RVs (no increase from maximum number tested for each type)	?	X			
4. Weight of RVs (no decrease from lightest test flown)		X			
5. RV performance (no increase in ballistic coefficient above maximum already tested and no maneuvering)	?	?			
B. Monitor limit on operational ballistic missile flight tests (6 or less per year)	X	X	X	X	
III. Nuclear Weapons Testing Freeze (CTB)					
A. Detect ambiguous seismic events					
B. Monitor activity/geography at potential test sites	X	?	?		
C. Detect evidence of nuclear explosions on land/in sea/air/space				X	X
D. Identify ambiguous events					
IV. Ballistic Missile/Strategic Bomber/SSBN Production Freeze*					
A. Monitor shut-down of existing main assembly plants and shipyard(s)	X				
B. Detect ambiguous activity at other facilities	X	X			
C. Identify ambiguous activity	X	?			
V. Nuclear Warhead Production Freeze					
A. Monitor shut-down of existing key nuclear component fabrication facilities	X	?			
B. Detect ambiguous activity at additional facilities	X	?			
C. Identify ambiguous activity at additional facilities	X	?			
VI. Weapons-Grade Nuclear Materials Cutoff					
A. Monitor military nuclear materials production facilities	X	?			
B. Detect ambiguous activity at civilian nuclear facilities	X	?			
C. Identify ambiguous activity	X	?			

*Comprehensive freeze could include a ban on replacement of these systems from new production.

X^i = *indirect* assistance in monitoring provision.

FREEZE: TASKS AND SYSTEMS

Detection Ground-based Seismic Sensors	Acoustic Underwater Surveillance	Ground Based Monitoring Posts	Test Observation Radars	Aircraft and Ships	HUMINT and Overt Collection	On Site Inspection	Overall Monitoring Confidence Level (estimate)
							high
							high moderate
							high
					Counting Rule		high
							high
		X		X	X		high moderate
	X			X	X		high moderate
		?		X	X		high moderate
		X	X	X	X		moderate-high moderate
		X	X	X	X		high moderate
		X	X	X	X		high
		X	X	X	X		high moderate
			X	X	X		high
	X^i	X	X	X	X		high
X	X						high moderate
		?		?			high moderate
X	X			X			high moderate
						X	moderate-high moderate
							high
		X^i	X^i	X^i	X		moderate
					X	X	low-high
		?		?	X		high
		?		?	X		low-moderate
		?		?	X	X	low-high
					X		high moderate-high
					X		low-moderate
					X	X	low-high

Source: C. Payne and T. Karas, Federation of American Scientists Public Interest Report 35, 6 (September, 1982).

SIPRI: CRITICAL MASSES OF VARIOUS FISSILE MATERIALS

Enrichment (% of U-235 or Pu-239)	Critical mass (kg) in metal form					
	Uranium-235		Plutonium		Uranium-233	
	Without reflector	With Be reflector	Without reflector	With Be reflector	Without reflector	With Be reflector
0	–	–	–	–	–	–
10	–	~1300	–	–	–	–
20	–	~250	–	–	–	–
30	–	–	–	–	–	–
40	–	75	–	–	–	–
50	145	50	–	9.6	–	–
60	105	37	–	7.8	–	–
70	82	–	23	6.7	–	–
80	66	21	–	5.6	–	–
90	54	–	–	5.0	–	–
100	50	15	15	4.4	17	4 to 5

Source: Stockholm International Peace Research Institute, Nuclear energy and nuclear weapons proliferation. London: Taylor and Francis Ltd., 1979, p.3.

NASAP: HISTORICAL OVERVIEW OF LEADING ENRICHMENT TECHNOLOGIES
FOR PRODUCTION OF LOW ENRICHED URANIUM

Technology	Period	Physical Principle	Present implementation
Gaseous diffusion	Late 1940's on	Differential U-235/U-238 rates collision with permeable walls.	United States: 21,000,000 SWU* year--1979; 28,000,000--1980's; EURODIF/COREDIF: 22,000,000; SWU/year--1980's.
Centrifuge	1950's R. & D. 1980's implementation.	Enhancement of centrifugal effects by countercurrent flow.	United States: 8,000,000 SWU/year --1980's' URENCO: 1,200,000; SWU/year--1980; 10,000,000-1980's.
Aerodynamic	1960's on	Centrifugal effects on UF$_6$ in small curved-wall chamber.	German R. & D.; Brazilian pilot plant 1980's; South African variant 1970's.
Calutron	1940's	Mass-dependent deflection of ions by strong magnetic field.	Oak Ridge 1944. Very low throughput.
Chemical exchange	Concept dates from World War II. Serious R. & D. in 1970's.	Exploits isotope-dependent differential equilibria in a system of organic and aqueous uranium compounds.	United States, French, and Japanese R. & D. No implementation yet.
Atomic vapor laser isotope separation (AVLIS).	1972 on	Multistep ionization of uranium metal vapor by optical laser.	R & D. prototype in 1980's (part of AIS program).**
Molecular vapor laser isotope separation (MLIS).	"	Laser chemistry on UF$_6$ (super-cooled through large expansion nozzle.)	"
Plasma separation process (PSP).	1975 on	RF reinforcement of orbits of uranium vapor ions in strong magnetic field.	"

* The capacity of an enrichment plant is given in separative work units (SWU's).
** Competition in the early 1980's will determine which of these prototypes of about 500,000 SWU/year will be continued in the United States.

Source: Department of Energy. Nuclear Proliferation and Civilian Nuclear Power. Final report of the nonproliferation alternative systems assessment program. (NASAP) vol. II. Proliferation Resistance. June 1980, pp.3-4. (NASAP report.)

APPENDIX G

HOW TO CALCULATE THE KILL PROBABILITY OF A SILO

Kosta Tsipis
Program in Science and Technology for International Security
Department of Physics
Massachusetts Institute of Technology
Cambridge, Massachusetts 02139

A reentry vehicle from a ballistic missile which upon repeated firings against an aim point would land in a normal distribution about that aim point with a standard deviation σ will have a single-shot kill probability against a target

$$P_k = 1 - P_s = 1 - \exp(-r_k^2/2\sigma^2) \tag{1}$$

where r_k is the distance from the point of impact, and explosion, beyond which the target would survive.

Now you would want to be able to re-write (1) in terms of the characteristics of the target, say a silo, and the attacking warhead. A silo is characterized by the amount of blast overpressure it can take, or as is commonly referred to its <u>hardness</u>. An attacking warhead is characterized by the energy yield of its explosive charge, its accuracy, its bias, and its reliability. We will assume in this calculation that the bias is zero.

The overpressure P_o (in psi) caused by the explosion of a warhead of yield Y (in Mtons of TNT) at a point r (in n.m.) from the point of detonation is given[1] by:

$$P_o = 14.7 \frac{Y}{r^3} + 12.8 \left(\frac{Y}{r^3}\right)^{1/2} \tag{2}$$

This is an empirical formula derived from early atmospheric tests.

Solving for $\left(\frac{Y}{r^3}\right)^{1/2}$

$$\left(\frac{Y}{r^3}\right)^{1/2} = \frac{-1 \pm (1 + .36P_o)^{1/2}}{2.3}$$

Now for $P_o \gg 300$ psi which is the case with silos, $.36P_o \gg 1$; therefore we can omit the 1 in the parenthesis. So

$$\left(\frac{Y}{r^3}\right)^{1/2} = -.432 \pm 0.26 P_o^{1/2}$$

0094-243X/83/1040353-04 $3.00 Copyright 1983 American Institute of Physics

or

$$r^3 = \frac{Y}{P_0 \cdot f(P_0)}$$

where $f(P_0)$ is a slowly varying function of P_0.

$$f(P_0) = .19P_0^{-1} - .23P_0^{-1/2} + .068$$

$$\approx .061 \text{ for } P_0 \text{ around } 1000 \text{ psi.}$$

Now if one substitutes the hardness H of the silo for P_0, r is equal to r_k, the distance at which the silo would survive a blast over-pressure P_0. Therefore

$$r_k = \frac{Y^{1/3}}{H^{1/3} f(H)^{1/3}} \quad \text{where } f(1000) = .061 \tag{3}$$

Substituting (3) into (1) we have

$$P_k = 1 - P_s = 1 - \exp(-Y^{2/3}/.3\sigma^2 \cdot H^{2/3})$$

This assumes somewhat artificially that all silos for which $r < r_k$ will be destroyed and all those for which $r > r_k$ will survive. This, known as the "cookie-cutter" approximation, is satisfactory for all purposes other than detailed targeting.

One would now like to describe σ in terms of the accuracy of the warhead expressed by its CEP or "circular error probable," a measure of accuracy used in characterizing missiles.

By definition CEP is the radius of the circle centered at the aim point and containing 50% of the projectiles launched against the aim point. Now assuming a normal distribution in r and θ we write

$$P_{(r)} = Ce^{-r^2/2\sigma^2}$$

so

$$1 = C \int_0^{2\pi} d\theta \int_0^{\infty} re^{-r^2/2\sigma^2} dr$$

$$1 = 2\pi C\sigma^2 \quad \text{so } C = \frac{1}{2\pi\sigma^2}$$

Then by definition of the CEP

$$.5 = \frac{1}{2\pi\sigma^2} \int_0^{2\pi} d\theta \int_0^{CEP} re^{-r^2/2\sigma^2} dr$$

$$.5 = 1 - e^{-(CEP)^2/2\sigma^2}$$

$$\ln .5 = -(CEP)^2/2\sigma^2$$

so $\qquad$ CEP = 1.17 σ $\qquad$ Therefore σ = .85 CEP $\qquad$ (4)

Now we substitute:

$$P_k = 1 - \exp\left(-Y^{2/3}/.22H^{2/3} \cdot CEP^2\right) \qquad \text{for perfect} \qquad (5)$$
$$\text{reliability}$$

Now define $\qquad\qquad$ $K = \dfrac{Y^{2/3}}{(CEP)^2}$ $\qquad\qquad\qquad$ (6)

a quality that describes the lethality of a missile warhead against a silo. (Remember it has to be a silo or something harder than 300 psi to use this formalism.)
So we can rewrite P_k

$$P_k = 1 - e^{-\frac{K}{.22H^{2/3}}} \qquad\qquad (7)$$

where the numerator describes the missile and the denominator describes the silo.
Note that K can become ridiculously large as CEP → 0. In reality K has a maximum numerical value beyond which its magnitude has no physical meaning. This limiting value comes about because when CEP ≤ r_c where r_c is the radius of the crater that the explosion excavates, $P_k \simeq 1$, i.e. the kill probability for the silo is unity. Making the missile more accurate, or the K larger, loses physical significance and operational importance because a missile inside the silo cannot survive if it is inside the crater.
So one defines:

$$K_{max} = \frac{Y^{2/3}}{r_c^2} \qquad\qquad \text{for CEP} = r_c$$

but $\qquad r_c = \alpha Y^{1/3}$ $\qquad\qquad\qquad$ where α is an excavation constant depending on the soil.

Therefore, $\qquad K_{max} = \dfrac{2/3}{\alpha^2 Y^{2/3}} = \dfrac{1}{\alpha^2} \simeq 100$.

So it is not meaningful to speak of lethality values above, say, K= 100-125.

We have assumed this far that the missile reliability is 1, or 100% reliable. Now what about the missile reliability ρ smaller than 1? There things are easy.

$$P_k(\rho<1) = \rho \cdot P_k(\rho=1)$$

And what about using n warheads against one silo? If all n come from the same launcher

$$P_k(\rho,n) = \rho \cdot (1 - P_s(\rho=1)^n) \qquad \text{for } \rho < 1$$

but if they come from different launchers

$$P_k(\rho,n) = 1 - (1 - \rho \cdot P_k(\rho = 1))^n .$$

1) H.L. Brode, Ann. Rev. Nuc. Sci. 18, 153 (1968).

APPENDIX H

Basic Physics of EMP, Beam Weapons, and ABM

D.W. Hafemeister
Physics Department
California Polytechnic University
San Luis Obispo, CA 93407

I: ELECTROMAGNETIC PULSE (EMP)

The command, control, communication, and intelligence (C^3I) systems which control the US strategic forces can be vulnerable to the large electromagnetic pulses[1] (E = 25,000 volts/meter) that are created by high altitude (100 to 500 km) or low altitude (0 to 2 km) bursts of nuclear weapons. This vulnerability has been perceived as creating a possible instability in the arms race. If a country perceives that the use of its strategic forces could be negated by the EMP so that it could not command and control its missiles, this country might be tempted to adopt a "launch on warning" policy so that it would use its weapons rather than lose them to a preemptive first strike. The situation is not as unstable as I have characterized it because both nations have hardened their systems to partially withstand the EMP and both nations have viable, second-strike missiles based on submarines that are not vulnerable to a first strike. However, the perception of vulnerability in the land-based leg of the strategic triad (land/sea/air) can create pressures for a "now, or never", launch-on-warning response. Since some of the C^3I facilities, such as the Air Force's Looking Glass command post in the sky, are more vulnerable than some of the strategic forces that they are intended to direct, and since the U.S. land-based missiles could be vulnerable to an EMP attack, it is clear that the EMP can affect military policy.

A brief discussion of the mechanism that creates the EMP is given below: In an explosion of a nuclear weapon, many of the fission fragments are created in an excited state. Since these excited states have energies of 1 MeV or more, their lifetimes are much less than the risetime of the explosion (about 10 nanoseconds); the de-excitation gamma rays are emitted promptly. These prompt gamma rays interact with air molecules and create forward scattered Compton electrons which constitute an energetic (about 1 MeV), negative current flowing radially outward from the weapon (Fig. 1). In order to have a large EMP, it is necessary to remove the spherical symmetry around the nuclear weapon. At low altitudes above the earth, a net dipole moment is created because

0094-243X/83/1040357-12 $3.00 Copyright 1983 American Institute of Physics

the Compton electrons are in the form of a hemisphere
rising above the earth; the asymmetry of the rising,
charged hemisphere creates the EMP. At high altitudes,
the symmetry is broken because the air density varies
exponentially with height above the earth so that the
Compton electrons are created asymmetrically and a net
dipole moment is created. The electrons created from the
very high altitude bursts (above the atmosphere) follow a
helical path about the earth's magnetic field lines in
the very thin upper atmosphere; the centripetal
acceleration of these Compton electrons creates the EMP.
In addition, the x-rays following a nuclear explosion can
also produce EMP effects by ionizing the atmosphere and
momentarily affecting the magnetic field configuration of
the Earth.

Figure 1. Electromagnetic Pulse (EMP). During a nuclear
explosion gamma rays are emitted promptly from the
excited states of the fission fragments; these prompt
gamma rays transfer their energy to electrons in air
molecules by the Compton scattering process. The EMP is
generated by these relativistic electrons by several
mechanisms which depend on the altitude (H) of the
nuclear explosion above the earth. A surface blast at
low altitudes (H = 0-2 km) causes a hemispherical, rising
Compton current which creates a large net vertical
dipole current. An air burst (H = 2-20 km) creates a
small net vertical dipole moment because the asymetrical

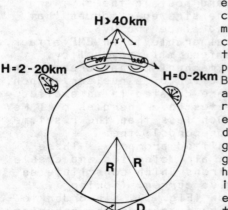

density of air which varies
exponentially with height. A
high altitude (H > 40 km)
explosion creates Compton
electrons that revolve in
circles around the earth's
magnetic field lines; the
centripetal acceleration of
the Compton electrons
produces EMP radiation.
Because the Compton electrons
are relativistic, the EMP
radiation from the individual
electrons have a partial
degree of coherence which can
give very large E fields of
greater than 25,000 V/m. The
high-altitude EMP pulse can
illuminate wide areas of the
earth's surface; the range of
the EMP is determined by its
line-of-sight distance, $D = (2RH)^{1/2}$ where R is the radius
of the Earth.

I.A. The Range of EMP. Assume that sufficient megatonnage is available for the EMP pulse, and that one must only consider line-of-sight geometry when considering covering the US with the EMP pulse. The height (H) of the blast that is necessary to cover 50% and 100% of the US is determined as follows: The extremum distance (D) from the nuclear detonation to the ground is obtained (Fig. 1) from a right triangle which has one side as the distance D, the other side the radius (R) of the earth at the location in question, and the hypotenuse extends from the center of the earth to the blast (H + R). It follows that

$$R^2 + D^2 = (H + R)^2 \cong R^2 + 2HR$$

when H << R. From this we obtain the height of the EMP blast to cover all the US

$$H = D^2/2R = (2500 \text{ km})^2/(2)(6400 \text{ km}) \cong 500 \text{ km} \quad .$$

For the case of 50% of the US, D = 1800 km and H = 250 km.

I.B. The Strength of the EMP Fields. The approximate average (rms) E field generated by the EMP pulse at a location D = 1000 km from the blast is determined by assuming the following: (1) A one megaton weapon contains 10^{15} calories or 4.2×10^{15} J; (2) The prompt gamma rays constitute 0.3% of the total energy; they have an average energy of about 1 MeV; and they are emitted over the 10 ns risetime of the blast; and (3) About 0.6% of the total prompt gamma ray energy is converted into forward-scattered Compton electrons near $\theta = 0$. (If each of the fission fragments emitted a 1 MeV gamma ray, and if 50% of these gamma rays were not absorbed by the materials of the weapon, then about 0.5% (1 MeV/200 MeV) of the energy of the weapon would be in the form of prompt gamma rays.) The amount of energy available for the EMP pulse from the 1 megaton weapons is about

$$\text{UEMP} = (4.2 \times 10^{15} \text{ J})(0.003)(0.006) = 7.6 \times 10^{10} \text{ J}.$$

This energy can be considered to approximately evenly distributed in a spherical shell with a volume $(4\pi D^2)(T)$ where D is the distance from the site to the blast. The thickness, T, of the EMP shell is the product of the speed of light and the time duration of the blast, or T = $(3 \times 10^8 \text{ m/s})(10^{-8} \text{ s}) = 3$ meters. Since 50% of the stored energy resides in the electric fields, we obtain

$$\text{UEMP}/2 = (\varepsilon_0 E^2/2)(4\pi D^2 T), \quad \text{or}$$

$$E = (\text{UEMP}/4\pi \varepsilon_0 T)^{1/2} /D = (7.6 \times 10^{10} \times 9 \times 10^9 /3)^{1/2}$$

$$/(10^6) = 15000 \text{ volts/meter.}$$

A more sophisticated calculation of the E field is carried out in I.C below, but this approximate estimate is sufficient to show that the E field can be quite large. The energy/cm^2 deposited on the semiconducting components used in C^3I is of the order of

$$(7.6 \times 10^{10} \text{ J})/(4\pi)(10^8 \text{ cm})^2 = 6 \times 10^{-7} \text{ J/cm}^2 \quad .$$

This flux of energy can be effectively multiplied by the area of powerlines, antennae or missile bodies, and the resulting electrical transients can be sufficient to destroy many of the semiconducting devices.

I.C Frequency Spectrum. A fundamental calculation of the frequency and field strength distributions from a high altitude burst involves a calculation of the magnitude of the Compton electronic charge and how this charge accelerates in the earth's magnetic field. In order to do this we must consider the relativistic expression for the radiative power,

$$P = e^2 a^2 \gamma^4 /6\pi \varepsilon_0 c \quad ,$$

from an accelerated charge where a is the acceleration and $\gamma = 1/(1 - v^2/c^2)^{1/2}$. Using this relationship, the approximate magnitude of the E field from a high altitude, 1 megaton explosion, and the maximum frequency of the EMP spectrum are determined as follows: Because the duration of the explosion is about 10^{-8} s, the maximum frequency possible for the EMP is about 100 MHz, but in practice it is considerably lower. Since the Compton electrons have a kinetic energy of about 1 MeV = 2 $m_0 c^2$; $\gamma=3$ and v/c = 0.94. The radius of gyration (r) of the Compton electrons in the earth's field of 0.6 Gauss is

$$r = \gamma m_0 v/eB = (3)(9.1 \times 10^{-31} \text{ kg})(0.94)(3 \times 10^8 \text{ m/s})/$$

$$(1.6 \times 10^{-19} \text{ C})(0.6 \times 10^{-4} \text{ T}) = 80 \text{ m} \quad .$$

The frequency of radiation associated with this motion is = v/r = (0.94)(3 $\times$ 10^8 m/s)/(80 m) = 3.5 Mrad/s, or f = ω /2π = 0.56 MHz.

The approximate number of prompt gamma rays emitted from the 1 megaton explosion is

$$N_\Gamma = (4.2 \times 10^{15} \text{ J})(0.003)(1 \text{ MeV}/1.6 \times 10^{-13} \text{ J})/$$

$$(1 \text{ MeV}/\Gamma) = 7.9 \times 10^{25} \Gamma/\text{Mton} \quad .$$

Since about 0.6% of this energy is converted to Compton

electrons, the approximate initial number of Compton electrons produced is $N_e = (0.006)(7.9 \times 10^{25}) = 4.7 \times 10^{23}$ electrons/Mton, or 75,000 coulombs/Mton. The initial centripetal acceleration for each Compton electron is about

$$a = r\omega^2 = (80 \text{ m})(3.5 \times 10^6/\text{s})^2 = 1.0 \times 10^{15} \text{ m/s} .$$

The initial power radiated by this electron is

$$P = e^2 a^2 \gamma^4/6\pi\varepsilon_0 c^3 = (1.6 \times 10^{-19} \text{ C})^2 (1.0 \times 10^{15} \text{ m/s}^2)^2 \times$$

$$(3^4)(6 \times 10^9)/(3 \times 10^8)^3 = 4.6 \times 10^{-22} \text{ Watts.}$$

If the E fields from the $N = 4.7 \times 10^{23}$ Compton electrons added incoherently, the EMP radiated power would be only about $(4.7 \times 10^{23})(4.6 \times 10^{-22} \text{ W}) = 220$ W which would not produce a very large E field. However, since the wavelength of this radiation, $\lambda = c/f = 540$ m, is much longer than the spatial thickness of the prompt radiation, $T = c/(10 \text{ ns}) = 3$ m, and since the velocity of the Compton electrons is approximately c, the EMP electric fields are partially additive, or coherent. Let us initially assume that all N_e electrons give coherent EMP radiation; a comparison of the field from the this completely coherent situation can then be compared to the given value of about 25,000 V/m to determine the degree of coherence.

If the Compton electrons radiated coherently, the effective radiative power would be $(N_e^2)P = 1.0 \times 10^{26}$ W which is similar to the power of the sun! The average value of the Poynting vector for this radiation (neglecting the $\sin^2(\theta)$ distribution) is $P/4\pi D^2$ where D is the distance from the closest Compton electrons to the site (perhaps an average effective value of 100 km). From this the average E field is given by

$$E \cong (P/4\pi\varepsilon_0 cD^2)^{1/2} = ((1.0 \times 10^{26} \text{ W})(9 \times 10^9)/$$

$$(3 \times 10^8 \text{ m/s})(10^{10} \text{ m}^2))^{1/2} = 5.5 \times 10^8 \text{ V/m.}$$

Since this result is about 10^4 times larger than the reported value of 2.5×10^4 V/m, we see that, on average, only about 0.01% of the forward-scattered Compton electrons must be totally coherent. As the Compton electrons loes energy, lower frequency, longer wavelength components are added to the EMP distribution, and the radiation becomes less coherent.

Rather large currents can be generated by the EMP. The voltage developed along the length of a 20 m missile could be as much as $V = (25,000 \text{ V/m})(20 \text{ m}) = 500,000$ V. Since the inductive impedance of a missile at 1 MHz is about $Z = 10$ ohms, the current flowing along the surface of the missile could be as high as

$$I = V/Z = (5 \times 10^5 \text{ V})/(10 \text{ ohms}) = 50,000 \text{ Amperes.}$$

By coupling through various inductive and capacitive transfer impedances, this current can produce voltage transients on the center conductors of the cables in the missile on the order of 1 to 100 volts. While this amount of voltage will only burn out the most sensitive semiconductors, it could upset the digital circuitry and nullify its computer logic. Much larger voltages could be developed on these logic elements if one considers the voltages developed on the powerlines to the missile sites, but it is possible to overcome these difficulties by shielding and hardening the electronics with filter circuits and fiber optics.

II: PARTICLE AND LASER-BEAM WEAPONS

The superpowers are currently developing weapons that would utilize the kinetic energy (in the GeV range) of particle beams or the energy content of laser beams to destroy incoming missiles, surface cruise missiles, ships, or satellites. The particle-beam and laser-beam weapons might be mounted on the surface of the earth, in the space shuttle, or on satellite stations in space. These weapons have to be described as futuristic since there are many technological obsticals that must be be overcome before they would become operational. In addition, these weapons would be vulnerable to countermeasures by the attacking force. These weapon systems have been described[3] in some detail by K. Tsipis, et al, and we will discuss only their more general[2] properties.

II.A. <u>Energy Deposition</u>. The flux of energy (J/cm^2) that would preignite the chemical explosives around a nuclear weapon in the incoming missile can be estimated. Following Tsipis, et al, assume that a temperature rise of 500 ^{0}C can preignite the explosive material which has a density of $\rho = 0.8$ g/cm^3 and a molecular weight of about M = 50. In addition, assume that the absorption length for 1 GeV electrons is about 10 cm; the beam loses about 10% of its energy in 1 cm. The energy density needed to predetonate the chemical explosive is $Q = C\rho\Delta T/M$ where the ΔT is the temperature rise and the specific heat $C = 3R = 25$ J/mole-^{0}C (since $T > \Theta_D$, the Debye temperature). We obtain

$$Q = (3R)\rho\Delta T/M = (25)(0.8)(500)/50 = 200 \text{ J/cm}^3 \ .$$

Since about 10% of the energy is desposited in 1 cm, the

energy intensity of the beam must be about 2000 J/cm^2 in order to deposit 200 J/cm^3 in the chemical explosive. This same flux of energy would raise the temperature, ΔT, of the aluminum missile:

$$Q = 200 \text{ J/cm}^3 = (3R)\rho\,\Delta T/M = (25)(2.7)(\Delta T)/27 \quad ,$$

which gives T = 80 ^{0}C. This modest temperature rise would begin to cause some internal stresses and misalignment in the missile. Doses about 8 times larger (16,000 J/cm^2 and 1500 J/cm^3) would raise aluminum to its melting temperature of 660 ^{0}C. In addition, the particles can cause havoc with the semiconducting components in the guidance systems; energy densities as low 25 J/cm^3 can cause shifts in the switching thresholds of the circuit elements and 1000 J/cm^3 can destroy the elements. The energetic particles would create centers for trapping, and recombination which would reduce the lifetime of the minority carriers.

II.B. Beam Current and Angular Resolution. What minimum beam current of 1 GEV electrons would be necessary to preignite the chemical explosive in the incoming missile? Assume that the size of the beam is dictated by the 1 meter diameter of the missile booster and that the pulse duration is determined by the velocity of the missile, v = 10^4 m/s. What angular accuracy ($\Delta\theta$) would be required to disable a missile which is 1000 km away and above the atmosphere?

Using the energy intensity of 2000 J/cm^2 (Sec. II.A), the energy delivered to an area of 1 m^2 (dictated by the diameter of the missile) is (2000 J/cm^2)(10^4 cm^2) = 2 x 10^7 J. In order to keep the area of the beam on the missile for a single shot (without continuous tracking), the time duration of the pulse must be less than t = (1 m/10^4 m/s) = 0.1 msec. The energy in the pulse is VIt = 2 x 10^7 J where V = 10^9 volts and I is the current in amperes. Solving for I, we get

$$I = (2 \times 10^7)/Vt = (2 \times 10^7)/(10^9)(10^{-4}) = 200 \text{ A}.$$

If the beam was intended to melt aluminum or destroy semiconducting components, the current and energy of the beam must be at least five times higher, or 1000 amperes and 10^8 J. It has been estimated that the efficiency of producing a pulse of particles is about 1/6, thus it takes about 6 x 10^8 J to make the 1000 ampere pulse. Since it takes about 0.9 pound of coal to generate a kwh = 3.6 x 10^6 J, it would take about 150 pounds of coal (or 1000 pounds of TNT) to generate one pulse; the large amount of any type of fuel would certainly complicate the logistics of placing either particle- or laser-beam weapons on a space station. For comparison sake, there

are already existing high-voltage, low-current
accelerators, such as the Fermilab's proton accelerator
which has a current of 0.1 milliamperes at an energy of
500 GeV, and there are low-voltage, high-current electron
accelerators used in the fusion program which have a
current of 10^6 amperes (per beam) at an energy of 2 MeV.
Because the earth's magnetic field will deflect
long-range, charged particle beams, a beam of neutral
hydrogen atoms (produced from H$^-$ beams passing through a
stripper gas) might be considered. The present
technology is not yet capable of these requirements as
the Los Alamos Meson Factory is capable of developing
currents of only 0.15 amperes of H$^-$ ions at 1 GeV.

In order to "shoot a bullet 1000 km away with a
bullet", the beam must be aligned to within an accuracy
of $\Delta\theta = 1$ m/10^6 m $= 10^{-6}$ radians. Since this kind of
accuracy would be difficult to carry out, it is likely
that the beam current would have to be correspondingly
increased to compensate for a larger spread in the beam
size at the target. If $\Delta\theta \simeq 10^{-5}$, the current and total
energy of the beam would have to be increased by a factor
of 100; if the target was considerably closer, this would
not be necessary.

II.C. <u>Burning a Hole in the Atmosphere.</u> Since
relativistic electrons and protons of about 1 GeV loes
about 0.2 GeV/km when they pass through air, they would
not be able to penetrate through the entire atmosphere.
Particle beam weapons that were located on the surface of
the earth would have to be able to "burn a hole in the
atmosphere" in order to reduce energy losses and make the
weapon viable over longer distances. If the air density
in the "hole" in the atmosphere was reduced by a factor
of ten, then the energy loss rate would be reduced to
0.02 GeV/km, and the beam would then loes about 0.2 Gev
(20% of the beam energy) to pass through the entire
atmosphere of 10 km. Approximately how much energy would
it take to reduce the density of the air by a factor of
ten in a "hole" that had an area of 1 cm^2 and was 1 km
long? The density of air is 1.3 kg/m^3 and its specific
heat at constant pressure is 1000 J/kg-^{0}C.

Using the perfect gas law, PV = nRT, we see that the
temperature must be raised from about 300 K to 3000 K in
order to allow a reduction in the density, n/V, by a
factor of 10. The mass of the heated air in the
atmospheric hole is (1.3 kg/m^3)(10^3 m)(10^{-4} m^2) = 0.13 kg.
The amount of energy required to heat the "hole" is

$$Q = mC\Delta T = (0.13)(1000)(2700) = 3.5 \times 10^5 \text{ J.}$$

Additional energy would be lost by the partial heating of
the air forced out of the hole, by the loss of scattered
radiation and secondary particles, and by turbulence.

Our value of 0.4×10^6 J/km is a lower bound value and it is about 25% of the more accurately calculated[3] value of 1.5×10^6 J/km. This energy loss is about 1%/km of the beams energy of 10^8 J; a 10 km path would use 10% of the beam's energy; this is similar to the energy that would be lost by the beam as it passed through the hole. At this time it is unclear how long the high energy beam would remain in the hole over these extended distances.

II.D. Laser Beam Weapons. Either a 25 kJ pulsed laser or a 1 MW continuous laser is capable of destroying nearby aircraft. However, since laser beams do not readily pass through fog or rain, the laser-beam weapons could not be counted on to defend against a surprise attack on a rainy day. For this reason, the laser-beam weapons (if they are ever deployed) would probably be used above the atmosphere. Let us determine the necessary energy for a laser pulse to destroy a missile 1000 km away using these assumptions: (1) It takes a pulse of about 1000 J/cm^2 (Sec. II.A) to create an impulsive failure in thin metal targets such as missiles; (2) The HF chemical laser has a wavelength λ = 2.7 μm; and (3) The beam extraction mirror of the laser has a diameter d = 4 meters. The finite diameter of the extraction mirror causes a diffraction broadening in the laser beam

$$\Delta\theta = 1.22\lambda/d = (1.22)(2.7 \times 10^{-6} \text{ m})/4 \text{ m} = 0.8 \times 10^{-6} .$$

This effect will broaden the radius of the laser beam 1000 km away to

$$r = (\Delta\theta)(1000 \text{ km}) = 0.8 \text{ m} .$$

Since it takes about 1000 J/cm^2 of absorbed energy to partially melt and crack the missile body, the energy in the laser pulse must be (1000 J/cm^2)(π)(82 cm)^2 = 2×10^7 J for the case of total absorption by the missile body; if the missile reflects 90% of the energy, ten times as much energy will be needed, or 2×10^8 J. This energy requirement is considerably larger than the presently contemplated 25 kJ pulsed CO_2 lasers (λ = 10.6 μm) being developed for the fusion program. If the weapon was a 1 MW continuous laser, then the laser would have to be focused on the target for t = (2×10^8 J)/(10^6 J/s) = 200 seconds = 3 minutes. The energy content of the H and F fuels is about 1.4×10^6 J/kg; if the laser had an efficiency of 10%, it would take about

$$(2 \times 10^9 \text{ J})/(1.4 \times 10^6 \text{ J/kg}) = 1400 \text{ kg}$$

of fuel for one pulse from the HF laser. Thus, lasers in the visible region (λ = 0.5 μ = $\lambda_{HF}/6$) are being studied.

III.   ABM SYSTEMS

There are a variety of systems, other than particle
and laser beams weapons (Sec. II), that are being
considered[4] to destroy incoming missiles. In this
section we will briefly consider the basic physics of
(III.A) the Low Altitude Defense System (LoADS), (III.B)
the high altitude Spartan ABM (1969), and (III.C) simple
and novel systems. This appendix will not discuss the
broader political, legal, and strategic aspects of the
ABM. (See the chapter by A. Carnesale in this book).

III.A.   Low Altitude Defense System (LoADS). This
system is similar to the Sprint system that was
considered in 1969, but it is intended to operate at
lower altitudes, below about 50,000 feet. As in the case
of Sprint, the fast neutrons from a 20 kiloton (kt)
explosion are used to destroy the solid-state electronics
and to disarm the nuclear warhead by the absorption of
neutrons in the fissile material of the incoming warhead.
Since 1 kt (4.2×10^{12} J) is the equivalent of the total
fission of about 55 grams of fissile material (0.23 gram
moles), the number of neutrons escaping from a 20 kt
fission warhead is about

$$(20 \text{ kt})(0.23 \text{ g moles})(6 \times 10^{23})(3 - 2) = 3 \times 10^{24} \text{ n}.$$

In this simple calculation, we have assumed that one
neutron/fission (3 - 2) escapes from the warhead. The
number of neutrons/cm^2 at a distance of 400 meters is
about

$$(3 \times 10^{24})/(4\pi)(4 \times 10^{4})^{2} = 1.5 \times 10^{14} \text{ n/cm}^2.$$

This neutron flux at 400 m is similar to value of 2×10^{14}
presented in Glasstone and Dolan[5] and it is capable of
disarming a nuclear warhead under certain conditions.

III.B   SPARTAN. The SPARTAN ABM was intended (in
1969) to interecept incoming missiles high above the
earth. Since the explosion would take place above the
atmosphere in a vacuum, the energy going into the shock
and blast mode would be essentially zero, and the
fraction of the energy delievered as thermal radiation
would be enhanced. This system is no longer being
considered because the very high altitude burst would
develop an EMP pulse (Sec. I) which could disarm our own
C^3I systems. Nevertheless, for historical reasons it is
of interest to discuss briefly this system.
    If one assumes that the average temperature of the
exploding warhead is similar to the core of the sun (18 $\times$
10^6 K), the most probable photon energy can be obtained
from Wien's law, and it is about 7 keV, or in the soft

x-ray region. If one assumes that the SPARTAN uses a 2 Mt warhead ($2000 \times 4.2 \times 10^{12}$ J), and that 75% of its energy goes into thermal radiation, the heat flux at a distance of one km is about

$$(0.75)(8.4 \times 10^{15} J)/(4\pi)(10^5 \text{ cm})^2 = 5000 \text{ J/cm}^2.$$

The 7 keV x-rays penetrate about 0.004 cm in aluminum and would cause severe heating at this flux of energy (Sec. II.A).

III.C Simple and Novel ABM Systems. This section will briefly discuss two systems, Swarmjet and Dust Defense which rely on the kinetic energy of impact to destroy an incoming re-entry vehicle (RV). A Swarmjet ABM system would work as follows: After radar has detected an attacking RV, a launcher containing many small, rod-like projectiles is sent to the neighborhood of the RV. When the Swarmjet is near the incoming RV, the rod like projectiles are spread into a pattern which would destroy the RV upon impact. In addition, others have consider developing a large cloud of dust particles by exploding a nuclear warhead at the surface of the Earth near the silo which is to be defended. Since the incoming RV has a velocity of the order of 10^4 m/s; the kinetic energy of one gram of dust (with respect to the RV) is about

$$(0.5)(0.001 \text{ kg})(10^4 \text{ m/s})^2 = 0.5 \times 10^5 \text{ J/g}$$

which is about 10 - 100 times greater than chemical explosives.

REFERENCES

1. S. Glasstone and P. Dolan, THE EFFECTS OF NUCLEAR
    WEAPONS, Departments of Energy and Defense,
    Washington, D.C., 1977, Chapters 8 and 11; W. Broad,
    Science 212, 1009, 1116, 1248 (1981); Ricketts,
    FUNDAMENTALS OF NUCLEAR HARDENING OF ELECTRONIC
    EQUIPMENT, Wiley, New York, 1972; L. Ricketts, J.
    Bridges, and J. Milletta, EMP RADIATION AND
    PROTECTIVE TECHNIQUES, Wiley, 1976; V. Gilinsky,
    Phys. Rev. 137, 1A, 50 (1965).

2. D. Hafemeister, Am. J. Phys. 51, 215 (1983).

3. G. Bekefi, B. Feld, J. Parmentola, and K. Tsipis,
    Nature 284, 219 (1980); M. Callaham and K. Tsipis,
    HIGH ENERGY LASER WEAPONS: A TECHNICAL ASSESSMENT,
    Program in Science and Technology in International
    Security, MIT, Cambridge, MA, 1980; K. Tsipis, Sci.
    Am. 245, 51 (Dec. 1981).

4. Chapter 3 in MX MISSILE BASING, Office of Technology
    Assessment, Washington, DC, 1981; ABM: AN EVALUATION
    OF THE DECISION TO DEPLOY AN ANTIBALLISTIC MISSILE
    SYSTEM, edited by A. Chayes and J. Wiesner, Harper
    and Row, New York, 1969; WHY ABM: POLICY ISSUES IN
    THE MISSILE DEFENSE CONTROVERSEY, edited by J. Holst
    and W. Schneider, Pergamon, New York, 1969; A.
    Carnesale, Chapter 12 of this book; D. Hafemeister,
    Am. J. Phys. 41, 1191 (1973).

5. S. Glasstone and P. Dolan, THE EFFECTS OF NUCLEAR
    WEAPONS, Deptartments of Energy and Defense,
    Washington, DC, 1977, page 366.

APPENDIX I

BIOGRAPHICAL NOTES ON THE AUTHORS

Albert Carnesale (Ph.D., North Carolina State, 1966):
Professor of Public Policy and Academic Dean of the John
F. Kennedy School of Government, Harvard University.
Former Senior Advisor in the negotiations of SALT I and
the International Nuclear Fuel Cycle Evaluation (INFCE);
presently serves as a consultant on nuclear weapons
policies to the Departments of Defense and State.
Co-author of LIVING WITH NUCLEAR WEAPONS.

Inés Cifuentes (Ph.D. candidate, Columbia): Research
Associate at the Lamont-Doherty Geological Observatory,
Columbia University. Former seismologist at the U.S.
Geological Survey, Menlo Park, CA.

Paul Craig (Ph.D., Cal Tech, 1959): Professor of Applied
Science, University of California at Davis. After
working in cryogenics at Los Alamos and Brookhaven, he
spent five years with the National Science Foundation in
Washington, mostly with the Office of Energy R&D Policy
providing support to the President's Science Advisor.
Co-editor of DECENTRALIZED ENERGY.

John Dowling (Ph.D., Arizona State, 1964): Professor and
Chairperson, Department of Physics, Mansfield State
College, PA. Film Editor for the Bulletin of Atomic
Scientists; editor of PHYSICS and SOCIETY (APS Forum on
Physics and Society). Co-author of the AAPT resource
letter, "Physics and the Nuclear Arms' Race" and author of
WAR PEACE FILM GUIDE.

Ralph Earle II (LL.B., Harvard, 1955): Partner, Earle
and Greene and Company. Former Director, U.S. Arms
Control and Disarmament Agency (1980-81); U.S. Chief
Negotiator for the SALT II Treaty (1978-80); and
Alternate Chief Negotiator for SALT II (1977-78). Former
Partner in Morgan, Lewis, and Bockius.

Jack Evernden (Ph.D., California-Berkeley, 1951):
Research Geophysicist at the U.S. Geological Survey,
Menlo Park, Ca. Former Professor of Geophysics at
University of California at Berkeley, and seismologist at
the Air Force Applications Center, ACDA, and ARPA. Has
published many papers on the issue of the "Verification
of a Comprehensive Test Ban" which appeared in the
Scientific American in 1982.

Richard Garwin (Ph.D., Chicago, 1949):   IBM Fellow at the
Thomas J. Watson Research Center; Adjunct Professor of
Physics, Columbia University; Adjunct Research Fellow,
Center for Science and International Affairs, Harvard
University; Andrew D. White Professor-at-Large, Cornell
University. Former member of the President's Science
Advisory Committee, and the Defense Science Board.
Co-author of NUCLEAR WEAPONS AND WORLD POLITICS, NUCLEAR
POWER: ISSUES AND CHOICES, ENERGY: THE NEXT TWENTY YEARS,
THE GENESIS OF WEAPONS SYSTEMS, and many articles on
national security and arms control.

G. Allen Greb (Ph.D., California-San Diego, 1978):
Associate Director of the Program in Science, Technology,
and Public Affairs, and Professor of International
Security Issues at the University of California at San
Diego. Author of numerous articles on nuclear weapons
systems, defense decisionmaking, and arms control policy.

David Hafemeister (Ph.D., Illinois, 1964): Professor of
Physics, California Polytechnic University. Former AAAS
Congressional Fellow and Science Advisor, U.S. Senate;
Special Assistant to the Under-Secretary of State for
Security Assistance, Science and Technology; and Science
Fellow, Los Alamos Scientific Laboratory. Author of
"Science and Society Tests" (American Journal of Physics)
and co-author of PHYSICS FOR MODERN ARCHITECTURE.

Warren Heckrotte (Ph.D., California-Berkeley, 1953):
Staff Physicist at Lawrence-Livermore Laboratory. Has
served since 1961 on U.S. delegations concerned with arms
control; in 1980 he was deputy head of the Comprehensive
Nuclear Test Ban negotiations in Geneva. Received the
Distinguished Associate Award from ERDA in 1976.
Co-editor of ARMS CONTROL IN TRANSITION: PROCEEDINGS OF
THE LIVERMORE ARMS CONTROL CONFERENCE.

Bobby R. Hunt (Ph.D., Arizona, 1967): Director of the
Digital Image Analysis Laboratory and Professor of
Electrical Engineering, University of Arizona. Chief
Scientist for signal processing, Science Applications
Incorporated, Tucson. Co-author of DIGITAL IMAGE
RESTORATION, and numerous papers on digital image
processing.

Gerald Johnson (Ph.D., California-Berkeley, 1947):
Member of Senior Technical Staff at TRW working on arms
control, energy, and national defense. Former Associate
Director, Lawrence Livermore Laboratory, for nuclear
testing for several years (1953-1966) except during the
Kennedy administration when he served as Assistant to the
Secretary of Defense for Atomic Affairs. From 1977-1979
he was the personal representative of the Secretary of
Defense to the Strategic Arms Limitations Talks and the
Comprehensive Test Ban negotiations.

Mark Levine (Ph.D., California-Berkeley, 1975): Staff
member at the Lawrence Berkeley Laboratory. Former staff
member with the Ford Energy Policy Project and Senior
Energy Analyst at Stanford Research Institute
International. Co-editor of DECENTRALIZED ENERGY.

Dietrich Schroeer (Ph.D., Ohio State, 1965): Professor
of Physics, University of North Carolina at Chapel Hill.
Author of PHYSICS AND ITS FIFTH DIMENSION: SOCIETY;
SCIENCE, TECHNOLOGY AND THE NUCLEAR ARMS RACE; and
co-author of the AAPT resource letter, "Physics and the
Nuclear Arms Race."

Peter Sharfman (Ph.D., Chicago, 1972): Program Manager
for International Security and Commerce, Office of
Technology Assessment, U.S. Congress. Directed OTA's
assessments of THE EFFECTS OF NUCLEAR WAR and MX MISSILE
BASING. Former Assistant Professor of Government at
Cornell University; Foreign Affairs Officer at the U.S.
Arms Control and Disarmament Agency; and Assistant
Director of National Assessment, Department of Defense.

Lynn R. Sykes (Ph.D., Columbia, 1964): Higgins Professor
of Geology and Head of the Seismology Group,
Lamont-Doherty Geological Observatory, Columbia
University. President-Elect of Section on Seismology,
American Geophysical Union (1982-84); President
(1984-86). Member of Panel on Earthquake Prediction,
National Academy of Science. Co-author of "Verification
of a Comprehensive Nuclear Test Ban" and numerous other
contributions that led to the development of the
hypothesis of plate techtonics.

Kosta Tsipis (Ph.D., Columbia, 1966): Co-director,
Program in Science and Technology for International
Security, and Professor of Physics, Massachusetts
Institute of Technology. Former Visiting Scientist at
the Stokholm Institute for Peace Research. Author of
three books and more than 60 articles on missile
accuracy, cruise missiles, particle and laser beam
weapons, and catastrophic releases of radioactivity.

Frank von Hippel (Ph.D., Oxford, 1962): Professor,
Woodrow Wilson School of Public and International
Affairs, Princeton University. Chairman, Federation of
American Scientists. Former Senior Research Physicst,
Center for Energy and Environmental Studies, Princeton;
member of the APS study on Reactor Safety; Sloan
Foundation Fellow; and Resident Fellow, National Academy
of Sciences. Co-Author of "Limited Nuclear War", "The
Consequences of a 'Limited' Nuclear War in East and West
Germany," and ADVICE AND DISSENT.

Peter Zimmerman (Ph.D., Stanford, 1969): Associate
Professor of Physics, Louisiana State University. Former
member of the Office of Technology Assessment study on MX
MISSILE BASING; Visiting Associate Research Professor,
Program on Science, Technology, and Public Affairs,
University of California at San Diego; Consultant to the
Institute for Defense Analysis, Applied Physics
Laboratory (Johns Hopkins), and Science Applications,
Inc. Author of numerous papers on technology and
national security.

FEB 1 5 1995
Rec'd UCB PHYS

OCT 0 5 2000

Rec'd UCB PHYS

RETURN PHYSICS LIBRARY
TO ➡ 351 LeConte Hall   642-3122

LOAN PERIOD 1 1-MONTH	2	3
4	5	6

ALL BOOKS MAY BE RECALLED AFTER 7 DAYS
Overdue books are subject to replacement bills

DUE AS STAMPED BELOW

MAY 1 4 1986	MAR 1 3 1995	
JAN 2 4 1987	FEB 1 5 1995	
APR 8 1987	Rec'd UCB PHYS	
	OCT 0 5 2000	
	MAY 1 5 2008	
MAY 2 7 1987		
JUL 2 4 1987		
NOV 2 3 1989		
APR 6 1990		
JUN 4 1991 Rec'd UCB PHYS		
MAY 1 4 1992		

FORM NO. DD 25

UNIVERSITY OF CALIFORNIA, BERKELEY
BERKELEY, CA 94720